Cracking the AP® BIOLOGY EXAM

2015 Edition

Kim Magliore

PrincetonReview.com

PENGUIN RANDOM HOUSE

The Princeton Review
24 Prime Parkway, Suite 201
Natick, MA 01760
E-mail: editorialsupport@review.com

Published in the United States by Random House LLC, New York, and simultaneously in Canada by Random House of Canada Limited, Toronto.
A Penguin Random House Company.

ISBN: 978-0-8041-2524-6
ISSN: 1092-0080

Editor: Calvin S. Cato
Production Editor: Jim Melloan
Production Artist: Althea Gladstone

Printed in the United States of America on partially recycled paper.

10 9 8 7 6 5 4 3 2 1

2015 Edition

ACKNOWLEDGMENTS

The author would like to give special thanks to her parents, who instilled in her a love of science. She would also like to thank the staff of The Princeton Review and Andrew Taggart for his work on previous editions of this title.

The Princeton Review would like to thank Sarah Woodruff, Deborah Silvestrini, Althea Gladstone, and Bree Porcelli for their hard work on previous revisions to this edition.

The Princeton Review would also like to thank Christopher Stobart for his hard work in updating this current edition of this title. Very special thanks to Andrea Kornstein for her work in editing this title.

CONTENTS

PART I

USING THIS BOOK TO IMPROVE YOUR AP SCORE

PREVIEW ACTIVITY: YOUR KNOWLEDGE, YOUR EXPECTATIONS

Your route to a high score on the AP Biology Exam depends a lot on how you plan to use this book. Respond to the following questions.

1. Rate your level of confidence about your knowledge of the content tested by the AP Biology Exam:

   A. Very confident—I know it all
   B. I'm pretty confident, but there are topics for which I could use help
   C. Not confident—I need quite a bit of support
   D. I'm not sure

2. If you have a goal score in mind, circle your goal score for the AP Biology Exam:

   5 4 3 2 1 I'm not sure yet

3. What do you expect to learn from this book? Circle all that apply to you.

   A. A general overview of the test and what to expect
   B. Strategies for how to approach the test
   C. The content tested by this exam
   D. I'm not sure yet

YOUR GUIDE TO USING THIS BOOK

This book is organized to provide as much—or as little—support as you need, so you can use this book in whatever way will be most helpful to improving your score on the AP Biology Exam.

- The remainder of **Part One** will provide guidance on how to use this book and help you determine your strengths and weaknesses

- **Part Two** of this book will
  - provide information about the structure, scoring, and content of the AP Biology Exam
  - help you to make a study plan
  - point you towards additional resources

- **Part Three** of this book will explore:
  - how to attack multiple choice questions
  - how to write high scoring free-response answers
  - how to manage your time to maximize the number of points available to you

- **Part Four** of this book covers the content you need for your exam.
- **Part Five** of this book contains practice tests.

You may choose the use some parts of this book over others, or you may work through the entire book. This will depend on your needs and how much time you have. Let's now look how to make this determination.

HOW TO BEGIN

1. **Take a Test**

   Before you can decide how to use this book, you need to take a practice test. Doing so will give you insight into your strengths and weaknesses, and the test will also help you make an effective study plan. If you're feeling test-phobic, remind yourself that a practice test is a tool for diagnosing yourself—it's not how well you do that matters but how you use information gleaned from your performance to guide your preparation.

   So, before you read further, take AP Biology Practice Test 1 starting at page 277 of this book. Be sure to do so in one sitting, following the instructions that appear before the test.

2. **Check Your Answers**

   Using the answer key on page 297, count how many multiple choice questions you got right and how many you missed. Don't worry about the explanations for now, and don't worry about why you missed questions. We'll get to that soon.

3. **Reflect on the Test**

   After you take your first test, respond to the following questions:

   - How much time did you spend on the multiple-choice questions?
   - How much time did you spend on each free-response question?
   - How many multiple-choice questions did you miss?
   - Do you feel you had the knowledge to address the subject matter of the essays?
   - Do you feel you wrote well organized, thoughtful responses to the free-response questions?

4. **Read Part Two of this Book and Complete the Self-Evaluation**

As discussed in the Guide section above, Part Two will provide information on how the test is structured and scored. It will also set out areas of content that are tested.

As you read Part Two, re-evaluate your answers to the questions above. At the end of Part Two, you will revisit and refine the questions you answer above. You will then be able to make a study plan, based on your needs and time available, that will allow you to use this book most effectively.

5. **Engage with Parts Three and Four as Needed**

Notice the word **engage**. You'll get more out of this book if you use it intentionally than if you read it passively, hoping for an improved score through osmosis.

The strategy chapters in Part Three will help you think about your approach to the question types on this exam. Part Three will open with a reminder to think about how you approach questions now and then close with a reflection section asking you to think about how and/or whether you will change your approach in the future.

The content chapters in Part Four are designed to provide a review of the content tested on the AP Biology Exam, including the level of detail you need to know and how the content is tested. You will have the opportunity to assess your mastery of the content of each chapter through test-appropriate questions.

6. **Take Test 2 and Assess Your Performance**

Once you feel you have developed the strategies you need and gained the knowledge you lacked, you should take Test 2, which starts at page 315 of this book. You should do so in one sitting, following the instructions at the beginning of the test.

When you are done, check your answers to the multiple-choice sections. See if a teacher will read your essays and provide feedback.

Once you have taken the test, reflect on what areas you still need to work on, and revisit the chapters in this book that address those deficiencies. Through this type of reflection and engagement, you will continue to improve. For extra practice, please check out our supplemental title *550 AP Biology Practice Questions*, which will be in stores soon.

7. Keep Working

After you have revisited certain chapters in this book, continue the process of testing, reflection, and engaging with the second practice test in this book. Consider what additional work you need to do and how you will change your strategic approach to different parts of the test.

As we will discuss in Part Two, there are other resources available to you, including a wealth of information at AP Central, the official site of the AP Exams. You can continue to explore areas that can stand to improve and engage in those areas right up to the day of the test.

PART II

ABOUT THE AP BIOLOGY EXAM

THE STRUCTURE OF THE TEST

The AP Biology Exam is three hours long and is divided into two sections: Section I (multiple-choice questions) and Section II (free-response questions).

Section I consists of 69 questions. These are broken down into Part A (63 multiple-choice questions) and Part B (6 grid-in questions). The multiple-choice questions are further broken down into two parts: regular multiple-choice questions and questions dealing with experiments or data.

Section II involves free-response questions. You'll be presented with two long-form free-response questions and six short-form free-response questions touching upon key issues in biology. You'll be given a 10-minute reading period followed by 80 minutes to answer all eight questions.

If you're thinking that this sounds like a heap of work to try to finish in three hours, you're absolutely right. Here's how it breaks down: You have roughly 75 seconds per multiple-choice or grid-in question and 21 minutes per free-response question. How can you possibly tackle so much science in so little time?

Fortunately, there's absolutely no need to. As you'll soon see, we're going to ask you to leave a small chunk of the test blank. Which part? The parts you don't like. This selective approach to the test, which we call "pacing," is probably the most important part of our overall strategy. But before we talk strategy, let's look at the topics that are covered by the AP Biology Exam.

The AP Biology Exam covers these four Big Ideas:

- Big Idea 1: The process of evolution drives the diversity and unity of life.
- Big Idea 2: Biological systems utilize free energy and molecular building blocks to grow, to reproduce, and to maintain dynamic homeostasis.
- Big Idea 3: Living systems store, retrieve, transmit, and respond to information essential to life processes.
- Big Idea 4: Biological systems interact, and these systems and their interactions possess complex properties.

These four areas are further subdivided into major topics. These topics include the following:

1. Chemistry of Life
    - Organic molecules in organisms
    - Water
    - Free-energy changes
    - Enzymes
2. Cells
    - Prokaryotic and eukaryotic cells
    - Membranes
    - Subcellular organization
    - Cell cycle and its regulation

3. Cellular Energetics
    - Coupled reactions
    - Fermentation and cellular respiration
    - Photosynthesis
4. Heredity
    - Meiosis and gametogenesis
    - Eukaryotic chromosomes
    - Inheritance patterns
5. Molecular Genetics
    - RNA and DNA structure and function
    - Gene regulation
    - Mutation
    - Viral structure and replication
    - Nucleic acid technology and applications
6. Evolutionary Biology
    - Early evolution of life
    - Evidence for evolution
    - Mechanism of evolution
7. Diversity of Organisms
    - Evolutionary patterns
    - Survey of the diversity of life
    - Phylogenetic classification
    - Evolutionary relationships
8. Structure and Function of Plants and Animals
    - Reproduction, growth, and development
    - Structural, physiological, and behavioral adaptation
    - Response to the environment
9. Ecology
    - Population dynamics
    - Communities and ecosystems
    - Global issues

This might seem like an awful lot of information. But for each topic, there are just a few key facts you'll need to know. Your biology textbooks may go into far greater detail about some of these topics than we do. That's because they're trying to teach you "correct science," whereas we're aiming to improve your scores. Our science is perfectly sound; it's just cut down to size. We've focused on crucial details and given you only what's important. Moreover, as you'll soon see, our treatment of these topics is far easier to handle.

HOW AP EXAMS ARE USED

Different colleges use AP Exams in different ways, so it is important that you go to a particular college's website to determine how it uses AP Exams. The three items below represent the main ways in which AP Exam scores can be used:

- **College Credit.** Some colleges will give you college credit if you score well on an AP Exam. These credits count towards your graduation requirements, meaning that you can take fewer courses while in college. Given the cost of college, this could be quite a benefit, indeed.

- **Satisfy Requirements.** Some colleges will allow you to "place out" of certain requirements if you do well on an AP Exam, even if they do not give you actual college credits. For example, you might not need to take an introductory-level course, or perhaps you might not need to take a class in a certain discipline at all.

- **Admissions Plus.** Even if your AP Exam will not result in college credit or allow you to place out of certain courses, most colleges will respect your decision to push yourself by taking an AP Course or an AP Exam outside of a course. A high score on an AP Exam shows mastery of more difficult content than is taught in many high school courses, and colleges may take that into account during the admissions process.

OTHER RESOURCES

There are many resources available to help you improve your score on the AP Biology Exam, not the least of which are your **teachers**. If you are taking an AP class, you may be able to get extra attention from your teacher, such as obtaining feedback on your essays. If you are not in an AP course, reach out to a teacher who teaches AP Biology and ask if the teacher will review your essays or otherwise help you with content.

Another wonderful resource is **AP Central**, the official site of the AP Exams. The scope of the information at this site is quite broad and includes:

- A course description, which includes details on what content is covered and sample questions
- Sample questions from the AP Biology exam
- Free-response question prompts and multiple-choice questions from previous years

The AP Central home page address is: http://apcentral.collegeboard.com/.

For up-to-date information about the ongoing changes to the AP Biology Exam Course, please visit: http://apcentral.collegeboard.com/apc/public/courses/teachers_corner/2117.html

Finally, **The Princeton Review** offers tutoring and small group instruction. Our expert instructors can help you refine your strategic approach and add to your content knowledge. For more information, call 1-800-2REVIEW.

DESIGNING YOUR STUDY PLAN

As part of the Introduction, you identified some areas of potential improvement. Let's now delve further into your performance on Test 1, with the goal of developing a study plan appropriate to your needs and time commitment.

Read the answers and explanations associated with the multiple-choice questions (starting at page 299). After you have done so, respond to the following questions:

- Review the bulleted list of topics on pages 8 and 9. Next to each topic, indicate your rank of the topic as follows: "1" means "I need a lot of work on this," "2" means "I need to beef up my knowledge," and "3" means "I know this topic well."
- How many days/weeks/months away is your exam?
- What time of day is your best, most focused study time?
- How much time per day/week/month will you devote to preparing for your exam?
- When will you do this preparation? (Be as specific as possible: Mondays and Wednesdays from 3 to 4 pm, for example)
- Based on the answers above, will you focus on strategy (Part Three) or content (Part Four) or both?
- What are your overall goals in using this book?

PART III

TEST-TAKING STRATEGIES FOR THE AP BIOLOGY EXAM

HOW TO USE THE CHAPTERS IN THIS PART

For the following Strategy chapters, think about what you are doing now before you read the chapters. As you read and engage in the directed practice, be sure to appreciate the ways you can change your approach. At the end of Part Three, you will have the opportunity to reflect on how you will change your approach.

How to Approach Multiple-Choice Questions

SECTION I

As we mentioned earlier, the multiple-choice section consists of the following two parts:

- Part A—contains regular multiple-choice questions and multiple choice questions dealing with an experiment or a set of data
- Part B—contains grid-in questions

PART A

Part A of the AP Biology Exam consists of 63 run-of-the-mill multiple-choice questions. These questions test your grasp of the fundamentals of biology and your ability to apply biological concepts to help problem-solve. Here's an example:

22. If a segment of DNA reads 5′-ATG-CCA-GCT-3′, the mRNA strand that results from the transcription of this segment will be

    (A) 3′-TAC-GGT-CGA-5′
    (B) 3′-UAC-AGT-CAA-5′
    (C) 3′-TAA-GGU-CGA-5′
    (D) 3′-TAC-GGT-CTA-5′

Don't worry about the answer to this question for now. By the end of this book, it will be a piece of cake. The majority of the questions in this section are presented in this format. A few questions may include a figure, a diagram, or a chart.

The second part of the first portion also consists of multiple-choice questions, yet here you're asked to think logically about different biological experiments or data. Here's a typical example:

Questions 60 and 61 refer to the following diagram and information.

To understand the workings of neurons, an experiment was conducted to study the neural pathway of a reflex arc in frogs. A diagram of a reflex arc is given below.

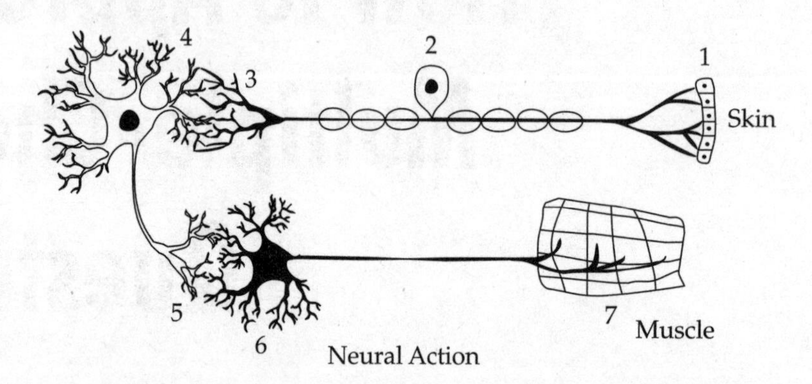

Neural Action

60. Which of the following represents the correct pathway taken by a nerve impulse as it travels from the spinal cord to effector cells?

    (A) 1-2-3-4
    (B) 6-5-4-3
    (C) 2-3-4-5
    (D) 4-5-6-7

61. The brain of the frog is destroyed. A piece of acid-soaked paper is applied to the frog's skin. Every time the piece of paper is placed on its skin, one leg moves upward. Which of the following conclusions is best supported by the experiment?

(A) Reflex actions are not automatic.
(B) Some reflex actions can be inhibited or facilitated.
(C) All behaviors in frogs are primarily reflex responses.
(D) This reflex action bypasses the central nervous system.

You'll notice that these particular questions refer to an experiment. Many of the questions in this portion test your ability to integrate information, interpret data, and draw conclusions from the results.

PART B

Part B consists of six grid-in questions where an answer needs to be calculated based on information presented in the question. That numeric response is then filled in on a grid and bubbled accordingly. Answers can be in the form of integers, decimals, or fractions. A four-function calculator can be used on these questions. Here's a typical example:

34. If the genotype frequencies of an insect population are AA = 0.49, Aa = 0.42, and aa = 0.09, what is the gene frequency of the dominant allele?

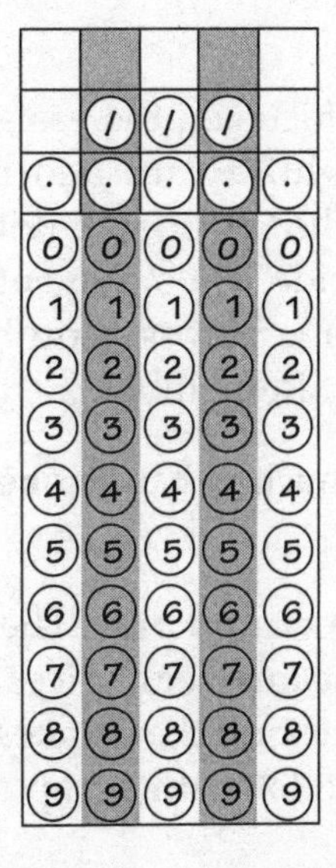

We mentioned earlier that our approach is strategy-based. As you're about to see, many of these strategies are based on common sense—for example, using mnemonics like "ROY G. BIV." (Remember that one? It's the mnemonic for red, orange, yellow, green, blue, indigo, violet—the colors of the spectrum.) Others do not make as much common sense. In fact, we're going to ask you to throw out much of what you've been taught when it comes to taking standardized tests.

PACE YOURSELF

When you take a test in school, how many questions do you answer? Naturally, you try to answer all of them. You do this for two reasons: (1) Your teacher told you to, and (2) if you left a question blank, your teacher would mark it wrong. However, that's not the case when it comes to the AP Biology Exam. In fact, finishing the test is the worst thing you can do. Before we explain why, let's talk about timing.

One of the main reasons that taking the AP Biology Exam is so stressful is the time constraint we discussed above—75 seconds per multiple-choice question and 21 minutes per essay. If you had all day, you would probably do much better. We can't give you all day, but we can do the next best thing: We can give you more time for each question. How? By having you slow down and answer fewer questions.

Slowing down, and doing well on the questions you do answer, is the best way to improve your score on the AP Biology Exam. Rushing through questions in order to finish, on the other hand, will always hurt your score. When you rush, you're far more likely to make careless errors, misread, and fall into traps. Keep in mind that blank answers are not counted against you.

THE THREE-PASS SYSTEM

The AP Biology Exam covers a broad range of topics. There's no way, even with our extensive review, that you will know everything about every topic in biology. So what should you do?

Do the Easiest Questions First

The best way to rack up points is to focus on the easiest questions first. Many of the questions asked on the test will be straightforward and require little effort. If you know the answer, nail it and move on. Others, however, will not be presented in such a clear, simple way. As you read each question, decide if it's easy, medium, or hard. During a first pass, do all the easy questions. If you come across a problem that seems time-consuming or completely incomprehensible, skip it. Remember:

> **Easier questions count just as much as harder ones, so your time is better spent on shorter, easier questions.**

Save the medium questions for the second pass. These questions are either time-consuming or require that you analyze all the answer choices (i.e., the correct answer doesn't pop off the page). If you come across a question that makes no sense from the outset, save it for the last pass. You're far more likely to fall into a trap or settle on a silly answer.

Watch Out for Those Bubbles!

Since you're skipping problems, you need to keep careful track of the bubbles on your answer sheet. One way to accomplish this is by answering all the questions on a page and then transferring your choices to the answer sheet. If you prefer to enter them one by one, make sure you double-check the number beside the ovals before filling them in. We'd hate to see you lose points because you forgot to skip a bubble!

So then, what about the questions you don't skip?

Process of Elimination (POE)

On most tests, you need to know your material backward and forward to get the right answer. In other words, if you don't know the answer beforehand, you probably won't answer the question correctly. This is particularly true of fill-in-the-blank and essay questions. We're taught to think that the only way to get a question right is by knowing the answer. However, that's not the case on Section I of the AP Biology Exam. You can get a perfect score on this portion of the test without knowing a single right answer, provided you know all the wrong answers!

What are we talking about? This is perhaps the single most important technique in terms of the multiple-choice section of the exam. Let's take a look at the example below.

1. The structures that act as the sites of gas exchange in a woody stem are the

(A) lungs
(B) lenticels
(C) ganglia
(D) lentil beans

Now if this were a fill-in-the-blank-style question, you might be in a heap of trouble. But let's take a look at what we've got. You see "woody stem" in the question, which leads you to conclude that we're talking about plants. Right away, you know the answer is not (A) or (C) because plants don't have lungs or ganglia. Now we've got it down to (B) and (D). Notice that (B) and (D) look very similar. Obviously, one of them is a trap. At this point, if you don't know what "lentil beans" are, you have to guess. However, even if we don't know precisely what they are, it's safe to say that most of us know that lentil beans have nothing to do with plant respiration. Therefore, the correct answer is (B), lenticels.

Although our example is a little goofy and doesn't look exactly like the questions you'll be seeing on the test, it illustrates an important point:

Process of Elimination (POE) is the best way to approach the multiple-choice questions.

Even if you don't know the answer right off the bat, you'll surely know that two or three of the answer choices are not correct. What then?

Aggressive Guessing

Educational Testing Service (ETS), the service that develops and administers the exam, tells you that random guessing will not affect your score. This is true. There is no guessing penalty on the AP Biology exam. For each correct answer you'll receive one point and you will not lose any points for each incorrect answer.

Although you won't lose any points for wrong answers, you should guess aggressively by getting rid of the incorrect answer choices. The moment you've eliminated a couple of answer choices, your odds of getting the question right, even if you guess, are far greater. If you can eliminate as many as two answer choices, your odds improve enough that it's in your best interest to guess.

Word Associations

Another way to rack up the points on the AP Biology Exam is by using word associations in tandem with your POE skills. Make sure that you memorize the words in the Key Words lists throughout this book. Know them backward and forward. As you learn them, make sure you group them by association, since ETS is bound to ask about them on the AP Biology Exam. What do we mean by "word associations"? Let's take the example of mitosis and meiosis.

You'll soon see from our review that there are several terms associated with mitosis and meiosis. *Synapsis*, *crossing-over*, and *tetrads*, for example, are words associated with meiosis but not mitosis. We'll explain what these words mean later in this book. For now, just take a look:

2. Which of the following typifies cytokinesis during mitosis?

   (A) Crossing-over
   (B) Formation of tetrads
   (C) Synapsis
   (D) Division of the cytoplasm

This might seem like a difficult problem. But let's think about the associations we just discussed. The question asks us about mitosis. However, answer choices (A), (B), and (C) all mention events that we've associated with meiosis. Therefore, they are out. Without even racking your brain, you've managed to find the correct answer choice: (D). Not bad!

Once again, don't worry about the science for now. We'll review it later. What is important to recognize is that by combining the associations we'll offer throughout this book and your aggressive POE techniques, you'll be able to rack up points on problems that might have seemed difficult at first.

MNEMONICS—OR THE BIOLOGY NAME GAME

One of the big keys to simplifying biology is the organization of terms into a handful of easily remembered packages. The best way to accomplish this is by using mnemonics. Biology is all about names: the names of chemical structures, processes, theories, etc. How are you going to keep them all straight? A mnemonic, as you may already know, is a convenient device for remembering something.

For example, one important issue in biology is taxonomy, that is, the classification of life forms, or organisms. Organisms are classified in a descending system of similarity, leading from domains (the broadest level) to species (the most specific level). The complete order runs: domain, kingdom, phylum, class, order, family, genus, and species. Don't freak out yet. Look how easy it becomes with a mnemonic:

King Philip of Germany decided to walk to America. What do you think happened?

Dumb → Domain
King → Kingdom
Philip → Phylum
Came → Class
Over → Order
From → Family
Germany → Genus
Soaked → Species

Learn the mnemonic and you'll never forget the science!

Mnemonics can be as goofy as you like, so long as they help you remember. Throughout this book, we'll give you mnemonics for many of the complicated terms we'll be seeing. Use ours, if you like them, or feel free to invent your own. Be creative! Remember: The important thing is that you remember the information, not how you remember it.

Identifying Except Questions

About 10 percent of the multiple-choice questions in Section I are EXCEPT/NOT/LEAST questions. With this type of question, you must remember that you're looking for the *wrong* (or the least correct) answer. The best way to go about these is by using POE.

More often than not, the correct answer is a true statement, but is wrong in the context of the question. However, the other three tend to be pretty straightforward. Cross off the three that apply and you're left with the one that does not. Here's a sample question:

17. All of the following are true statements about gametes EXCEPT:

(A) They are haploid cells.
(B) They are produced only in the reproductive structures.
(C) They bring about genetic variation among offspring.
(D) They develop from polar bodies.

If you don't remember anything about gametes and gametogenesis, or the production of gametes, this might be a particularly difficult problem. We'll see these again later on, but for now, remember that gametes are the "sex cells" of sexually reproducing organisms. As such, we know that they are haploid and are produced in the sexual organs. We also know that they come together to create offspring.

From this very basic review, we know immediately that (A) and (B) are not our answers. Both of these are accurate statements, so we eliminate them. That leaves us with (C) and (D). If you have no idea what (D) means, focus on (C). In sexual reproduction, each parent contributes one gamete, or half the genetic complement of the offspring. This definitely helps vary the genetic makeup of the offspring. Answer choice (C) is a true statement, so it can be eliminated. The correct answer is (D).

Don't sweat it if you don't recall the biology. We'll be reviewing it in detail soon enough. For now, remember that the best way to answer these types of questions is: Spot all the right statements and cross them off. You'll wind up with the wrong statement, which happens to be the correct answer

Solving Grid-In Questions

Part B of the multiple-choice section requires you to answer six grid-in questions. Unlike the questions in Part A, you will not be able to use process of elimination. However, there are strategies you can use to increase your chances of getting this questions correct efficiently. First, identify what the question is actually asking for. Are you being asked to find a value on a graph or chart? Perhaps they are having you calculate the frequency of an allele using Hardy-Weinberg equilibrium. If you have no idea how to find out what they are asking you, take a quick and educated guess and move on. For instance, if they are asking for a frequency, try guessing a decimal between 0 and 1, such as 0.25. However, if you do know how to find the answer, then write down the question number and any work (e.g. notes, calculations), which you would need to answer the question neatly. Before filling in your answer, make sure that you are correct and haven't made any mistakes on interpreting data or in making calculations. Be sure not to forget to bubble in your answers.

REFLECT

Respond to the following questions:

- How long will you spend on multiple-choice questions?
- How will you change your approach to multiple-choice questions?
- What is your multiple-choice guessing strategy?
- Will you seek further help, outside of this book (such as a teacher, tutor, or AP Central), on how to approach the questions that you will see on the AP Biology Exam?

How to Approach Free-Response Questions

THE ART OF THE ETS ESSAY

You are given two essay questions and six short-form free-response questions to answer in 80 minutes. The best way to rack up points on this section is to give the essay readers what they're looking for. Fortunately, we know precisely what that is.

The ETS essay reviewers have a checklist of key terms and concepts that they use to assign points. We like to call these "hot button" terms. Simply put, for each hot button that you include in your essay, you will receive a predetermined number of points. For example, if the essay question deals with the function of enzymes, the ETS graders are instructed to give 2 points for a mention of the "lock-and-key theory of enzyme specificity."

Naturally, you can't just compose a "laundry list" of scientific terms. Otherwise, it wouldn't be an essay. What you can do, however, is organize your essay around a handful of these key, or hot button, points. The most effective and efficient way to do this is by using the 10-minute reading period to brainstorm and come up with the scientific terms. Then outline your essay before you begin to write, using your hot buttons as your guide.

READ THE QUESTIONS CAREFULLY

ETS gives you 10 minutes to read the questions and organize your thoughts before you begin writing. If you use these 10 minutes wisely, you can breeze your way through the essays. The first thing you should do is take less than a minute to skim all of the questions and put them into your own personal order of difficulty from easiest to toughest. Once you've decided the order in which you will answer the questions (easiest first, hardest last), you can begin to formulate your responses. Your first step should be a more detailed assessment of each question.

The most important advice we can give you is to read each question at least twice. As you read the question, focus on key words, especially "direction words." Almost every essay question begins with a direction word. Some examples of direction words are *discuss*, *define*, *explain*, *describe*, *compare*, and *contrast*. If a question asks you to discuss a particular topic, you should give a viewpoint and support it with examples. If a question asks you to compare two things, you should discuss how the two things are similar. On the other hand, if the question asks you to contrast two things, you need to show how these things are different.

Many students lose points on their essays because they either misread the question or fail to do what's asked of them.

BRAINSTORM

Your next objective during the 10-minute preview should be to organize your thoughts.

Once you've read the questions, you need to brainstorm. Jot down as many key terms and concepts as you can. Don't forget, the test reviewers assign points on the basis of these key concepts. For each one that you mention and/or explain, you get a point.

How many do you need? You won't need all of them, that's for sure. However, you will need enough to get you the maximum number of points for that question. Most of these you can pull directly from your reading in this book. At the end of each chapter is a "Key Words" list, which is full of terms that you should incorporate into your free responses. In addition to this list, you can feel free to create your own thorough lists of terms for your own use.

Don't spend more than about 2 minutes per question brainstorming. Once your 10-minute reading period is up you should be ready to start writing your essays.

OUTLINE YOUR ESSAY

Have you ever written yourself into a corner? You're halfway through your essay when you suddenly realize that you have no idea whatsoever where you're going with your train of thought. To avoid this (and the panic that accompanies it), take a few minutes to draft an outline.

Your outline should incorporate as many of the hot buttons as you need in order to maximize your score. In other words, a question asks for two examples, choose only those two with which you are most comfortable. In your outline, make notes about the crucial points to mention with regard to each topic or key word. Once your outline is complete, you're ready to move on to writing the actual essay.

All of this preparation may seem time-consuming. However, it should take no more than four or five minutes per essay. What's more, it will greatly simplify the whole essay-writing process. So while you lose a little time at the outset, you'll more than make up for it when it comes time to actually write your answer.

DEVELOP YOUR IDEAS IN EACH ESSAY

Now you can use your outline to write your essay. Most students can come up with key terms or phrases that concern a particular biology topic. What separates a high-scoring student from a low-scoring student is *how* the student develops his or her thoughts on each essay. Besides giving the hot buttons, you'll need to elaborate on your thoughts and ideas. For example, don't just throw out a list of terms that pertain to meiosis and mitosis (such as *synapsis*, *crossing-over*, and *gametes*). Go one step further. Make sure you mention the *significance* of meiosis (i.e., it produces genetic variability). This extra piece of information will earn you an extra point.

Generally, you'll need to write about two to four paragraphs, depending on the number of parts contained in each essay question. In addition, be sure to give the appropriate number of examples for each essay. If the question asks for three examples, give only three examples. If you present more than is required, the test reviewers won't even read them or count them toward your score. The bottom line is this: Stick to the question.

ANSWER EACH PART OF THE ESSAY QUESTION SEPARATELY

The more parts there are to an essay question, the more important it is to pace yourself. On each essay, you're better off writing a little bit for each part than you are spending all your time on any one part of a question. Why? Even if you were to write the perfect answer to one part of a question, there's a limit to the number of points the test reviewers can assign to that part. Moreover, by writing a separate paragraph for each section, you make the test grader's job that much easier. When test readers have an easy time reading your essay, they're more likely to award points: It comes across as clearer and more organized. Readers have also requested that students label each part of their essay answers, so write "Part a," "Part b," and so on, accordingly in your response.

Finally, don't spend too much time writing a fancy introduction. You won't get brownie points for beautifully written openings like, "It was the best of experiments, it was the worst of experiments." Just leap right into the essay. And don't worry too much about grammar or spelling errors. Your grammar can hurt only if it's so bad that it seriously impairs your ability to communicate.

Incorporate Elements of an Experimental Design

Since one of the essay questions will be experimentally based, you'll need to know how to design an experiment. Most of these questions require that you present an appropriately labeled diagram or graph. Otherwise, you'll only get partial credit for your work.

There are two things you must remember when designing experiments on the AP Biology Exam: (1) Always label your figures, and (2) include controls in all experiments. Let's take a closer look at these two points.

Know How to Label Diagrams and Figures

Let's briefly discuss the important elements in setting up a graph. The favorite type of graph on the AP Biology Exam is the *coordinate graph*. The coordinate graph has a horizontal axis (x-axis) and a vertical axis (y-axis).

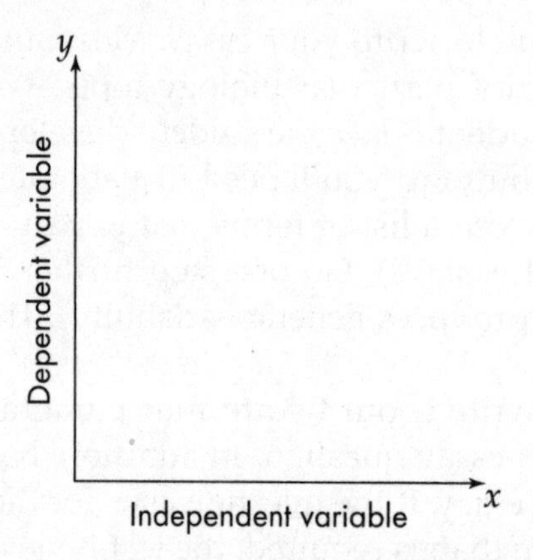

The x-axis usually contains the *independent variable*—the thing that's being manipulated or changed. The y-axis contains the *dependent variable*—the thing that's affected when the independent variable is changed.

Now let's look at what happens when you put some points on the graph. Every point on the graph represents both an independent variable and a dependent variable.

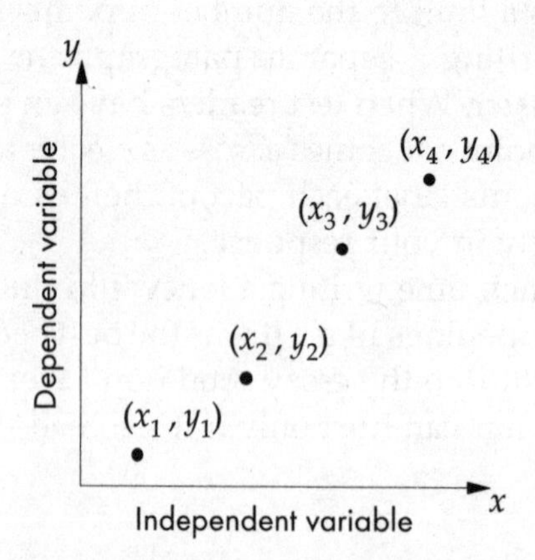

Once you draw both axes and label the axes as x and y, you can plot the points on the graph. Let's look at the following question.

1. Enzymes are biological catalysts.
   a. Relate the chemical structure of an enzyme to its catalytic activity and specificity.
   b. Design an experiment that investigates the influence of temperature, substrate concentration, or pH on the activity of an enzyme.
   c. Describe what information concerning enzyme structure could be inferred from the experiment you have designed.

For now, let's discuss only part b, which asks us to design an experiment. Let's set up a graph that shows the results of an experiment examining the relationship between pH and enzyme activity. Notice that we've chosen only one factor here, pH. We could have chosen any of the three. Why did we choose pH and not temperature or substrate concentration? Well, perhaps it's the one we know the most about.

What is the independent variable? It is pH. In other words, pH is being manipulated in the experiment. We'll therefore label the *x*-axis with pH values from 0 through 14.

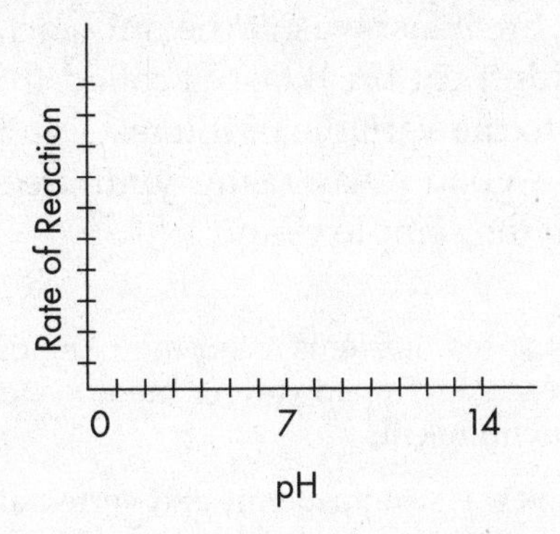

What is the dependent variable? It's the enzyme activity—the thing that's affected by pH. Let's label the *y*-axis "Rate of Reaction." Now we're ready to plot the values on the graph. Based on our knowledge of enzymes, we know that for most enzymes the functional range of pH is narrow, with optimal performance occurring at or around a pH of 7.

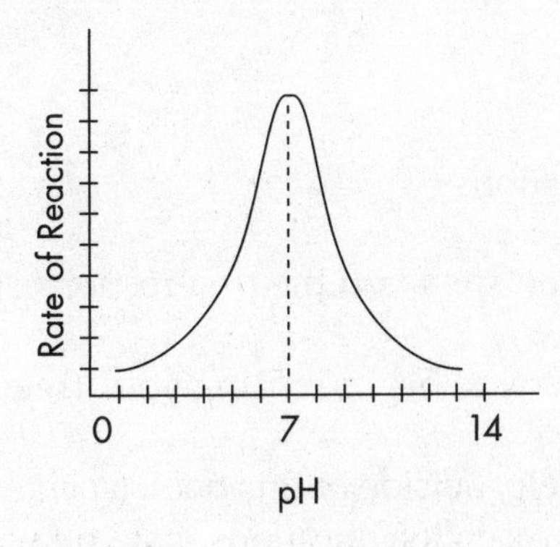

Now you should interpret your graph. If the pH level decreases from a neutral pH of 7, the reaction rate of the enzyme will decrease. If the pH level increases, the rate of reaction will also decrease. Don't forget to include a simple explanation of your graph.

Include Controls In Your Experiments

Almost every experiment will have at least one variable that remains constant throughout the study. This is called the *control*. A control is simply a standard of comparison. What does a control do? It enables the biologist to verify that the outcome of the study is due to changes in the independent variable and nothing else.

Let's say the principal of your school thinks that students who eat breakfast do better on the AP Biology Exam than those who don't eat breakfast. He gives a group of 10 students from your class free breakfast every day for a year. When the school year is over, he administers the AP Biology Exam and they all score brilliantly! Did they do well because they ate breakfast every day? We don't know for sure. Maybe the principal handpicked the smartest kids in the class to participate in the study.

In this case, the best way to be sure that eating breakfast made a difference is to have a control group. In other words, he would need to pick students in the class who *never* eat breakfast and follow them for a year. At the end of that year, he could send them in to take the AP Biology Exam. If they do just as well as the group that ate breakfast, then we can probably conclude that eating breakfast wasn't the only factor leading to higher AP scores. The group of students that didn't eat breakfast is called the control group because those students were not "exposed" to the variable of interest—in this case, breakfast.

Let's see if you can write a good essay using your free-response techniques. Take 21 minutes to write a response to the sample essay.

1. All organisms need nutrients to survive. Angiosperms and vertebrates have each developed various methods to obtain nutrients from their environment.
   a. Discuss the ways angiosperms and vertebrates procure their nutrients.
   b. Discuss two structures used for obtaining nutrients among angiosperms. Relate structure to function.
   c. Discuss two examples of symbiotic relationships that have evolved between organisms to obtain nutrients.

REFLECT

Respond to the following questions:

- How much time will you spend on the free-response questions?
- What will you do before you begin writing your free-response answers?
- Will you seek further help, outside of this book (such as a teacher, tutor, or AP Central), on how to approach the questions that you will see on the AP Biology Exam?

PART IV

CONTENT REVIEW FOR THE AP BIOLOGY EXAM

3

The Chemistry of Life

ELEMENTS

Although organisms exist in many diverse forms, they all have one thing in common: They are all made up of matter. Matter is made up of elements. **Elements**, by definition, are substances that cannot be broken down into simpler substances by chemical means.

THE ESSENTIAL ELEMENTS OF LIFE

Although there are 92 natural elements, 96 percent of the mass of all organisms is made up of just four of them: **oxygen** (O), **carbon** (C), **hydrogen** (H), and **nitrogen** (N). Other elements such as calcium (Ca), phosphorus (P), potassium (K), and magnesium (Mg) are also present, but in smaller quantities. These elements make up most of the remaining 4 percent of an organism's weight. Some elements are known as **trace elements** because they are only required by an organism in very small quantities. Trace elements include iron (Fe), iodine (I), and copper (Cu).

SUBATOMIC PARTICLES

If you break down an element into smaller pieces, you'll eventually come to the **atom**—the smallest unit of an element that retains its characteristic properties. Atoms are the building blocks of the physical world.

Within atoms, there are even smaller subatomic particles called **protons, neutrons**, and **electrons.** Let's take a look at a typical atom:

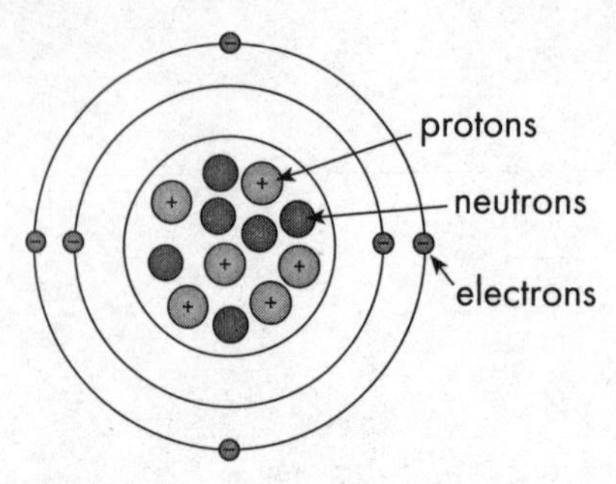

Protons and neutrons are particles that are packed together in the core of an atom called the **nucleus**. You'll notice that protons are positively charged (+) particles, whereas neutrons are uncharged particles.

Electrons, on the other hand, are negatively charged (–) particles that spin around the nucleus. Electrons are pretty small compared to protons and neutrons. In fact, for our purposes, electrons are considered massless. Most atoms have the same number of protons and electrons, making them electrically neutral. Some atoms have the same number of protons but differ in the number of neutrons in the nucleus. These atoms are called **isotopes**. Ancient artifacts can be dated by examining the rate of decay in carbon 14 isotopes.

COMPOUNDS

When two or more different types of atoms are combined in a fixed ratio, they form a chemical **compound**. You'll sometimes find that a compound has different properties from those of its elements. For instance, hydrogen and oxygen exist in nature as gases. Yet when they combine to make water, they often pass into a liquid state. When hydrogen atoms get together with oxygen atoms to form water, we've got a **chemical reaction:**

$$2H_2\,(g) + O_2\,(g) \rightarrow 2H_2O\,(l)$$

The atoms of a compound are held together by **chemical bonds**, which may be ionic bonds, covalent bonds, or hydrogen bonds.

An **ionic bond** is formed between two atoms when one or more electrons are transferred from one atom to the other. In this reaction, one atom *loses* electrons and becomes positively charged and the other atom *gains* electrons and becomes negatively charged. The charged forms of the atoms are called ions. For example, when Na reacts with Cl, charged ions, Na^+ and Cl^-, are formed.

A **covalent bond** is formed when electrons are *shared* between atoms. If the electrons are shared equally between the atoms, the bond is called **nonpolar covalent.** If the electrons are shared unequally, the bond is called **polar covalent.** When one pair of electrons is shared between two atoms, the result is a single covalent bond. When two pairs of electrons are shared, the result is a double covalent bond. When three pairs of electrons are shared, the result is a triple covalent bond.

WATER: THE VERSATILE MOLECULE

One of the most important substances in nature is water. Did you know that more than 60 percent of your body weight consists of water? Water is considered a unique molecule because it plays an important role in chemical reactions.

Let's take a look at one of the properties of water. Water has two hydrogen atoms joined to an oxygen atom:

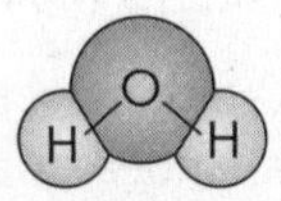

In water molecules, the hydrogen atoms have a partial positive charge and the oxygen atom has a partial negative charge. Molecules that have partially positive and partially negative charges are said to be **polar**. Water is therefore a polar molecule. The positively-charged ends of the water molecules strongly attract the negatively-charged ends of other polar compounds. Likewise, the negatively-charged ends strongly attract the positively-charged ends of neighboring compounds. These forces are most readily apparent in the tendency of water molecules to stick together, as in the formation of water beads or raindrops.

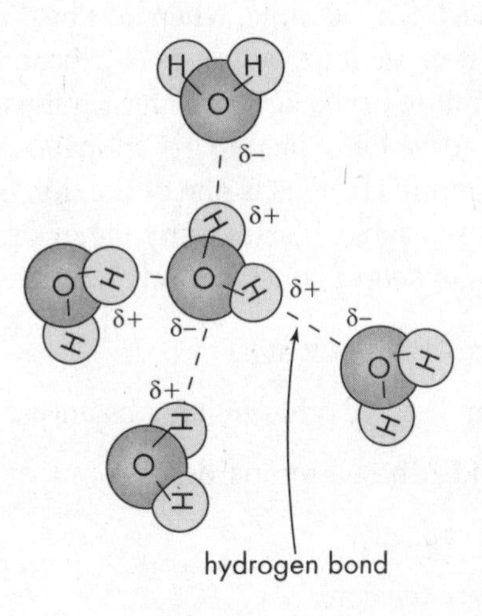

Another type of bond important in organisms is a hydrogen bond. **Hydrogen bonds** are weak chemical bonds that form when a hydrogen atom that is covalently bonded to one electronegative atom is also attracted to another electronegative atom. Water molecules are held together by hydrogen bonds. Although hydrogen bonds are individually weak, collectively, they are strong when present in large numbers. Because it can react with other polar substances, water makes a great solvent; it can dissolve many kinds of substances. The hydrogen bonds that hold water molecules together contribute to a number of special properties including the following:

- As mentioned above, water molecules have a strong tendency to stick together. That is, water exhibits *cohesive forces*. These forces are extremely important to life. For instance, during transpiration water molecules evaporate from a leaf "pulling" on neighboring water molecules. These, in turn, draw up the molecules immediately behind them, and so on, all the way down the plant vessels. The resulting chain of water molecules enables water to move up the stem. Water also has a high surface tension because of the cohesiveness of its molecules.
- Water molecules also like to stick to other substances—that is, they're *adhesive*. Have you ever tried to separate two glass slides stuck together by a film of water? They're difficult to separate because of the water sticking to the glass surfaces. These two forces taken together—**cohesion** and **adhesion**—account for the ability of water to rise up the roots, trunks, and branches of trees. Since this phenomenon occurs in thin vessels, it's called capillary action.
- Another remarkable property of water is its high **heat capacity**. What's heat capacity? Your textbook will give you a definition something like this: "Heat capacity is the quantity of heat required to change the temperature of a substance by 1 degree." What does that mean? In plain English, heat capacity refers to the ability of a substance to store heat. For example, when you heat up an iron kettle, it gets hot pretty quickly. Why? Because it has a *low* specific heat. It doesn't take much heat to increase the temperature of the kettle. Water, on the other hand, has a high heat capacity. You have to add a lot of heat to get an increase in temperature. Water's ability to resist temperature changes is one of the things that helps keep the temperature in our oceans fairly stable. It's also why organisms that are mainly made up of water, like us, are able to keep a constant body temperature.

So let's review the unique properties of water.

- Water is polar and can dissolve other polar substances.
- Water has cohesive and adhesive properties.
- Water has a high heat capacity.
- Water has a high surface tension.

THE ACID TEST

We just said that water is important because most reactions occur in watery solutions. Well, there's one more thing to remember: Reactions are also influenced by whether the solution in which they occur is **acidic**, **basic**, or **neutral**.

What makes a solution acidic or basic? A solution is acidic if it contains a lot of *hydrogen ions* (H^+). That is, if you dissolve an acid in water, it will release a lot of hydrogen ions. When you think about acids, you usually think of substances with a sour taste, like lemons. For example, if you squeeze a little lemon juice into a glass of water, the solution will become acidic. That's because lemons contain citric acid.

Bases, on the other hand, do not release hydrogen ions when added to water. They release a lot of *hydroxide ions* (OH^-). These solutions are said to be **alkaline**. Bases usually have a slippery consistency. Common soap, for example, is composed largely of bases.

The acidity or alkalinity of a solution can be measured using a **pH scale**. The pH scale is numbered from 1 to 14. The midpoint, 7, is considered neutral pH. The concentration of hydrogen ions in a solution will indicate whether it is acidic, basic, or neutral. If a solution contains a lot of hydrogen ions, then it will be acidic and have a low pH. Here's the trend:

An increase in H^+ ions causes a decrease in the pH.

You'll notice from the scale that stronger acids have lower pHs. If a solution has a low concentration of hydrogen ions, it will have a high pH.

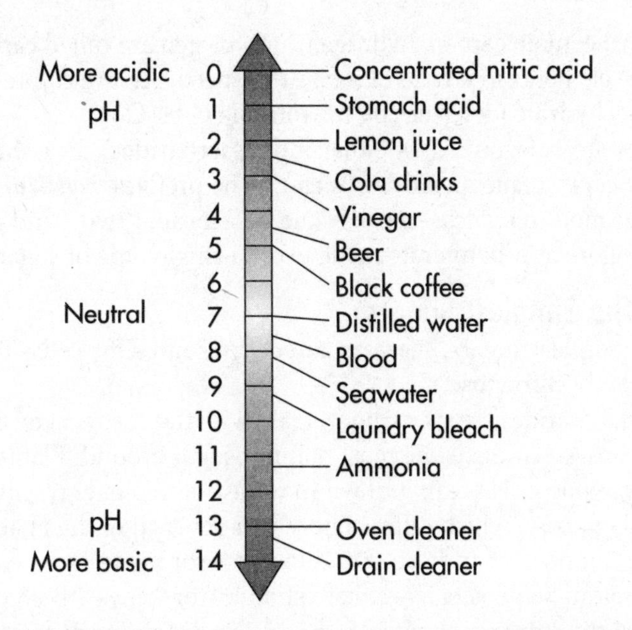

One more thing to remember: The pH scale is not a linear scale—it's logarithmic. That is, a change of *one* pH number actually represents a *tenfold* change in hydrogen ion concentration. For example, a pH of 3 is actually ten times more acidic than a pH of 4. This is also true in the reverse direction: A pH of 4 represents a tenfold decrease in acidity compared to a pH of 3.

ORGANIC MOLECULES

Now that we've discussed chemical compounds in general, let's talk about a special group of compounds. Most of the chemical compounds in living organisms contain a skeleton of carbon atoms. These molecules are known as **organic compounds**. By contrast, molecules that do not contain carbon atoms are called **inorganic compounds**. For example, salt (NaCl) is an inorganic compound.

Carbon is important for life because it is a versatile atom, meaning that it has the ability to bind with other carbons but with a number of other atoms including nitrogen, oxygen, and hydrogen. The resulting molecules are key in carrying out the activities necessary for life.

To recap:

- Organic compounds contain carbon atoms.
- Inorganic compounds do not contain carbon atoms (except carbon dioxide).

Now let's focus on four classes of organic compounds central to life on earth:

- Carbohydrates
- Proteins
- Lipids
- Nucleic acids

CARBOHYDRATES

Organic compounds that contain carbon, hydrogen, and oxygen are called **carbohydrates**. They usually contain these three elements in a ratio of 1 : 2 : 1, respectively. We can represent the proportion of elements within carbohydrate molecules by the formula $C_nH_{2n}O_n$.

Most carbohydrates are categorized as either **monosaccharides, disaccharides**, or **polysaccharides**. The term *saccharides* is a fancy word for "sugar." The prefixes *mono-*, *di-*, and *poly-* refer to the number of sugars in the molecule. *Mono-* means "one," *di-* means "two," and *poly-* means "many." A monosaccharide is therefore a carbohydrate made up of a single type of sugar molecule.

Monosaccharides: The Simplest Sugars

Monosaccharides, the simplest sugars, serve as an energy source for cells. The two most common sugars are (1) **glucose** and (2) **fructose**.

Both of these monosaccharides are six-carbon sugars with the chemical formula $C_6H_{12}O_6$. Glucose, the most abundant monosaccharide, is the most popular sugar around. Plants produce it by capturing sunlight for energy, while cells break it down to release stored energy. Glucose can come in two forms: α-glucose and β–glucose, which differ simply by a reversal of the H and OH on the first carbon. Fructose, the other monosaccharide you need to know for the test, is a common sugar in fruits.

Glucose and fructose can be depicted as either "straight" or "rings." Both of them are pretty easy to spot; just look for the six carbon molecules. Here are the two different forms:

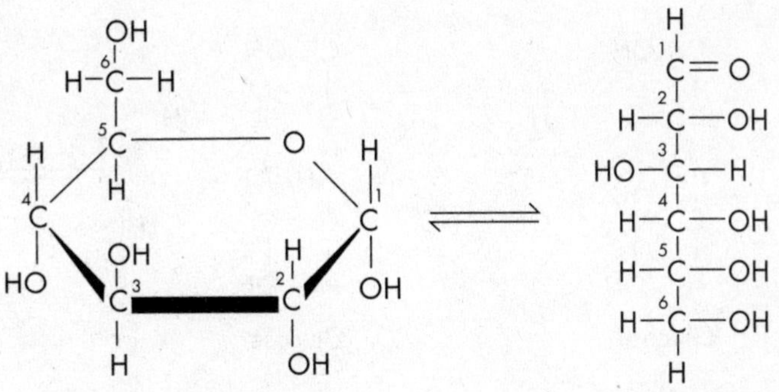

OH
6
H–C–H
5
C
H
H
O
H
4
C
1
C
OH
H
HO
3
C
2
C
OH
H
OH

Ring form of glucose

H
1
C=O
2
H–C–OH
3
HO–C–H
4
H–C–OH
5
H–C–OH
6
H–C–OH
H

Straight-chain form of glucose

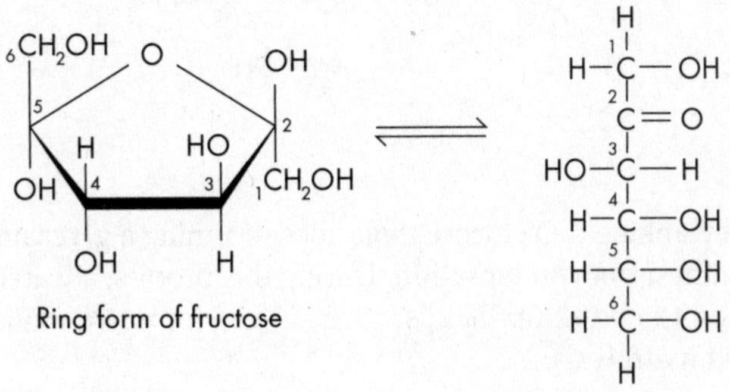

$_6CH_2OH$
O
OH
5
2
H
HO
OH
4
3
$_1CH_2OH$
OH
H

Ring form of fructose

H
1
H–C–OH
2
C=O
3
HO–C–H
4
H–C–OH
5
H–C–OH
6
H–C–OH
H

Straight-chain form of fructose

Disaccharides

What happens when two monosaccharides are brought together? The hydrogen (–H) from one sugar molecule combines with the hydroxyl group (–OH) of another sugar molecule. What do H and OH add up to? Water (H_2O)! So a water molecule is removed from the two sugars. The two molecules of monosaccharides are chemically linked and form a disaccharide.

Maltose is an example of a disaccharide:

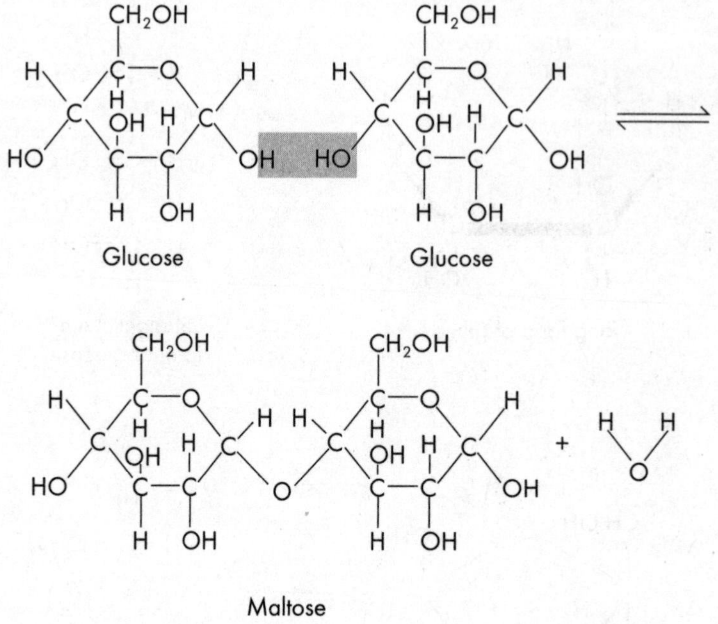

Glucose Glucose

Maltose

Maltose is formed by linking two glucose molecules—forming a **glycosidic bond**. This process is called **dehydration synthesis**, or **condensation**. During this process, a water molecule is lost.

Now what if you want to break up the disaccharide and form two monosaccharides again? Just add water. That's called **hydrolysis**.

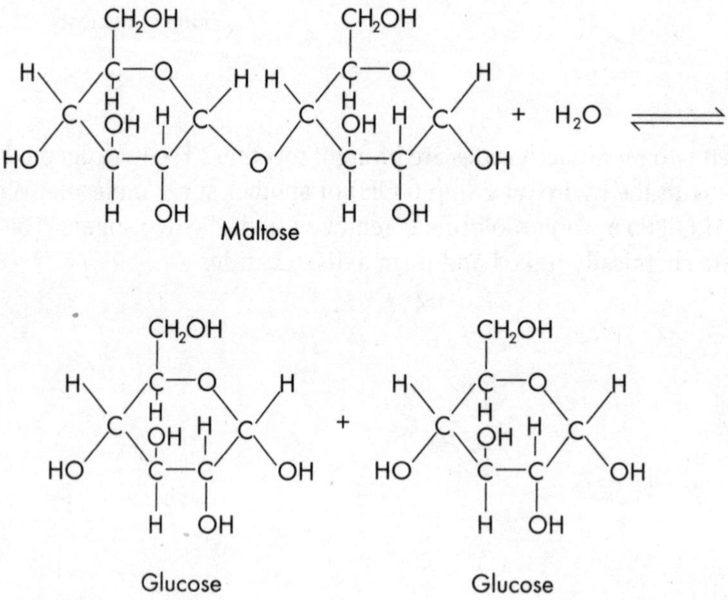

Maltose + H_2O

Glucose + Glucose

Polysaccharides

Polysaccharides are made up of many repeated units of monosaccharides. Therefore, a polysaccharide is a kind of **polymer**, a molecule with repeating subunits of the same general type. The most common polysaccharides you'll need to know for the test are **starch**, **cellulose**, and **glycogen**. Polysaccharides are often storage forms of sugar or structural components of cells. For instance, animals store glucose molecules in the form of glycogen in the liver and muscle cells. Plants "stockpile" α–glucose in the form of starch in structures called **plastids**. Cellulose, on the other hand, made up of β–glucose, is a major part of the cell wall in plants. Its function is to lend structural support. Chitin, a polymer of β–glucose molecules, serves as a structural molecule in the walls of fungus and in the exoskeletons of arthropods.

Here's an AP question that may come up on the test: Why can't humans digest cellulose? The glycosidic bond in polymers that have α–glucose can easily be broken down by humans but glycoside bonds in polymers containing β–glucose polymers, such as cellulose, cannot. This is because the bonds joining the glucose subunits in cellulose are different than those in starch. Starch is composed of α–glucose subunits held together by 1–4 glycoside linkages while cellulose contain β–glucose subunits held together by 1–4 linkages.

PROTEINS

Amino acids are organic molecules that serve as the building blocks of proteins. They contain carbon, hydrogen, oxygen, and nitrogen atoms. There are 20 different amino acids commonly found in proteins. Fortunately, you don't have to memorize the 20 amino acids. But you do have to remember that every amino acid has four important parts: an **amino group** (–NH_2), a **carboxyl group** (–COOH), a hydrogen, and an **R group**.

Here's a typical amino acid:

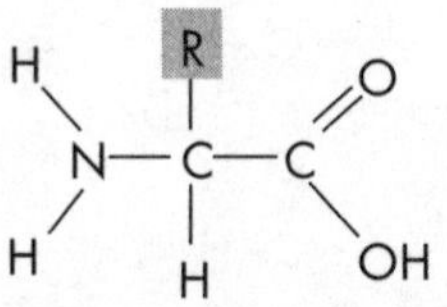

Amino acids differ only in the R group, which is also called the **side chain**. The R group associated with an amino acid could be as simple as a hydrogen atom (as in the amino acid *glycine*) or as complex as a carbon skeleton (as in the amino acid *arginine*).

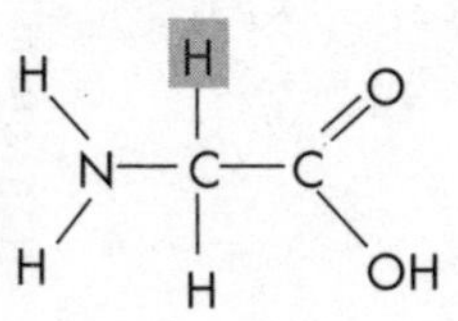

Glycine

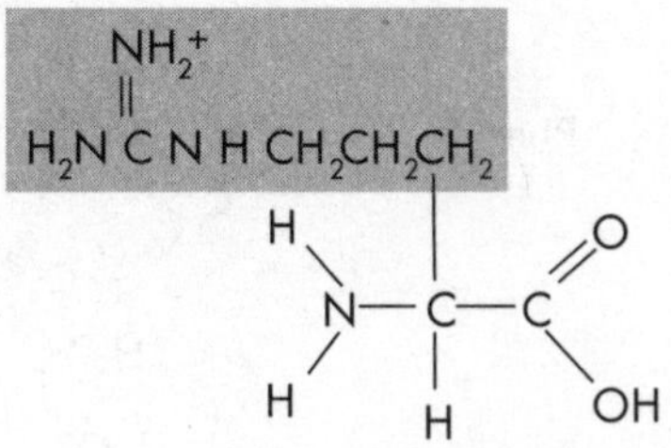

Arginine

Since ETS will probably test you on chemical diagrams, it's a good idea to identify the functional groups of the different structures. **Functional groups** are the distinctive groups of atoms that play a large role in determining the chemical behavior of the compound they are a part of. For example, an organic acid has a functional group, the carboxyl group (–COOH) that releases hydrogen ions in water. This makes the solution acidic. When it comes to spotting an amino acid, simply keep an eye out for the amino group (NH_2), then look for the carboxyl molecule (COOH). The most common functional groups of organic compounds are listed on the next page.

Functional Group	Structural Formula	Molecular Formula
Amino	$-\overset{H}{\overset{\mid}{N}}-H$	$-NH_2$
Alkyl	$-\underset{H}{\overset{H}{C}}-\underset{H}{\overset{H}{C}}-\cdots-\underset{H}{\overset{H}{C}}-H$	$-C_nH_{2n+1}$
Methyl	$-\underset{H}{\overset{H}{C}}-H$	$-CH_3$
Ethyl	$-\underset{H}{\overset{H}{C}}-\underset{H}{\overset{H}{C}}-H$	$-C_2H_5$
Propyl	$-\underset{H}{\overset{H}{C}}-\underset{H}{\overset{H}{C}}-\underset{H}{\overset{H}{C}}-H$	$-C_3H_7$
Carboxyl	$-\overset{O}{\overset{\Vert}{C}}-O-H$	$-COOH$
Hydroxyl	$-O-H$	$-OH$
Aldehyde	$-\overset{O}{\overset{\Vert}{C}}-H$	$-CHO$
Keto (carbonyl)	$-\overset{O}{\overset{\Vert}{C}}-$	$-CO$
Sulfhydryl (thiol)	$-S-H$	$-SH$
Phenyl	ring: H–C–C–H (top), C, C–H, C=C with H, H (bottom)	$-C_6H_5$
Phosphate	$-O-\underset{O}{\overset{O}{\overset{\Vert}{P}}}-O$	$-PO_4$

Polypeptides

When two amino acids join they form a **dipeptide**. The carboxyl group of one amino acid combines with the amino group of another amino acid. Here's an example:

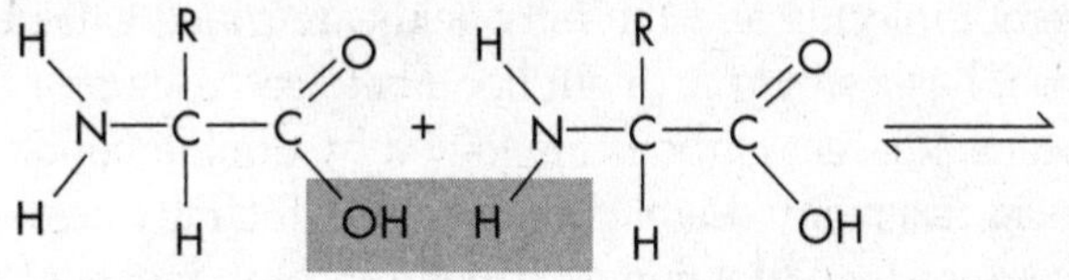

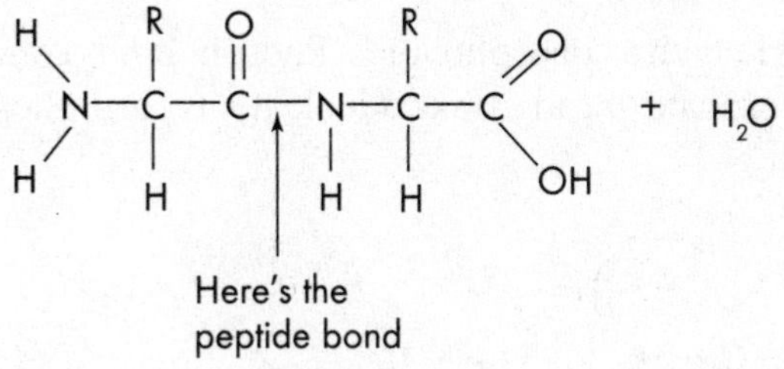

Here's the peptide bond

This is the same process we saw earlier: dehydration synthesis. Why? Because a water molecule is removed to form a bond. By the way, the bond between two amino acids has a special name—a **peptide bond**. If a group of amino acids are joined together in a "string," the resulting organic compound is called a **polypeptide**. Once a polypeptide chain twists and folds on itself, it forms a three-dimensional structure called a **protein**.

LIPIDS

Like carbohydrates, **lipids** consist of carbon, hydrogen, and oxygen atoms, but not in the 1:2:1 ratio typical of carbohydrates. The most common examples of lipids are **fats**, **oils**, **phospholipids**, and **steroids**. Let's talk about the simple lipids—**neutral fats**. A typical fat consists of three fatty acids and one molecule of **glycerol**. If you see the word *triglyceride* on the test, it's just a fancy word for "fat." Let's take a look:

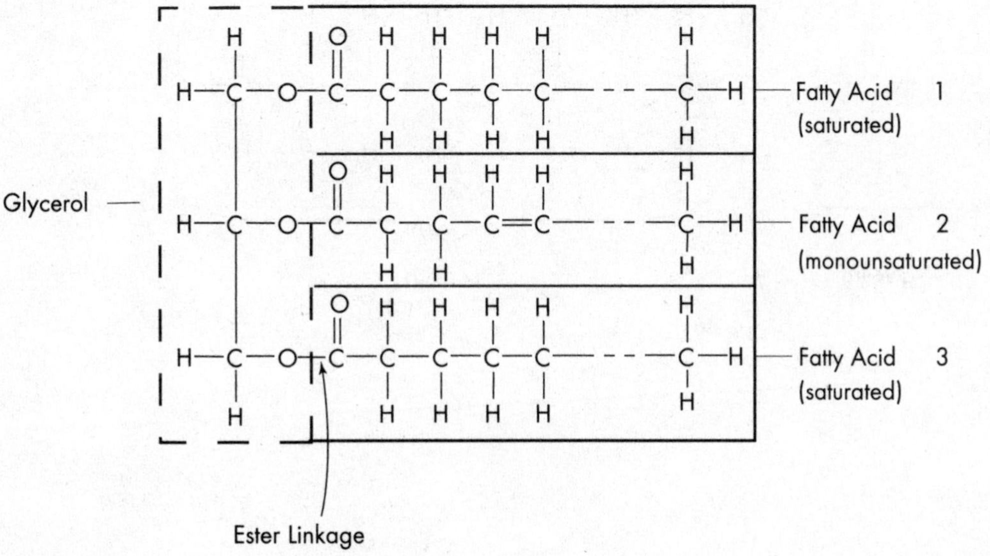

Glycerol

Fatty Acid 1 (saturated)

Fatty Acid 2 (monounsaturated)

Fatty Acid 3 (saturated)

Ester Linkage

To make a triglyceride, each of the carboxyl groups (–COOH) of the three fatty acids must react with one of the three hydroxyl groups (–OH) of the glycerol molecule. This happens by the removal of a water molecule. So, the creation of a fat requires the removal of three molecules of water. Once again, what have we got? You probably already guessed it—dehydration synthesis! The linkage now formed between the glycerol molecule and the fatty acids are called **ester linkage**. A fatty acid can be **saturated**, which means it has a single covalent bond between each pair of carbon atoms or it can be **unsaturated**, which means adjacent carbons are joined by double bonds instead of single bonds. A **polyunsaturated** fatty acid has many double bonds within the fatty acid.

Lipids are important because they function as structural components of cell membranes, sources of insulation, and a means of energy storage.

Phospholipids

Another special class of lipids is known as phospholipids. Phospholipids contain two fatty acid "tails" and one negatively charged phosphate "head." Take a look at a typical phospholipid:

```
        H         O
        |         ||
    H — C — O — P — O — R   } phosphate
        |         |
        |         O
        |
        |       H   H   H   H   H   H   H   H   H   H
        |       |   |   |   |   |   |   |   |   |   |
    H — C — O — C — C — C — C — C — C — C — C — C — C — ... etc.
        |       |   |   |   |   |   |   |   |   |   |
        |       H   H   H   H   H   H   H   H   H   H
        |
        |       H   H   H   H   H   H   H   H   H   H
        |       |   |   |   |   |   |   |   |   |   |
    H — C — O — C — C — C — C — C — C — C — C — C — C — ... etc.
        |       |   |   |   |   |   |   |   |   |   |
        H       H   H   H   H   H   H   H   H   H   H

     glycerol                  fatty acids
```

Phospholipids are extremely important, mainly because of some unique properties they possess, particularly with regard to water.

Interestingly enough, the two fatty acid tails are **hydrophobic** ("water–hating"). In other words, just like oil and vinegar, fatty acids and water don't mix. The reason for this is that fatty acid tails are nonpolar, and nonpolar substances don't mix well with polar ones, such as water.

On the other hand, the phosphate "head" of the lipid is **hydrophilic** ("water-loving"), meaning that it does mix well with water. Why? It carries a negative charge, and this charge draws it to the positively charged end of a water molecule. A molecule is **amphipathic** if it has both a hydrophilic region and a hydrophobic region. A phospholipid is amphipathic.

This arrangement of the fatty acid tails and the phosphate group head provides phospholipids with a unique shape. The two fatty acid chains orient themselves away from water, while the phosphate portion orients itself toward the water. Keep these properties in mind. We'll see later how this orientation of phospholipids in water relates to the structure and function of cell membranes.

One class of lipids is known as steroids. All steroids have a basic structure of four linked carbon rings. This category includes cholesterol, vitamin D, and a variety of hormones. Take a look at a typical steroid:

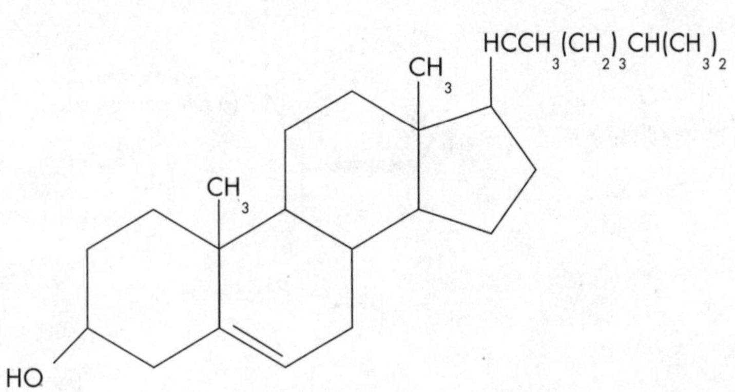

$HCCH_3(CH_2)_3CH(CH_3)_2$

CH_3

CH_3

HO

NUCLEIC ACIDS

The fourth class of organic compounds are the **nucleic acids**. Like proteins, nucleic acids contain carbon, hydrogen, oxygen, and nitrogen, but nucleic acids also contain phosphorus. Nucleic acids are molecules that are made up of simple units called **nucleotides**. For the AP Biology Exam, you'll need to know about two kinds of nucleic acids: **deoxyribonucleic acid (DNA)** and **ribonucleic acid (RNA)**.

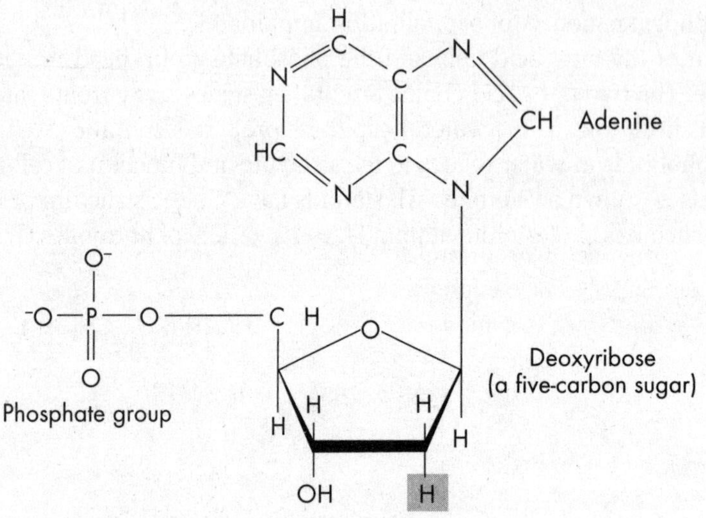

DNA is important because it contains genes, the hereditary "blueprints" of all life. RNA is important because it's essential for protein synthesis. We'll discuss DNA and RNA in greater detail when we discuss heredity.

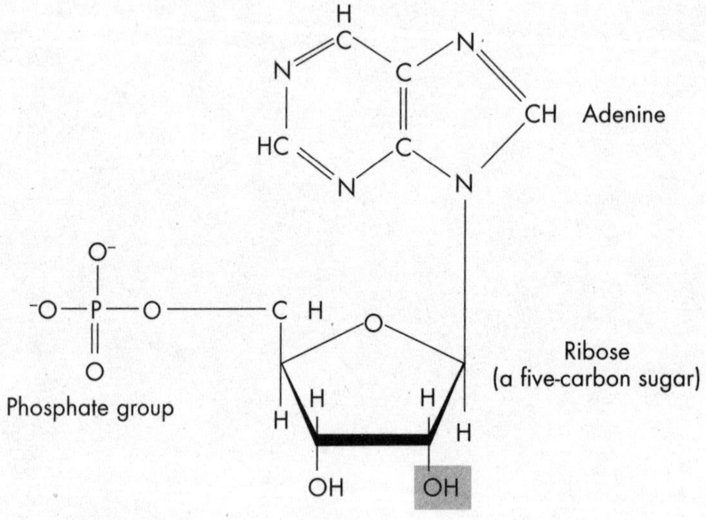

THE HETEROTROPH HYPOTHESIS

We've just seen the organic compounds that are essential for life. But where did they come from in the first place? This is still a hotly debated topic among scientists. Most scientists believe that the earliest precursors of life arose from nonliving matter (basically gases) in the primitive oceans of the earth. But this theory didn't take shape until the 1920s. Two scientists, **Alexander Oparin** and **J. B. S. Haldane**, proposed that the primitive atmosphere contained the following gases: methane (CH_4), ammonia (NH_3), hydrogen (H_2), and water (H_2O). Interestingly enough, there was almost no free oxygen (O_2) in this early atmosphere. They believed that these gases collided, producing chemical reactions that eventually led to the organic molecules we know today.

This theory didn't receive any substantial support until 1953. In that year, **Stanley Miller** and **Harold Urey** simulated the conditions of primitive Earth in a laboratory. They put the gases theorized to be abundant in the early atmosphere into a flask, struck them with electrical charges in order to mimic lightning, and organic compounds similar to amino acids appeared!

But how do we make the leap from simple organic molecules to more complex compounds and life as we know it? Since no one was around to witness the process, no one knows for sure how (or when) it occurred. Complex organic compounds (such as proteins) must have formed via dehydration synthesis. Simple cells then used organic molecules as their source of food. Over time, simple cells evolved into complex cells.

Now let's throw in a few new terms. Living organisms that rely on organic molecules for food are called **heterotrophs** (or consumers). For example, we're heterotrophs. Eventually, some organisms (such as plants) found a way to make their own food. These organisms are called **autotrophs** (or producers). Early autotrophs are responsible for Earth's oxygenated atmosphere.

KEY WORDS

elements
oxygen
carbon
hydrogen
nitrogen
trace elements
atom
protons
neutrons
electrons
nucleus
isotopes
compound
chemical reaction
chemical bond
ionic bond
covalent bond
nonpolar covalent
polar covalent
polar
hydrogen bonds
cohesion
adhesion
surface tension
capillary action
heat capacity
acidic
basic
neutral
alkaline
pH scale
organic compounds
inorganic compounds
carbohydrates
monosaccharides
disaccharides
polysaccharides
glucose
fructose
glycosidic bond
dehydration synthesis
(or condensation)
hydrolysis
polymer
starch
cellulose
glycogen
plastids
amino acids
amino group
carboxyl group
R group
side chain
functional groups
dipeptide
peptide bond
polypeptide
protein
lipids
fats
oils
phospholipids
steroids
neutral fats
glycerol
ester linkage
saturated
unsaturated
polyunsaturated
hydrophobic
hydrophilic
amphipathic
nucleic acids
nucleotides
deoxyribonucleic acid (DNA)
ribonucleic acid (RNA)
Oparin and Haldane
Stanley Miller
Harold Urey
heterotrophs (or consumers)
autotrophs (or producers)

CHAPTER 3 REVIEW QUESTIONS

Answers can be found in Chapter 15.

1. Water is a critical component of life due to its unique structural and chemical properties. Which of the following does NOT describe a way that the exceptional characteristics of water are used in nature for life?

   (A) The high heat capacity of water prevents lakes and streams from rapidly changing temperature and freezing completely solid in the winter.

   (B) The high surface tension and cohesiveness of water facilitates capillary action in plants.

   (C) The low polarity of water prevents dissolution of cells and compounds.

   (D) The high intermolecular forces of water such as hydrogen bonding results in a boiling point which exceeds the tolerance of most life on the planet.

2. The intracellular pH of human cells is approximately 7.4. Yet, the pH within the lumen (inside) of the human stomach averages 1.5. Which of the following accurately describes the difference between the acidity of the cellular and gastric pH?

   (A) Gastric juices contain approximately 5 times more H^+ ions than the intracellular cytoplasm of cells and are more acidic.

   (B) Gastric juices contain approximately 100,000-fold more H^+ ions than the intracellular cytoplasm of cells and are more acidic.

   (C) The intracellular cytoplasm of cells contain approximately 5 times more H^+ ions than gastric juices and is more acidic.

   (D) The intracellular cytoplasm of cells contains approximately 100,000-fold more H^+ ions than gastric juices and is more acidic.

3. Amino acids are the basic molecular units, which comprise proteins. All life on the planet forms proteins by forming chains of amino acids. Which labeled component of the amino acid structure of phenylalanine shown below will vary from amino acid to amino acid?

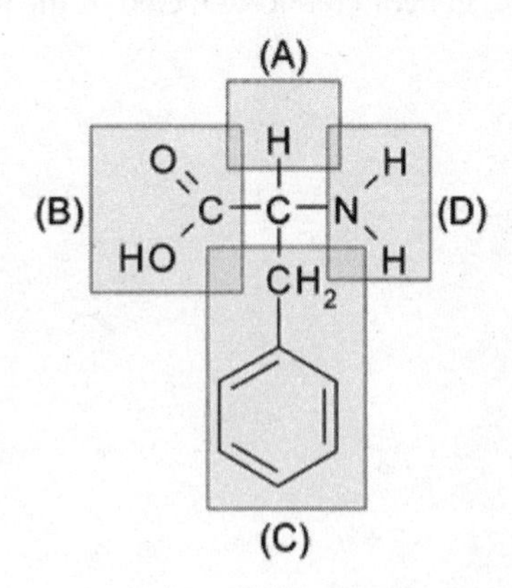

Questions 4 - 6 refer to the following paragraph and diagram.

In 1953, Stanley Miller and Harold Urey performed an experiment at the University of Chicago to test the hypothesis that the conditions of the early Earth would have favored the formation of larger more complex organic molecules from basic precursors. The experiment as shown below consisted of sealing basic organic chemicals (representing the atmosphere of the primitive Earth) in a flask, which was exposed to electric sparks (to simulate lightning) and water vapor.

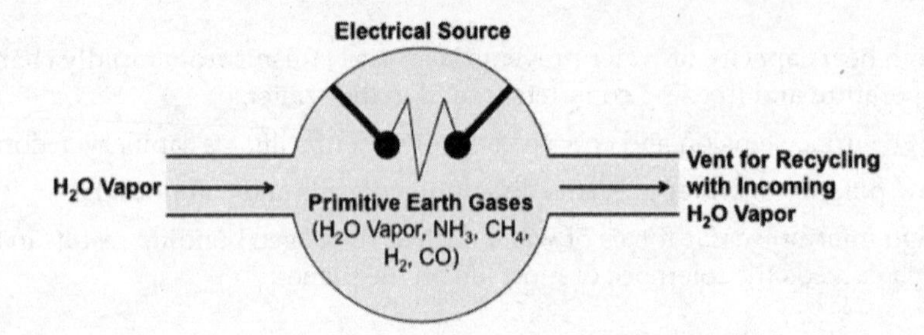

After one day of exposure, the mixture in the flask had turned pink in color and later analysis showed that at least 10% of the carbon had been transformed into simple and complex organic compounds including at least 11 different amino acids and some basic sugars. No nucleic acids were detected in the mixture.

4. Which of the following contradicts the hypothesis of the experiment that life may have arisen from the formation of complex molecules in the conditions of the primitive Earth?

(A) Complex carbon-based compounds were generated after only one day of exposure to simulated primitive Earth conditions.

(B) Nucleic acid compounds such as DNA and RNA were not detected in the mixture during the experiment.

(C) Over half of known amino acids involved in life were detected in the mixture during the experiment.

(D) Basic sugar molecules were generated and detected in the mixture during the experiment.

5. Some amino acids such as cysteine (shown below) and methionine could not be formed in this experiment. Which of the following best explains why these molecules could not be detected?

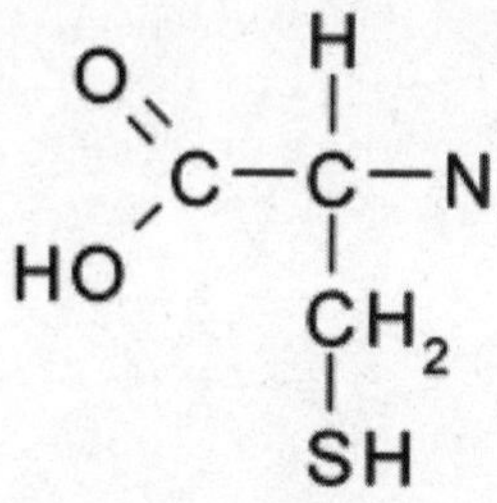

(A) The chemical reactions necessary to create amino acids such as cysteine and methionine requires more energy than the simulated lightning provided in the experiment.
(B) The chemical reactions necessary to create amino acids such as cysteine and methionine require enzymes for catalysis to occur, which were not included in the experiment.
(C) Sulfur-based compounds were not included in the experiment.
(D) Nitrogen-based compounds were not included in the experiment.

6. A scientist believes that the Miller-Urey experiment was unable to capture the remaining amino acids and the formation of nucleic acids was due to the absence of critical chemical substrates that would have existed on the primordial Earth due to volcanism. All of the following basic compounds which are associated with volcanism should be included by the scientist EXCEPT:

(A) H_2S gas
(B) SiO_2 (silica)
(C) SO_2
(D) H_3PO_4 (phosphoric acid)

Cells

LIVING THINGS

All living things are composed of **cells**. According to the cell theory, the cell is life's basic unit of structure and function. This simply means that the cell is the smallest unit of living material that can carry out all the activities necessary for life.

Cells are studied using different types of microscopes. **Light microscopes**, also known as compound microscopes, are used to study stained or living cells. They can magnify the size of an organism up to 1,000 times. **Electron microscopes** are used to study detailed structures of a cell that cannot be easily seen or observed by light microscopy. They are capable of resolving structures as small as a few nanometers in length such as individual virus particles or the pores on the surface of the nucleus.

WHAT ARE THE DIFFERENT TYPES OF CELLS?

For centuries, scientists have known about cells. However, it wasn't until the development of the electron microscope that scientists were able to figure out what cells do. We now know that there are two distinct types of cells: **eukaryotic cells** and **prokaryotic cells**. A eukaryotic cell contains a membrane-bound structure called a nucleus and **cytoplasm**, filled with tiny structures called **organelles** (literally "little organs"). Examples of eukaryotic cells are *fungi, protists, plant cells,* and *animal cells.*

A prokaryotic cell, which is a lot smaller than a eukaryotic cell, lacks both a nucleus and membrane-bound organelles. Bacteria, for example, are prokaryotic cells. The genetic material in a prokaryote is one continuous, circular DNA molecule that lies free in the cell in an area called the **nucleoid**. In addition to a plasma membrane, most prokaryotes have a cell wall composed of peptidoglycan. Prokaryotes may also have ribosomes (although smaller than those found in eukaryotic cells) as well as one or more **flagella**, which are long projections used for motility (movement).

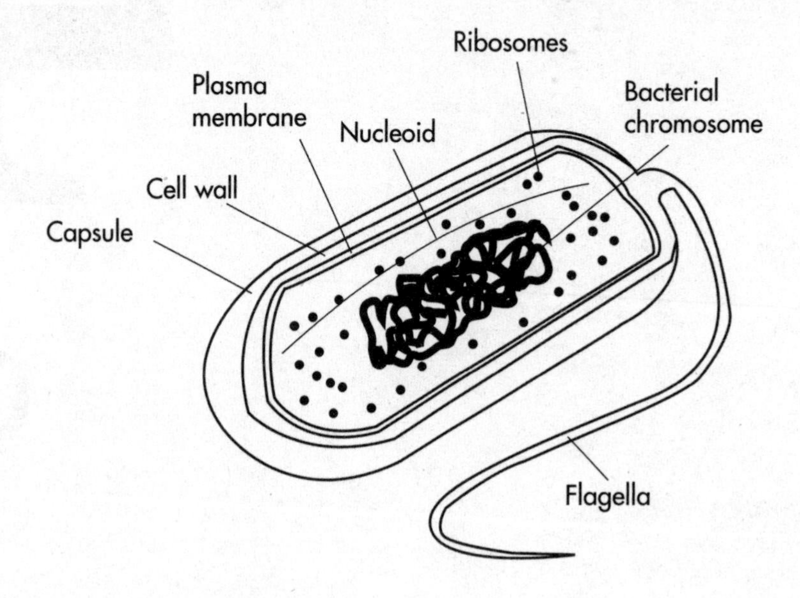

ORGANELLES

A eukaryotic cell is like a microscopic factory. It's filled with organelles, each of which has its own special tasks. Let's take a tour of a eukaryotic cell and focus on the structure and function of each organelle. Here's a picture of a typical animal cell and its principal organelles:

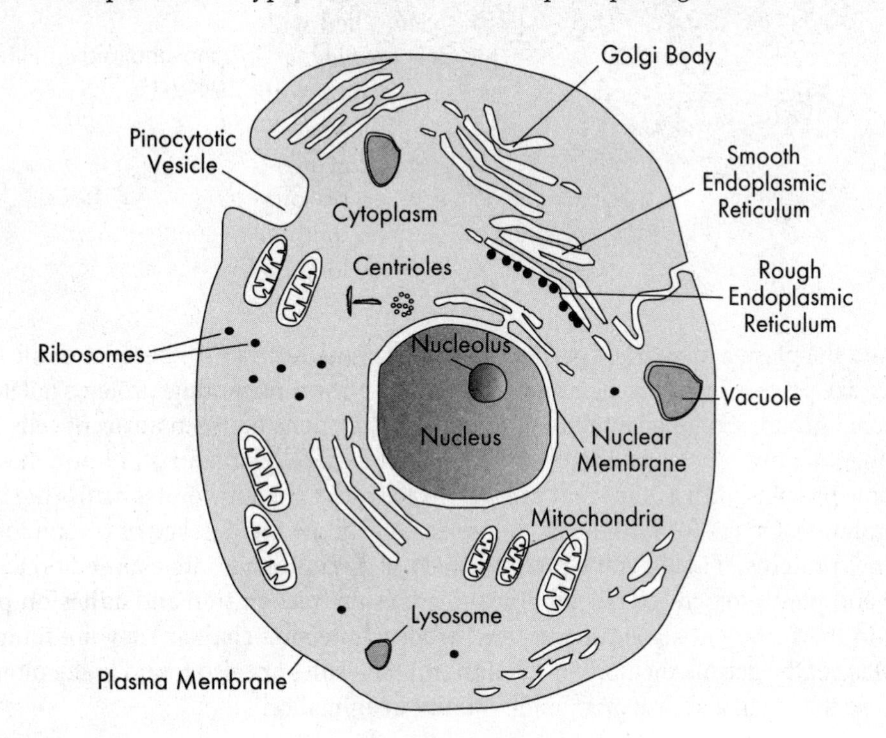

Plasma Membrane

The cell has an outer envelope known as the **plasma membrane**. Although the plasma membrane appears to be a simple, thin layer surrounding the cell, it's actually a complex double-layered structure made up of phospholipids and proteins. The *hydrophobic* fatty acid tails face inward and the *hydrophilic* phosphate heads face outward.

The plasma membrane is important because it regulates the movement of substances into and out of the cell. The membrane itself is semipermeable, meaning that only certain substances, namely small hydrophobic molecules, pass through it unaided. Many proteins are associated with the cell membrane. Some of these proteins are loosely associated with the lipid bilayer (**peripheral proteins**). They are located on the inner or outer surface of the membrane. Others are firmly bound to the plasma membrane (**integral proteins**). These proteins are amphipathic, which means that their hydrophilic regions extend out of the cell or into the cytoplasm while their hydrophobic regions interact with the tails of the membrane phospholipids. Some integral proteins do not extend all the way through the membrane (**transmembrane proteins**). This arrangement of phospholipids and proteins is known as the **fluid-mosaic model**.

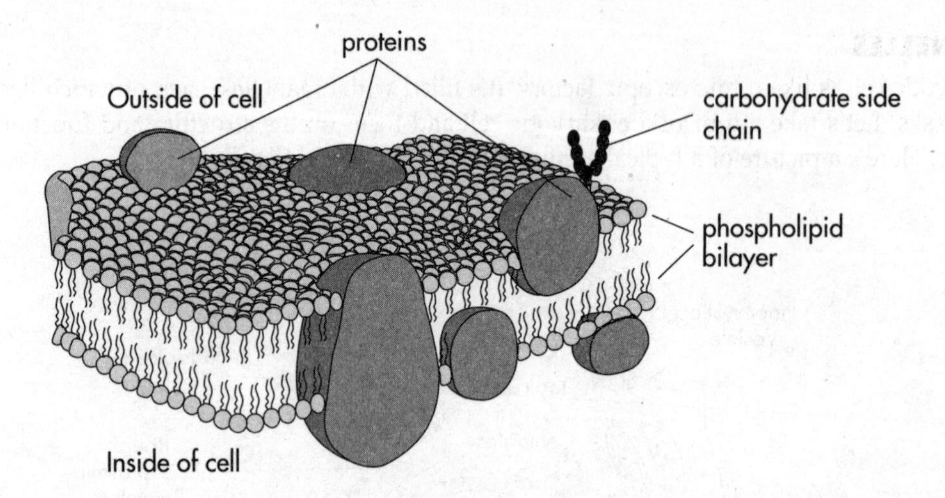

Why should the plasma membrane need so many different proteins? It's because of the number of activities that take place in or on the membrane. Generally, plasma membrane proteins fall into several broad functional groups. Some membrane proteins form junctions between adjacent cells (**adhesion proteins**). Others serve as docking sites for proteins of the extracellular matrix or hormones (**receptor proteins**). Some proteins form pumps that use ATP to actively transport solutes across the membrane (**transport proteins**). Others form channels that selectively allow the passage of certain ions or molecules (**channel proteins**). Finally, some proteins, such as glycoproteins, are exposed on the extracellular surface and play a role in cell recognition and adhesion (**recognition and adhesion proteins**).

Attached to the surface of some proteins are **carbohydrate side chains.** They are found only on the outer surface of the plasma membrane. **Cholesterol** molecules are also found in the **phospholipid bilayer** because they help stabilize membrane fluidity in animal cells.

The Nucleus

The nucleus, which is usually the largest organelle, is the control center of the cell. The nucleus not only directs what goes on in the cell, it is also responsible for the cell's ability to reproduce. It's the home of the hereditary information—DNA—which is organized into large structures called **chromosomes**. The most visible structure within the nucleus is the **nucleolus**, which is where rRNA is made and ribosomes are assembled.

Ribosomes

The **ribosomes** are the sites of protein synthesis. Their job is to manufacture all the proteins required by the cell or secreted by the cell. Ribosomes are round structures composed of RNA and two subunits of proteins. They can be either free floating in the cell or attached to another structure called the **endoplasmic reticulum (ER)**.

Endoplasmic Reticulum (ER)

The endoplasmic reticulum (ER) is a continuous channel that extends into many regions of the cytoplasm. The region of the ER that is "studded" with ribosomes is called the *rough ER* (RER). Proteins generated in the rough ER are trafficked to or across the plasma membrane. The region of the ER that lacks ribosomes is called the *smooth ER* (SER). The smooth ER makes lipids, hormones, and steroids and breaks down toxic chemicals.

Golgi Bodies

The **Golgi bodies**, which look like stacks of flattened sacs, also participate in the processing of proteins. Once the ribosomes on the rough ER have completed synthesizing proteins, the Golgi bodies modify, process, and sort the products. They're the packaging and distribution centers for materials destined to be sent out of the cell. They package the final products in little sacs called **vesicles**, which carry the products to the plasma membrane. Golgi bodies are also involved in the production of lysosomes.

Mitochondria: The Powerhouses of the Cell

Another important organelle is the *mitochondrion*. The **mitochondria** are often referred to as the "powerhouses" of the cell. They're power stations responsible for converting the energy from organic molecules into useful energy for the cell. The energy molecule in the cell is **adenosine triphosphate (ATP)**.

The mitochondrion is usually an easy organelle to recognize because it has a unique oblong shape and a characteristic double membrane consisting of an inner portion and an outer portion. The inner mitochondrial membrane forms folds known as *cristae*. As we'll see later, most of the production of ATP is done on the cristae.

Since mitochondria are the cell's powerhouses, you're most likely to find more of them in cells that require a lot of energy. Muscle cells, for example, are rich in mitochondria.

Lysosomes

Throughout the cell are small, membrane-bound structures called **lysosomes**. These tiny sacs carry digestive enzymes, which they use to break down old, worn-out organelles, debris, or large ingested particles. The lysosomes make up the cell's cleanup crew, helping to keep the cytoplasm clear of unwanted flotsam.

Centrioles

The **centrioles** are small, paired, cylindrical structures that are found within **microtubule organizing centers (MTOCs)**. Centrioles are most active during cellular division. When a cell is ready to divide, the centrioles produce *microtubules*, which pull the replicated chromosomes apart and move them to opposite ends of the cell. Although centrioles are common in animal cells, they are not found in plant cells.

Vacuoles

In Latin, the term *vacuole* means "empty cavity." But **vacuoles** are far from empty. They are fluid-filled sacs that store water, food, wastes, salts, or pigments.

Peroxisomes

Peroxisomes are organelles that detoxify various substances, producing hydrogen peroxide as a byproduct. They also contain enzymes that break down hydrogen peroxide (H_2O_2) into oxygen and water. In animals, they are common in the liver and kidney cells.

Cytoskeleton

Have you ever wondered what actually holds the cell together and enables it to keep its shape? The shape of a cell is determined by a network of fibers called the **cytoskeleton**. The most important fibers you'll need to know are **microtubules** and **microfilaments**.

Microtubules, which are made up of the protein **tubulin**, participate in cellular division and movement. These small fibers are an integral part of three structures: *centrioles*, *cilia*, and *flagella*. We've already mentioned that centrioles help chromosomes separate during cell division. **Cilia** and flagella are thread-like structures best known for their locomotive properties in single-celled organisms. The beating motion of cilia and flagella structures propels these organisms through their watery environments.

The two classic examples of organisms with these structures are the **euglena**, which gets about using its whiplike flagellum, and the **paramecium**, which is covered in cilia. The rhythmic beating of the paramecium's cilia enables it to motor about in waterways, ponds, and microscope slides in your biology lab. You've probably already checked these out in lab, but here's what they look like:

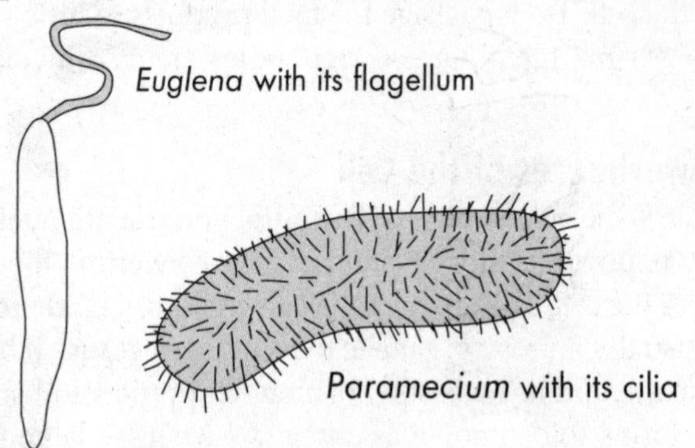

Though we usually associate such structures with microscopic organisms, they aren't the only ones with cilia and flagella. As you probably know, these structures are also found in certain human cells. For example, the cells lining your respiratory tract possess cilia that sweep constantly back and forth (beating up to 20 times per second), helping to keep dust and unwanted debris from descending into your lungs. And every sperm cell has a flagellum, which enables it to swim through the female reproductive organs to fertilize the waiting ovum.

Microfilaments, like microtubules, are important for movement. These thin, rodlike structures are composed of the protein actin, they are involved in cell mobility, and play a central role in muscle contraction.

Plant Cells Versus Animal Cells

Plant cells contain most of the same organelles and structures seen in animal cells, with several key exceptions. Plant cells, unlike animal cells, have a protective outer covering called the **cell wall** (made of cellulose). A cell wall is a rigid layer just outside of the plasma membrane that provides support for the cell. It is found in plants, protists, fungi, and bacteria. (In fungi, the cell wall is usually made of **chitin**, a modified polysaccharide. Chitin is also a principle component of an arthropod's exoskeleton.) In addition, plant cells possess **chloroplasts** (organelles involved in photosynthesis). Chloroplasts contain chlorophyll, the light-capturing pigment that gives plants their characteristic green color. Another difference between plant and animal cells is that most of the cytoplasm within a plant cell is usually taken up by a large vacuole that crowds the other organelles. In mature plants, this vacuole contains the **cell sap**. Plant cells also differ from animal cells in that plant cells do not contain centrioles.

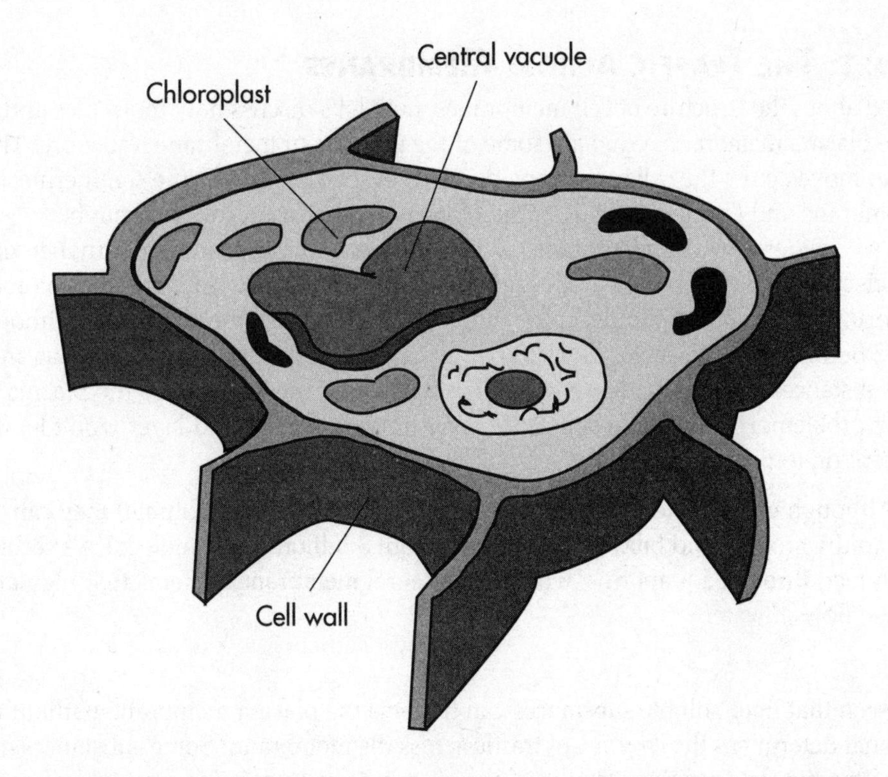

To help you remember the differences among prokaryotes, plant cells, and animal cells, we've put together this simple table. Make sure you learn it! ETS is bound to ask you which cells contain which structures:

STRUCTURAL CHARACTERISTICS OF DIFFERENT CELL TYPES			
Structure	**Prokaryote**	**Plant Cell**	**Animal Cell**
Cell Wall	Yes	Yes	No
Plasma Membrane	Yes	Yes	Yes
Organelles	No	Yes	Yes
Nucleus	No	Yes	Yes
Centrioles	No	No	Yes
Ribosomes	Yes	Yes	Yes

Why do we need to know about the structure of cells? Because biological structure is often closely related to function. (Watch out for this connection: It's a favorite theme for the AP Biology Exam.) And, more important, because ETS likes to test you on it!

TRANSPORT: THE TRAFFIC ACROSS MEMBRANES

We've talked about the structure of cell membranes, now let's discuss how molecules and fluids pass through the plasma membrane. What are some of the patterns of membrane transport? The ability of molecules to move across the cell membrane depends on two things: (1) the semipermeability of the plasma membrane and (2) the size and charge of particles that want to get through.

First let's consider how cell membranes work. For a cell to maintain its internal environment, it has to be selective in the materials it allows to cross its membrane. Since the plasma membrane is composed primarily of phospholipids, lipid-soluble substances cross the membrane without any resistance. Why? Because "like dissolves like." Generally speaking, the lipid membrane has an open-door policy for substances that are made up of lipids. These substances can cross the plasma membrane without any problem. However, if a substance is hydrophilic, the bilipid layer won't let it in.

One exception to the rule is water:

> Although water molecules are polar (and therefore not lipid-soluble) they can rapidly cross a lipid bilayer (at a rate of about 3 billion water molecules a second, in fact) through aquaporins, which are integral membrane proteins that regulate the flow of water.

Diffusion

We've just seen that lipid-soluble substances can traverse the plasma membrane without much difficulty. But what determines the *direction* of traffic across the membrane? Some substances move across a membrane by **simple diffusion**. That is, if there's a high concentration of a substance outside the cell and a low concentration inside the cell, the substance will move into the cell. In other words, the substance moves *down* a concentration gradient:

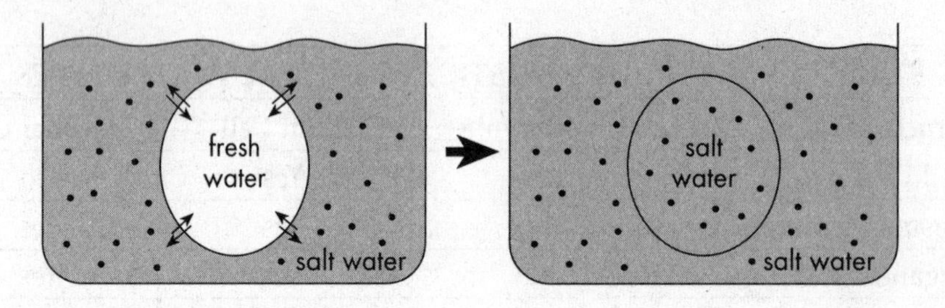

It's like riding a bicycle downhill. The bike "goes with the natural flow." Another name for this type of transport is **passive transport**. Here's one more thing you must remember:

Simple diffusion does not require energy.

A special kind of diffusion that involves the movement of water is called **osmosis**.

Facilitated Transport

How do *lipid-insoluble* substances, which are dissolved in the fluid on either side of the cell membrane, get in and out of the cell? These dissolved substances, or **solutes**, rely on the proteins embedded in the plasma membrane. Special proteins—called **channel proteins**—can help lipid-insoluble substances get in or out:

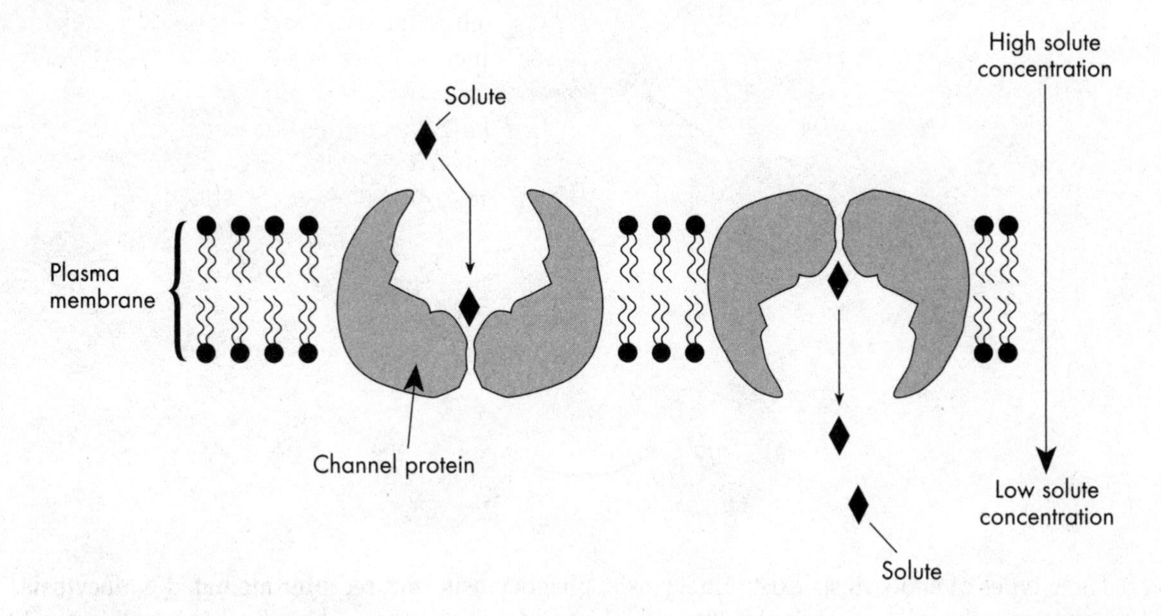

These proteins pick up the substance from one side of the membrane and carry it across to the other. This type of transport is known as **facilitated transport**, or facilitated diffusion. Facilitated diffusion is just like simple diffusion in one respect: The flow of the substance is *down* the concentration gradient. Therefore, it doesn't require any energy.

Active Transport

Suppose a substance wants to move in the opposite direction—from a region of lower concentration to a region of higher concentration. A transport protein can help usher the substance across the plasma membrane, but it's going to need energy to accomplish this. This time it's like riding a bicycle uphill. Compared with riding downhill, riding uphill takes a lot more work. Movement against the natural flow is called **active transport**.

But where does the protein get this energy? Some proteins in the plasma membrane are powered by ATP. The best example of active transport is a special protein called the **sodium-potassium pump**. It ushers out sodium ions (Na^+) and brings in potassium ions (K^+) across the cell membrane. These pumps depend on ATP to get ions across that would otherwise remain in regions of higher concentration. Where do we usually find these proteins? In vertebrates, they're found in neurons and skeletal muscle fibers.

We've now seen that small substances can cross the cell membrane by:

- Simple diffusion
- Facilitated transport
- Active transport

Endocytosis

When the particles that want to enter a cell are just too large, the cell uses a portion of the cell membrane to engulf the substance. The cell membrane forms a pocket, pinches in, and eventually forms either a vacuole or a vesicle. This is called **endocytosis.**

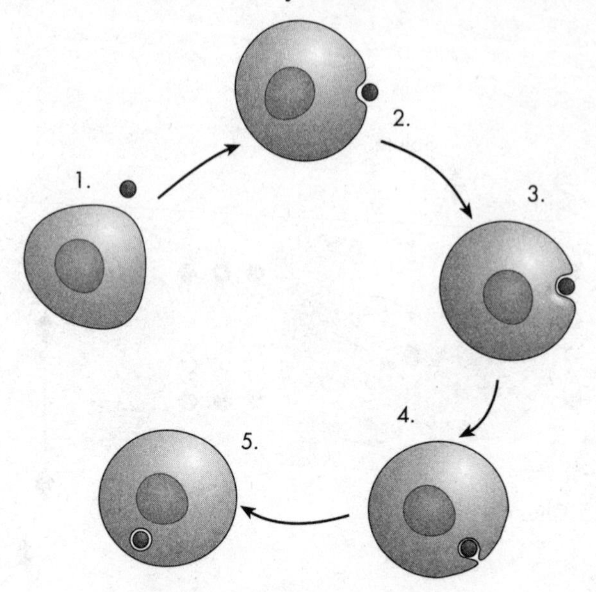

Three types of endocytosis exist: **pinocytosis**, **phagocytosis**, and **receptor-mediated endocytosis.** In pinocytosis, the cell ingests liquids ("cell-drinking"). In phagocytosis, the cell takes in solids ("cell-eating"). A special type of endocytosis, receptor-mediated endocytosis, involves cell surface receptors that are covered in clathrin-coated pits. (Clathrin is a kind of protein.) When a particle, or ligand, binds to one of these receptors, the ligand is brought into the cell by the invagination or "folding in" of the cell membrane. A vesicle then forms around the incoming ligand and carries it into the cell's interior.

Bulk Flow

Other substances move by **bulk flow**. Bulk flow is the one-way movement of fluids brought about by *pressure*. For instance, the movement of blood through a blood vessel or movement of fluids in xylem and phloem of plants are examples of bulk flow.

Dialysis

Dialysis is the diffusion of *solutes* across a selectively permeable membrane. For example, a cellophane bag is often used as an artificial membrane to separate small molecules from large molecules.

Exocytosis

Sometimes large particles are transported *out* of the cell. In **exocytosis**, a cell ejects waste products or specific secretion products such as hormones by the fusion of a vesicle with the plasma membrane.

Cell Junctions

When cells come in close contact with each other, they develop specialized **intercellular junctions** that involve their plasma membranes as well as other components. These structures may allow neighboring cells to form strong connections with each other, prevent passage of materials, or establish rapid communication between adjacent cells. There are three types of intercellular contact in animal cells: **desmosomes, gap junctions**, and **tight junctions**.

Desmosomes hold adjacent animal cells tightly to each other, like a rivet. They consist of a pair of discs associated with the plasma membrane of adjacent cells, plus the intercellular protein filaments that cross the small space between them. Intermediate filaments within the cells are also attached to the discs (see figure below).

Gap junctions are protein complexes that form channels in membranes and allow communication between the cytoplasm of adjacent animal cells or the transfer of small molecules and ions.

Tight junctions are tight connections between the membranes of adjacent animal cells. They're so tight that there is no space between the cells. Cells connected by tight junctions seal off body cavities and prevent leaks.

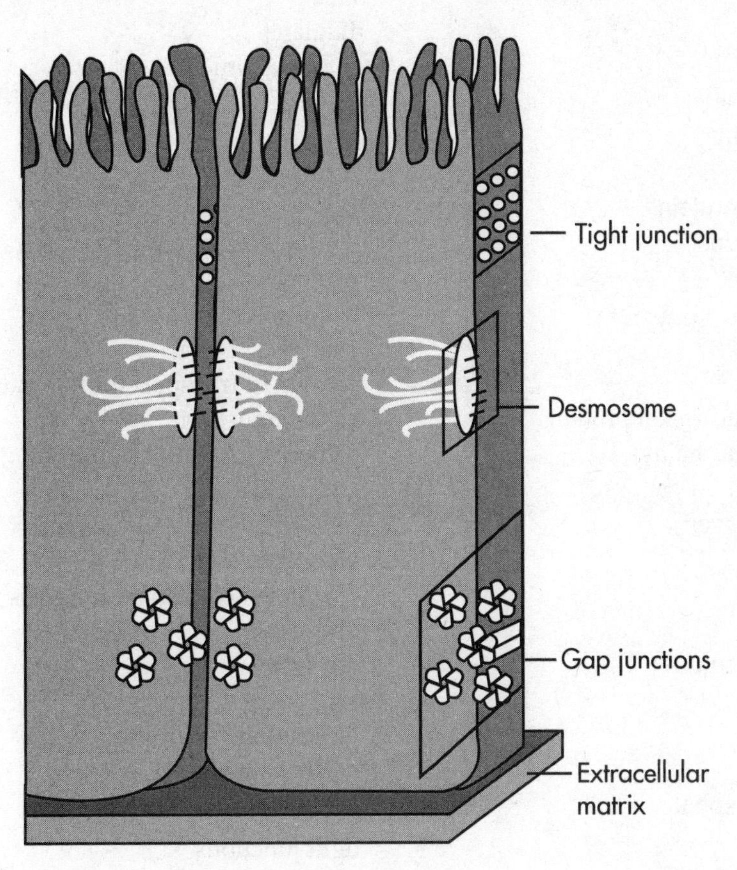

KEY WORDS

cells
light microscopes
electron microscopes
eukaryotic cells
prokaryotic cells
cytoplasm
organelles
nucleoid
flagellum
plasma membrane
peripheral proteins
integral proteins
transmembrane proteins
fluid-mosaic model
adhesion proteins
receptor proteins
transport proteins
channel proteins
recognition and adhesion proteins
carbohydrate side chains
cholesterol
phospholipid bilayer
chromosomes
nucleolus
ribosomes
endoplasmic reticulum (ER)
Golgi bodies
vesicles
mitochondria
adenosine triphosphate (ATP)
lysosomes
centrioles
microtubule organizing centers (MTOCs)
vacuoles
cytoskeleton
microtubules
microfilaments
tubulin
cilia
Euglena
Paramecium
cell wall
chitin
chloroplasts
cell sap
simple diffusion (or passive transport)
osmosis
solutes
channel proteins
facilitated transport (or facilitated diffusion)
active transport
sodium-potassium pump
endocytosis
pinocytosis
phagocytosis
receptor-mediated endocytosis
bulk flow
dialysis
exocytosis
intercellular junctions
desmosomes
gap junctions
tight junctions

CHAPTER 4 REVIEW QUESTIONS

Answers can be found in Chapter 15.

1. Movement of substances into the cell is largely dependent upon the size,,polarity, and concentration gradient of the substance. Which of the following represents an example of active transport of a substance into a cell?

   (A) Diffusion of oxygen into erythrocytes (red blood cells) in the alveolar capillaries of the lungs.
   (B) Influx of sodium ions through a voltage-gated ion channel in a neuron cell during an action potential.
   (C) The sodium-potassium pump which restores resting membrane potentials in neurons through the use of ATP.
   (D) Osmosis of water into an epithelial cell lining the lumen of the small intestine.

2. The development of electron microscopy has provided key insights into many aspects of cellular structure and function which had previously too small to be seen. All of the following would require the use of electron microscopy for visualization EXCEPT

   (A) the structure of a bacteriophage
   (B) the matrix structure of a mitochondrion
   (C) the shape and arrangement of bacterial cells
   (D) the pores on the nuclear membrane

3. *Vibrio cholerae* (shown below) are highly pathogenic bacteria that are associated with severe gastrointestinal illness and are the causative agent of cholera. In extreme cases, antibiotics are prescribed that target bacterial structures that are absent in animal cells. Which of the following structures is most likely targeted by antibiotic treatment?

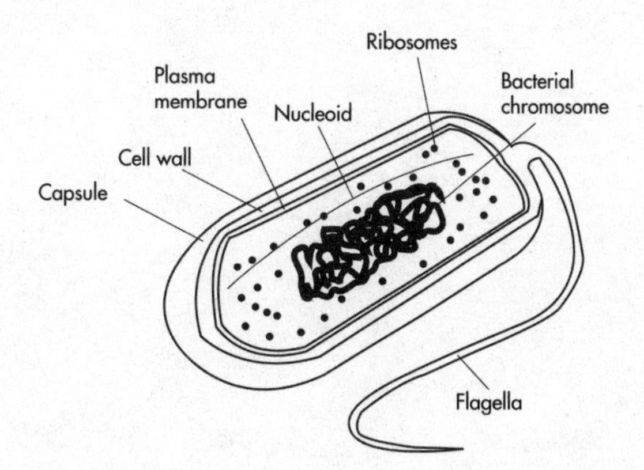

   (A) Flagella
   (B) Plasma membrane
   (C) Ribosomes
   (D) Cell wall

A new unicellular organism has recently been identified living in thermal pools in Yellowstone National Park. The thermal pools have average temperatures of 45 C, a pH of approximately 3.2, and have high concentrations of sulfur-containing compounds. To identify the organism, a microbiologist performs a series of tests to evaluate its structural organization. The table below summarizes the microscopy data of the newly identified organism.

Cellular Structure	**Analysis**
Plasma Membrane	Present
Cell Wall	Present, Very Thick
Mitochondria	Absent
Ribosomes	Present, Highly Abundant
Flagella	Present, Peritrichous Organization

4. This organism is most likely a new species of which of the following?

(A) Algae

(B) Protozoa

(C) Bacteria

(D) Fungi

Question 5 represents a question requiring a numeric answer.

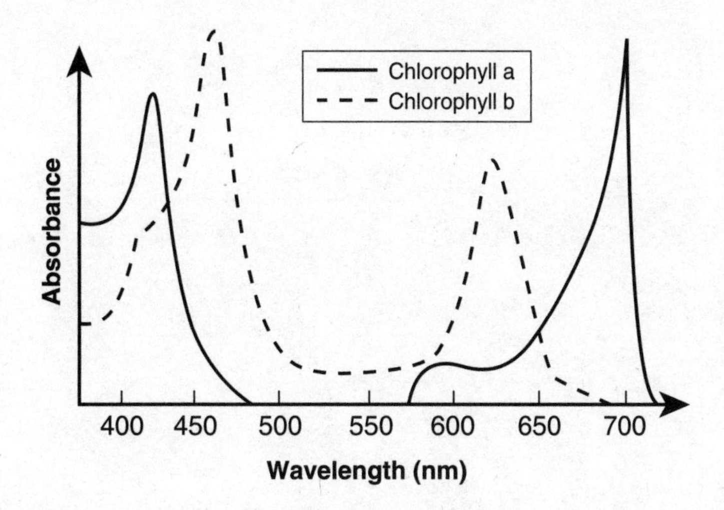

5. Chlorophyll is a green pigment present in the chloroplasts of algae and plants. It is essential for catalyzing the light-dependent cycles in photosynthesis. A scientist purifies both forms of chlorophyll (a and b) from plant chloroplasts and evaluates them for light absorption using a spectrophotometer. Using the spectrophotometer data provided above, at what wavelength is the absorbance of chlorophyll a at its maximum? Give your answer to the nearest whole number.

	/	/	/	
.	.	.	.	.
0	0	0	0	0
1	1	1	1	1
2	2	2	2	2
3	3	3	3	3
4	4	4	4	4
5	5	5	5	5
6	6	6	6	6
7	7	7	7	7
8	8	8	8	8
9	9	9	9	9

5

Cellular Energetics

BIOENERGETICS

In Chapter 3, we discussed some of the more important organic molecules. But what makes these molecules so important? Glucose, starch, and fat are all energy-rich. However, the energy is packed in the chemical bonds holding the molecules together. To carry out the processes necessary for life, cells must find a way to release the energy in these bonds when they need it and store it away when they don't. The study of how cells accomplish this is called **bioenergetics**. Generally, bioenergetics is the study of how energy from the sun is transformed into energy in living things.

All the energy for life comes from chemical bonds. During chemical reactions, such bonds are either broken or formed. This process involves energy, no matter in which direction we go. Every chemical reaction involves a change in energy.

Energy is invested in the formation of bonds, whereas energy is released when bonds are broken. However, the breaking apart of chemical bonds requires the input of energy. The way that cells are able to break these bonds without spending large amounts of energy is through the use of special molecules called **enzymes**.

ENZYMES

Most chemical reactions do not occur haphazardly in the cell. To help control the chemical reactions essential for life, cells rely on enzymes. These proteins help "kick-start" reactions and speed them up once they get rolling, enabling cells to get the most out of the energy sources available to them. Enzymes are **organic catalysts**; they speed up the rate of a reaction without altering the reaction itself.

That is, they *catalyze* them without being changed in the reaction themselves. Before we discuss precisely how this works, let's review some of the different types of reactions that can occur within a cell.

TYPES OF REACTIONS

Exergonic reactions are those in which the products have *less* energy than the reactants. Simply put, energy is given off during the reaction.

Let's look at an example. The course of a reaction can be represented by an energy diagram. Here's an energy diagram for an exergonic reaction:

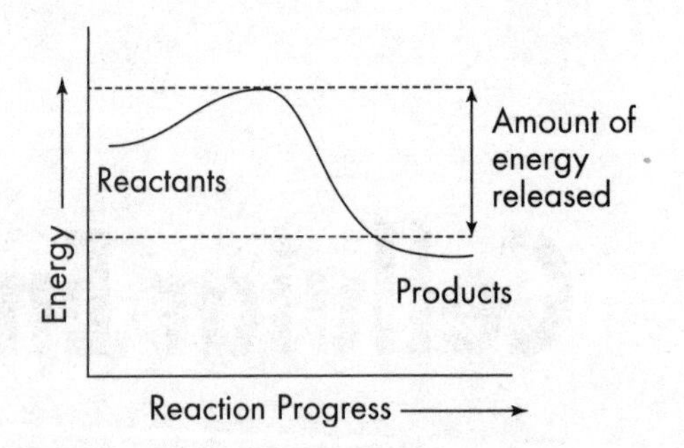

You'll notice that energy is represented along the *y*-axis. Based on the diagram, our reaction released energy. An example of an exergonic reaction is when food is oxidized in mitochondria of cells and then releases the energy stored in the chemical bonds.

Reactions that require an input of energy are called **endergonic reactions**. You'll notice that the products have *more* energy than the reactants.

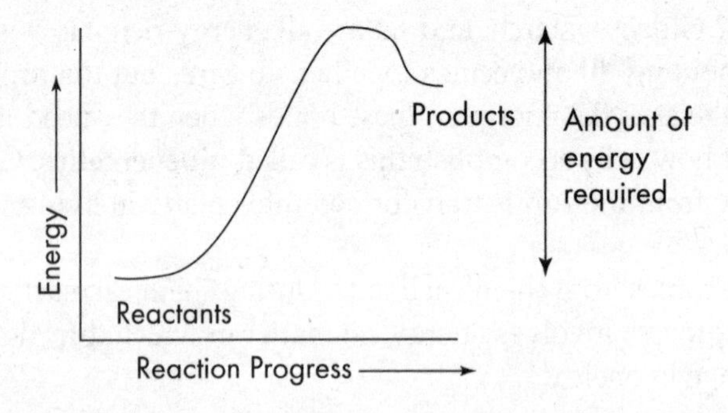

The products gained energy in the form of heat. An example is when plants use carbon dioxide and water to form sugars.

Activation Energy

Although we said that exergonic reactions release energy, this does not mean that they do not require any energy to get started. Take a look at this energy diagram of a typical exergonic reaction:

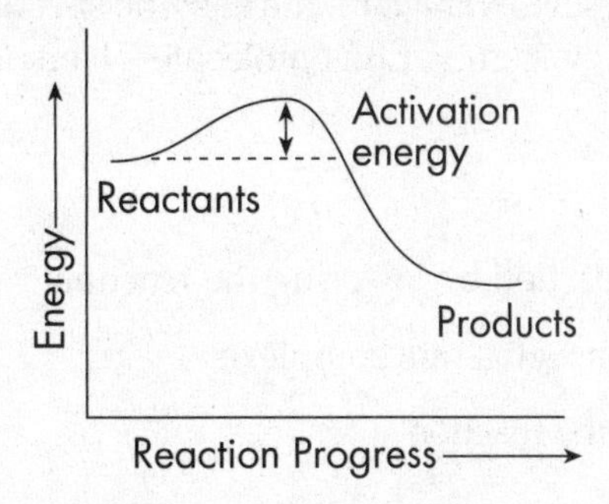

You'll notice that we needed a little energy to get us going. That's because chemical bonds must be broken before new bonds can form. This energy barrier—the hump in the graph—is called the **activation energy**. Once a set of reactants has reached its activation energy, the reaction can occur much faster than it would in the absence of the enzyme.

Getting Back to Enzymes

Why are enzymes so important in biology? They're important because many reactions would never occur in the cell if it weren't for the help of enzymes. As we saw earlier, enzymes, by definition, catalyze reactions: They activate them. What this means in chemical terms is that they *lower* the activation energy of a reaction, enabling the reaction to occur much faster than it would in the absence of the enzyme.

Enzyme Specificity

Most of the crucial reactions that occur in the cell require enzymes. Yet enzymes themselves are highly specific—in fact, each enzyme catalyzes only one kind of reaction. This is known as **enzyme specificity**. Since this is true, enzymes are usually named after the molecules they target. In enzymatic reactions, the targeted molecules are known as **substrates**. For example, maltose, a disaccharide, can be broken down into two glucose molecules. Our substrate, maltose, gives its name to the enzyme that catalyzes this reaction: *maltase*.

Many enzymes are named simply by replacing the suffix of the substrate with *-ase*. Using this nomenclature, malt*ose* becomes malt*ase*.

Enzyme-Substrate Complex

Enzymes have a unique way of helping reactions along. As we just saw, the reactants in an enzyme-assisted reaction are known as substrates. During a reaction, the enzyme's job is to bring the substrates together. It accomplishes this due to a special region on the enzyme known as an **active site**.

The enzyme temporarily binds the substrates to its active site, and forms an **enzyme-substrate complex**. Let's take a look:

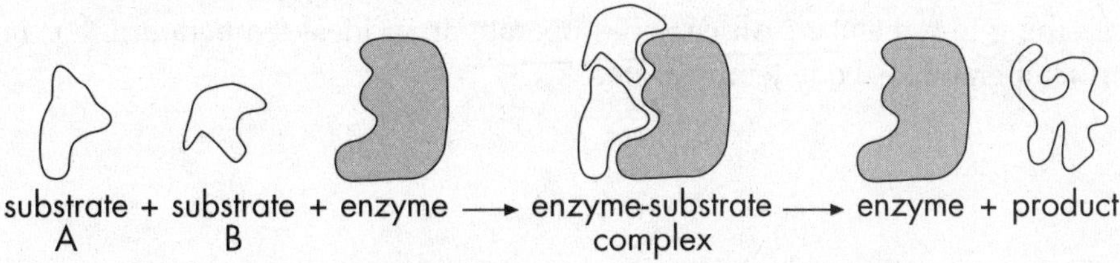

Once the reaction has occurred and the product is formed, the enzyme is released from the complex and restored to its original state. Now, the enzyme is free to react again with another bunch of substrates.

By binding and releasing over and over again, the enzyme speeds the reaction along, enabling the cell to release much-needed energy from various molecules. Here is a quick review on the function of enzymes.

Enzymes Do

- Increase the rate of a reaction by lowering the reaction's activation energy
- Form temporary enzyme-substrate complexes
- Remain unaffected by the reaction

Enzymes Don't

- Change the reaction
- Make reactions occur that would otherwise not occur at all

Induced Fit

However, scientists have discovered that enzymes and substrates don't fit together quite so seamlessly. It appears that the enzyme has to change its shape slightly to accommodate the shape of the substrates. This is called **induced fit**.

Enzymes Don't Always Work Alone

Enzymes sometimes need a little help in catalyzing a reaction. Those factors are known as **coenzymes**. Vitamins are examples of organic coenzymes. Your daily dose of vitamins is important for just this reason: The vitamins are "active and necessary participants" in crucial chemical reactions. The function of coenzymes is to accept electrons and pass them along to another substrate. Two examples of such enzymes are NAD^+ and $NADP^+$.

In addition to organic coenzymes, inorganic elements—called **cofactors**—help catalyze reactions. These elements are usually metal ions, such as Fe^{+2}.

Factors Affecting Reaction Rates

Enzymatic reactions can be influenced by a number of factors, such as temperature, pH, and the relative amounts of enzyme and substrate.

Temperature

The rate of a reaction increases with increasing temperature, up to a point, because an increase in the temperature of a reaction increases the chance of collisions among the molecules. But too much heat can damage an enzyme. If a reaction is conducted at an excessively high temperature (above 42°C), the enzyme loses its three-dimensional shape and becomes inactivated. Enzymes damaged by heat and deprived of their ability to catalyze reactions are said to be *denatured*.

Here's one thing to remember: All enzymes operate at an ideal temperature. For most human enzymes, this temperature is body temperature, 37°C.

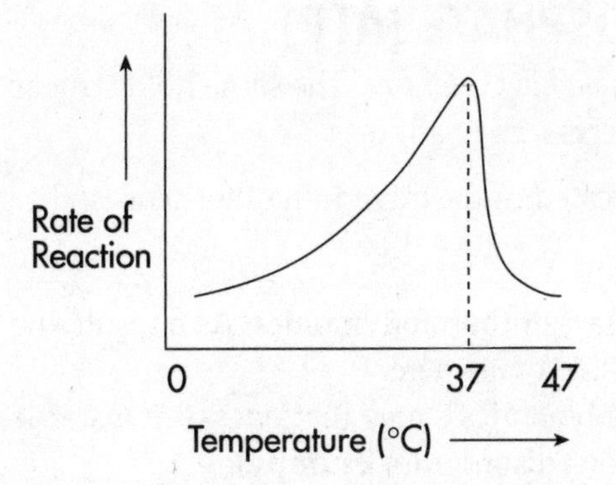

pH

Enzymes also function best at a particular pH. For most enzymes, the optimal pH is at or near a pH of 7:

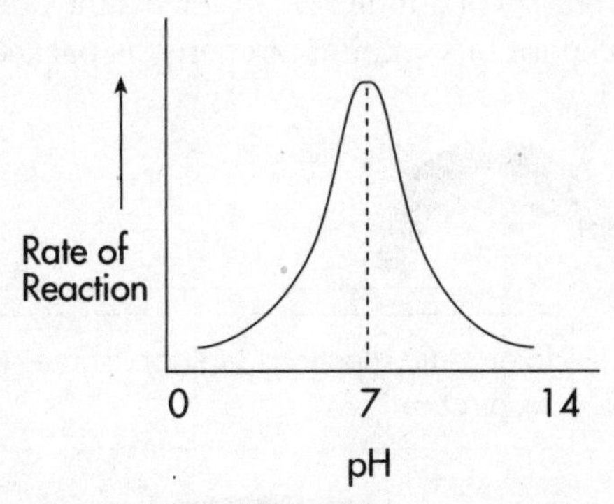

Other enzymes operate at a low pH. For instance, pepsin, the digestive enzyme found in the stomach, is most effective at an extremely acidic pH of 2.

Here's another important piece of information: Most enzymes are active only over a narrow range of pH.

ENZYME REGULATION

We know that enzymes control the rates of chemical reactions. But what regulates the activity of enzymes? It turns out that a cell can control enzymatic activity by regulating the conditions that influence the shape of the enzyme. For example, some enzymes have **allosteric sites**, a region of the enzyme other than the active site to which a substance can bind. Substances called **allosteric regulators** can either inhibit or activate enzymes. An **allosteric inhibitor** will bind to an allosteric site and keep the enzyme in its inactive form while an **allosteric activator** will bind to an enzyme and induce its active form. Allosteric enzymes are subject to **feedback inhibition** in which the formation of an end product inhibits an earlier reaction in the sequence.

Most enzymes can be inhibited by certain chemical substances. If the substance has a shape that fits the active site of an enzyme, it can compete with the substrate and effectively inactivate the enzyme. This is called **competitive inhibition**. Usually a competitive inhibitor is structurally similar to the normal substrate. In **noncompetitive inhibition**, the inhibitor binds with the enzyme at a site other than the active site and inactivates the enzyme by altering its shape. This prevents the enzyme from binding with the substrate at the active site.

ADENOSINE TRIPHOSPHATE (ATP)

We've all heard the expression *nothing is for free*. The same holds true in nature. Here's a fundamental principle of energy that it is necessary to address:

> Energy cannot be created or destroyed. In other words, the sum of energy in the universe is constant.

This rule is called the **first law of thermodynamics**. As a result, the cell cannot take energy out of thin air. Rather, it must harvest it somewhere.

The **second law of thermodynamics** states that energy transfer leads to less organization. That means the universe tends toward disorder (or **entropy**).

As we just saw, almost everything an organism does requires energy. How, then, can the cell acquire the energy it needs without becoming a major mess? Fortunately, it's through adenosine triphosphate (ATP).

ATP, as the name indicates, consists of a molecule of adenosine bonded to three phosphates. The great thing about ATP is that an enormous amount of energy is packed into those phosphate bonds, particularly the third bond.

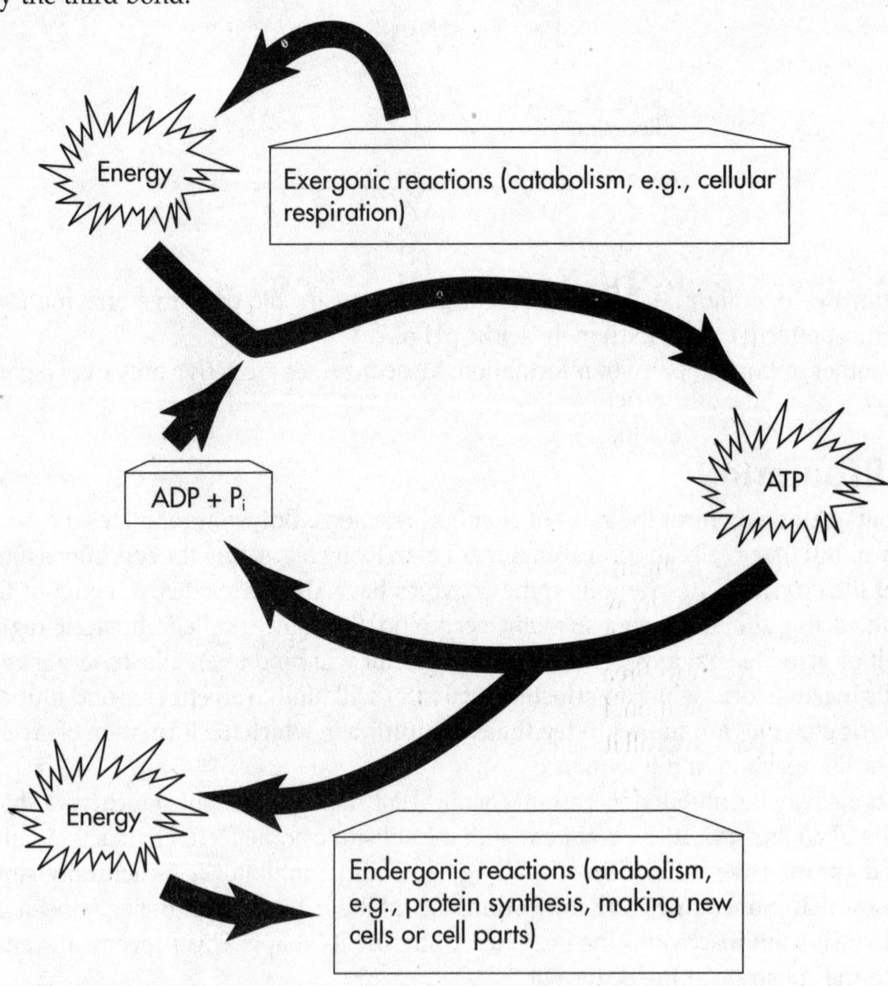

When a cell needs energy, it takes one of these potential-packed molecules of ATP and splits off the third phosphate, forming adenosine diphosphate (ADP) and one loose phosphate (P_i), while releasing energy in the process:

$$ATP \rightarrow ADP + P_i + energy$$

The energy released from this reaction can then be put to whatever use the cell so pleases. Of course, this doesn't mean that the cell is above the laws of thermodynamics. But within those constraints, ATP is the best source of energy the cell has available. It is relatively neat (only one bond needs to be broken to release that energy) and relatively easy to form. Organisms can use exergonic processes that increase energy, like breaking down ATP, to power endergonic reactions, like building organic macromolecules.

Sources of ATP

But where does all this ATP come from? It is produced in one of two ways: (1) through photosynthesis, or (2) through cellular respiration.

Photosynthesis involves the transformation of solar energy into chemical energy. Plants take carbon dioxide, water, and energy (in the form of sunlight) and use them to produce glucose. The overall reaction is:

$$6CO_2 + 6H_2O + sunlight \rightarrow C_6H_{12}O_6 + 6O_2$$

We'll hold off on a full discussion of photosynthesis until Chapter 6. For now, let's focus on the other means of ATP production—**cellular respiration.**

Cellular Respiration: The Shorthand Version

In cellular respiration, which is performed by all organisms, ATP is produced through the breakdown of nutrients. You'll recall from the beginning of this chapter that many organic molecules are important to cells because they are energy-rich. This is where that energy comes into play.

In the shorthand version, cellular respiration looks something like this:

$$C_6H_{12}O_6 + 6O_2 \rightarrow 6CO_2 + 6H_2O + ATP$$

Notice that we've taken a sugar, perhaps a molecule of glucose, and combined it with oxygen and water to produce carbon dioxide, water, and energy in the form of our old friend, ATP. However, as you probably already know, the actual picture of what really happens is far more complicated.

Generally speaking, we can break cellular respiration down to two different approaches: aerobic respiration or anaerobic respiration. If ATP is made in the presence of oxygen, we call it **aerobic respiration**. If oxygen isn't present, we call it **anaerobic respiration**. Let's jump right in with aerobic respiration.

AEROBIC RESPIRATION

Aerobic respiration consists of four stages:

1. Glycolysis
2. Formation of acetyl CoA
3. The Krebs cycle
4. Oxidative phosphorylation

There are so many steps within each stage that some students find this topic too confusing to follow. Don't sweat it. We've come up with a simple method to keep all the stages of cellular respiration in order: Just keep track of the number of carbons at each stage.

STAGE 1: GLYCOLYSIS

The first stage begins with **glycolysis**, the splitting (*-lysis*) of glucose (*glyco-*). Glucose is a six-carbon molecule that is broken into two three-carbon molecules, called **pyruvic acid**. This breakdown of glucose also results in the net production of two molecules of ATP:

$$\text{Glucose} + 2\text{ ATP} + 2\text{ NAD}^+ \rightarrow 2\text{ Pyruvic acid} + 4\text{ ATP} + 2\text{ NADH}$$

Although we've written glycolysis as if it were a single reaction, this process doesn't occur in one step. In fact, it requires a sequence of enzyme-catalyzed reactions!

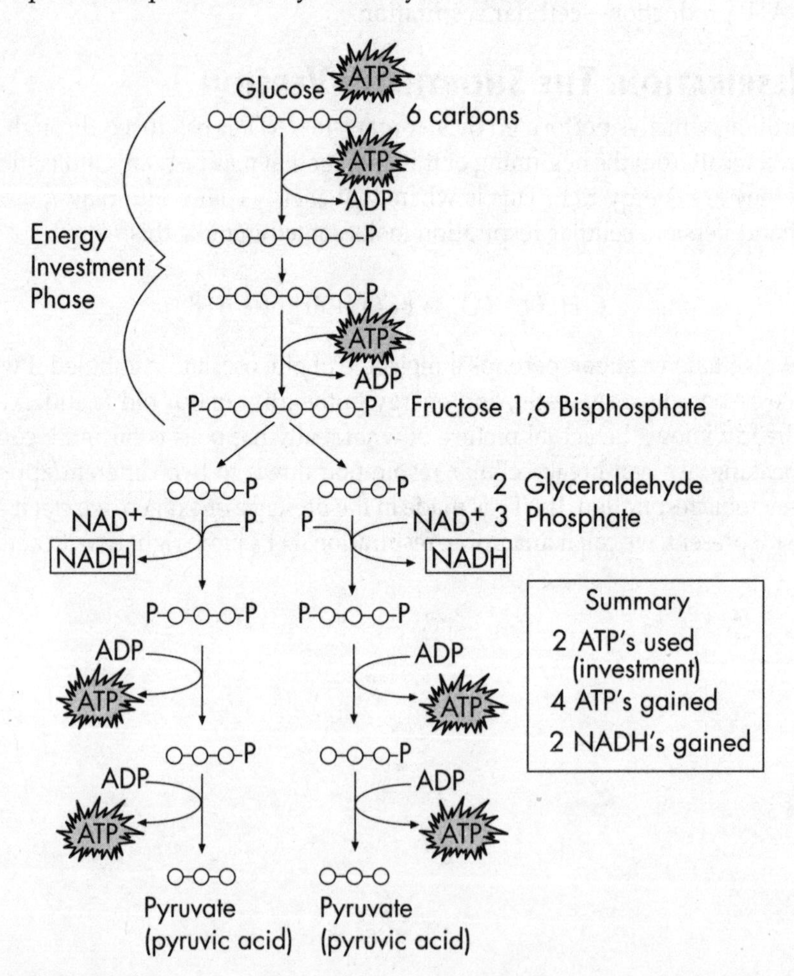

Fortunately, you don't need to memorize these steps for the test. What you do need to know is that glucose doesn't *automatically* generate ATP. It has to be activated. Once glucose is phosphorylated, it eventually splits into pyruvate.

If you take a good look at the reaction above, you'll see two ATPs are needed to produce four ATPs. You've probably heard the expression, "You have to spend money to make money." In biology, you have to invest ATP to make ATP: Our investment of two ATPs yielded four ATPs, for a net gain of two.

A second product in glycolysis is 2 NADH, which results from the transfer of H^+ to the hydrogen carrier NAD^+. NADH will be used elsewhere in respiration to make additional ATP.

There are four important tidbits to remember regarding glycolysis:

- Occurs in the cytoplasm
- Net of 2 ATPs produced
- 2 pyruvic acids formed
- 2 NADH produced

Once the cell has undergone glycolysis, it has two options: It can continue anaerobically, or it can switch to true aerobic respiration. As we'll soon see, the cell's decision has a lot to do with the environment in which it finds itself. If oxygen is present, many cells switch directly to aerobic respiration. If no oxygen is present, those same cells may carry out anaerobic respiration. Still others have no choice, and carry out only anaerobic respiration, with or without oxygen.

Since ETS is more likely to ask you about aerobic respiration, we'll look closely at the remaining steps.

However, before we do so, let's jump back to those important organelles, the mitochondria. We already know from our discussion in Chapter 4 that the mitochondria are the sites of cellular respiration. Now it's time to see exactly *where* they manufacture ATP.

The double membrane of the mitochondria divides the organelle into four regions:

- The **matrix**
- The **inner mitochondrial membrane**
- The **intermembrane space**
- The **outer membrane**

Let's have a look:

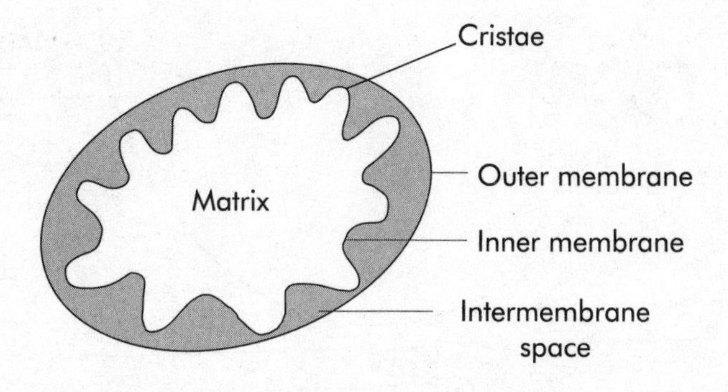

Why do you need to know about the different regions within a mitochondrion? Because we'll soon see that several of the stages of aerobic respiration occur within these regions of the mitochondria—and ETS loves to ask you questions about where things occur! Keep them in mind. We'll be discussing them below.

STAGE 2: FORMATION OF ACETYL COA

When oxygen is present, pyruvic acid is transported to the mitochondrion. Each pyruvic acid (a three-carbon molecule) is converted to **acetyl coenzyme A** (a two-carbon molecule) and CO_2 is released:

$$2 \text{ Pyruvic acid} + 2 \text{ Coenzyme A} + 2 \text{ NAD}^+ \rightarrow 2 \text{ Acetyl CoA} + 2 \text{ CO}_2 + 2 \text{ NADH}$$

Are you keeping track of our carbons? We've now gone from two three-carbon molecules to two two-carbon molecules. The extra carbons leave the cell in the form of CO_2. Once again, two molecules of NADH are also produced.

STAGE 3: THE KREBS CYCLE

The next stage is the **Krebs cycle**, also known as the **citric acid cycle**. Each of the two acetyl coenzyme A molecules will enter the Krebs cycle, one at a time, and all the carbons will ultimately be converted to CO_2. This stage occurs in the matrix of the mitochondria.

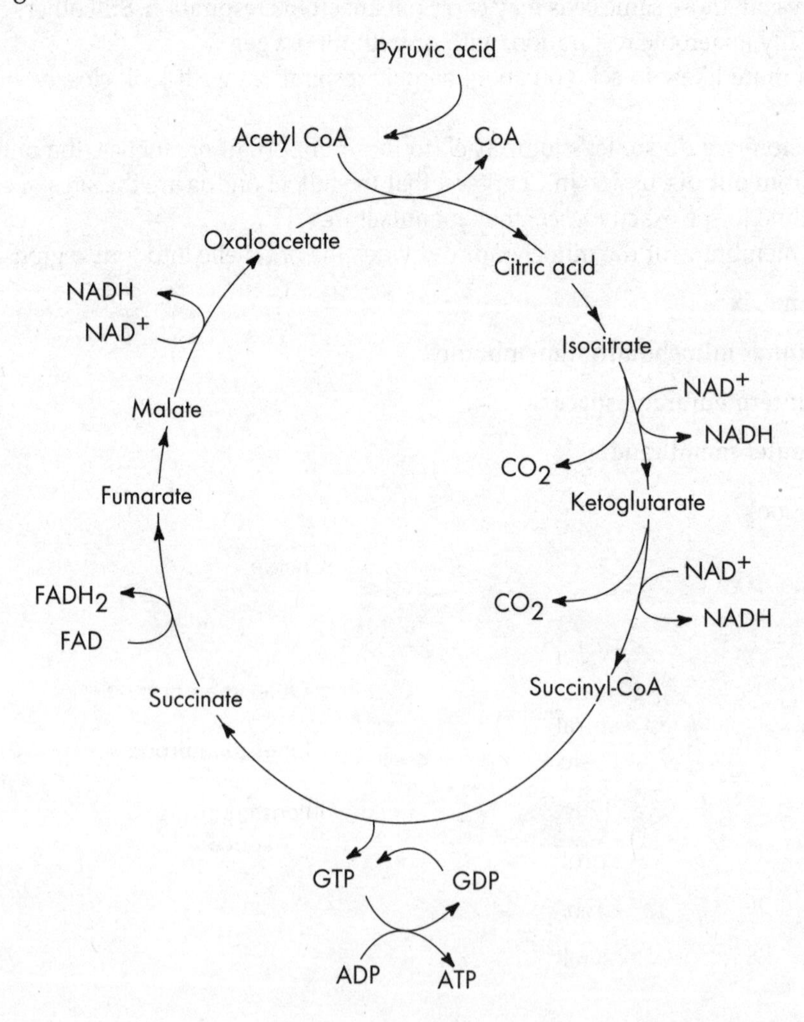

Let's track the carbons again. Each molecule of acetyl CoA produced from the second stage of aerobic respiration combines with **oxaloacetate**, a four-carbon molecule, to form a six-carbon molecule, **citric acid** or citrate:

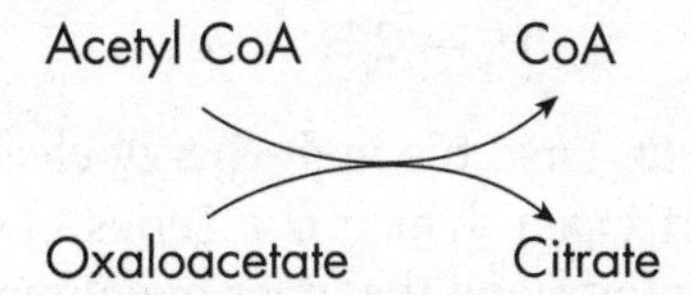

Since the cycle begins with a four-carbon molecule, oxaloacetate, it also has to end with a four-carbon molecule to maintain the cycle. So how many carbons do we have to lose to keep the cycle going? Two carbons, both of which will be released as CO_2. Now the cycle is ready for another turn with the second acetyl CoA.

With each turn of the cycle, three additional types of molecules are produced:

- 1 ATP
- 3 NADH
- 1 $FADH_2$

To figure out the total number of products per molecule of glucose, we simply double the number of products—after all, we started off the Krebs cycle with two molecules of acetyl CoA for each molecule of glucose!

Now we're ready to tally up the number of ATP produced.

After the Krebs cycle, we've made only four ATP—two ATP from glycolysis and two ATP from the Krebs cycle.

Although that seems like a lot of work for only four ATP, we have also produced hydrogen carriers in the form of NADH and $FADH_2$. These molecules will in turn produce lots of ATP.

Stage 4: Oxidative Phosphorylation

We said earlier that ATP is the energy currency of the cell. While this is true, ATP is not the only molecule that stores energy. Sometimes energy is stored by electron carriers like NAD^+ and FAD. (These electron carriers are also called hydrogen carriers because most electron carriers also carry hydrogen atoms.) Electrons are transferred from electron carriers to oxygen, resulting in ATP synthesis. This process is called oxidative phosphorylation.

Electron Transport Chain

As electrons (and the hydrogen atoms to which they belong) are removed from a molecule of glucose, they carry with them much of the energy that was originally stored in their chemical bonds. These electrons—and their accompanying energy—are then transferred to readied hydrogen carrier molecules. In the case of cellular respiration, these charged carriers are NADH and $FADH_2$.

Let's see how many "loaded" electron carriers we've produced. We now have:

- Two NADH molecules from glycolysis
- Two NADH from the production of acetyl CoA
- Six NADH from the Krebs cycle
- Two $FADH_2$ from the Krebs cycle

That gives us 12 altogether.

Now let's consider the fate of all the electrons removed from the breakdown of glucose. Here's what happens. The electron carriers—NADH and $FADH_2$—"shuttle" electrons to the electron transport chain, and the hydrogen atoms are split into hydrogen ions and electrons:

$$H_2 \rightarrow 2\,H^+ + 2e^-$$

Then, two interesting things occur. First, the high-energy electrons from NADH and $FADH_2$ are passed down the electron transport chain, which is a series of protein carrier molecules that are embedded in the cristae, the membrane along the inner membrane of a mitochondrion. Some of the carrier molecules in the electron transport chain are iron-containing carriers called **cytochromes**. Take a look at the entire chain below.

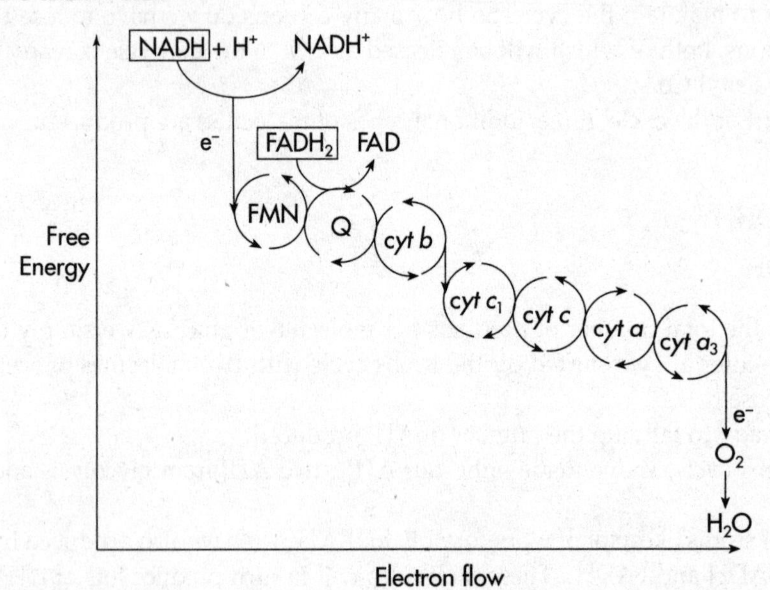

Each carrier molecule hands down the electrons to the next molecule in the chain. The electrons travel down the electron transport chain until they reach the final electron acceptor, oxygen. Oxygen combines with these electrons (and some hydrogens) to form water. This explains the "aerobic" in aerobic respiration. If oxygen weren't available to accept the electrons, they wouldn't move down the chain at all, thereby shutting down the whole process of ATP production.

CHEMIOSMOSIS

At the same time that electrons are being passed down the electron transport chain, another mechanism is at work. Remember those hydrogen ions (also called *protons*) that split off from the original hydrogen atom? Some of the energy released from the electron transport chain is used to pump hydrogen ions across the inner mitochondrial membrane to the intermembrane space. The pumping of hydrogen ions into the intermembrane space creates a **pH gradient**, or **proton gradient**. The potential energy established in this gradient is responsible for the production of ATP.

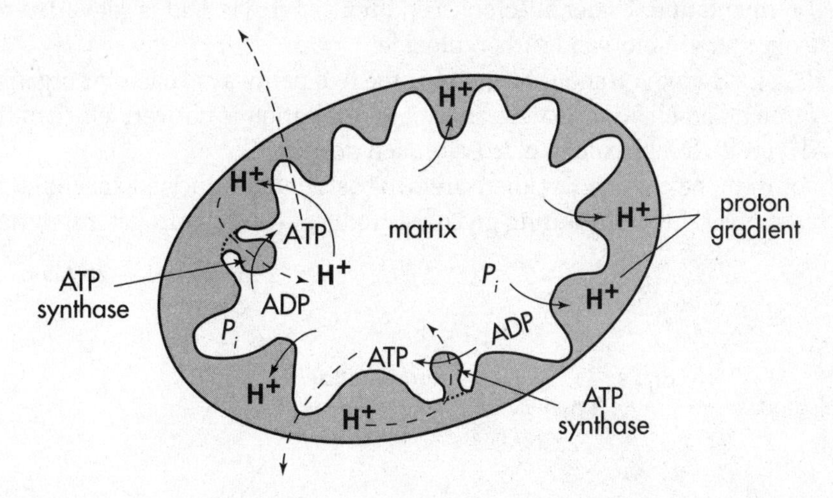

These hydrogen ions can diffuse across the inner membrane only by passing through channels called **ATP synthase**. Meanwhile, ADP and P_i are on the other side of these protein channels. The flow of protons through these channels produces ATP by combining ADP and P_i on the matrix side of the channel. This process is called **oxidative phosphorylation**.

Three other things you're expected to know for the AP Biology Exam:

- Every NADH yields 3 ATP (except NADH from glycolysis produces 2 ATP).
- Every $FADH_2$ yields 2 ATP.
- The total number of ATP produced in Stage 4 is 32 ATP.

Don't concern yourself with the exact number of NADH, $FADH_2$, CO_2, and ATP produced at each stage. For the AP Biology Exam, all you need to remember are the big steps and the overall outcome:

STAGES OF AEROBIC RESPIRATION		
Step in Respiration	**Takes Place in the . . .**	**Net Result**
Glycolysis	Cytoplasm	2 ATP (+2 NADH)
Split to acetyl CoA	Cytoplasm	0 ATP (+2 NADH)
Krebs cycle	Mitochondrial matrix	2 ATP (6 NADH + 2 $FADH_2$)
Oxidative phosphorylation	Inner mitochondrial membrane	32 ATP
		Net: 36 ADP

ANAEROBIC RESPIRATION

Some organisms can't undergo aerobic respiration. They're anaerobic. They can't use oxygen to make ATP. We just learned that oxygen is important because it's the final electron acceptor in the electron transport chain.

How do anaerobic organisms derive energy? Since glycolysis is an anaerobic process, they can make 2 ATP from this stage. However, instead of carrying out the other stages of aerobic respiration (the Krebs cycle, electron transport chain, and oxidative phosphorylation), these organisms carry out a process called **fermentation**. Under anaerobic conditions, pyruvic acid is converted to either **lactic acid** or **ethyl alcohol** (or **ethanol**) and carbon dioxide.

For the AP Biology Exam, you should remember the two pathways anaerobic organisms undergo: glycolysis and fermentation. Unfortunately, anaerobic respiration is not very efficient. It only results in a gain of 2 ATP for each molecule of glucose broken down.

As you can see from the chart below, there are two basic end products in anaerobic respiration. In both pathways, the NADH formed during glycolysis reduces (adds hydrogen to) pyruvate.

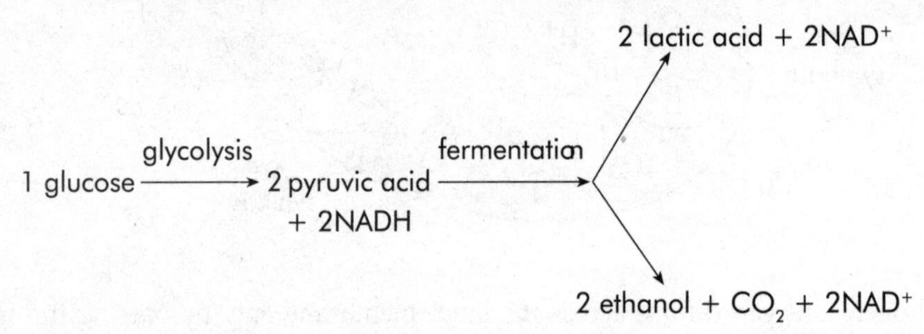

What types of organisms undergo fermentation?

- Yeast cells and some bacteria make ethanol and carbon dioxide.
- Other bacteria produce lactic acid.

YOUR MUSCLE CELLS CAN FERMENT

Although human beings are aerobic organisms, they can actually carry out fermentation in their muscle cells. Have you ever had a cramp? If so, that cramp was the consequence of anaerobic respiration.

When you exercise, your muscles require a lot of energy. To get this energy, they convert enormous amounts of glucose to ATP. But as you continue to exercise, your body doesn't get enough oxygen to keep up with the demand in your muscles. This creates an oxygen debt. What do your muscle cells do? They switch over to anaerobic respiration. Pyruvic acid produced from glycolysis is converted to lactic acid. As a consequence, the lactic acid causes your muscles to ache.

KEY WORDS

bioenergetics
enzymes
organic catalysts
exergonic reactions
endergonic reactions
activation energy
enzyme specificity
substrates
active site
enzyme-substrate complex
induced fit
coenzymes
cofactors
allosteric sites
allosteric regulators
allosteric inhibitor
allosteric activator
feedback inhibition
competitive inhibition
noncompetitive inhibition
first law of thermodynamics
second law of thermodynamics
entropy
photosynthesis
cellular respiration
aerobic respiration
anaerobic respiration
glycolysis
pyruvic acid
matrix
inner mitochondrial membrane
intermembrane space
outer membrane
acetyl coenzyme A
Krebs cycle (or citric acid cycle)
oxaloacetate
citric acid
cytochromes
pH gradient
(or proton gradient)
ATP synthase
oxidative phosphorylation
fermentation
lactic acid
ethanol (or ethyl alcohol)

CHAPTER 5 REVIEW QUESTIONS

Answers can be found in Chapter 15.

1. The mitochondrion is a critical organelle structure involved in cellular respiration. Below is a simple schematic of the structure of a mitochondrion. Which of the structural components labeled below in the mitochondrion is the primary location of the Krebs cycle?

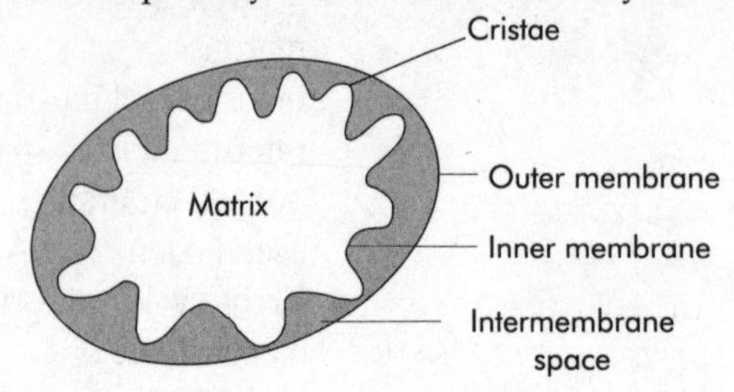

(A) Inner membrane
(B) Matrix
(C) Intermembrane space
(D) Outer membrane

2. Binding of Inhibitor Y as shown below inhibits a key catalytic enzyme by inducing a structural conformation change. Which of the following accurately describes the role of Inhibitor Y?

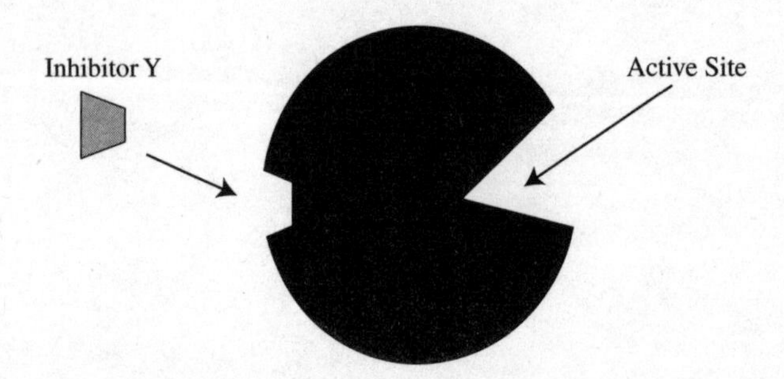

(A) Inhibitor Y competes with substrates for binding in the active site and functions as a competitive inhibitor.
(B) Inhibitor Y binds allosterically and functions as a competitive inhibitor.
(C) Inhibitor Y competes with substrates for binding in the active site and functions as a non-competitive inhibitor.
(D) Inhibitor Y binds allosterically and functions as a non-competitive inhibitor.

3. A single step chemical reaction is catalyzed by the addition of an enzyme. Which of the reaction coordinate diagrams accurately shows the effect of the added enzyme (represented by the dashed line) to the reaction?

(A)

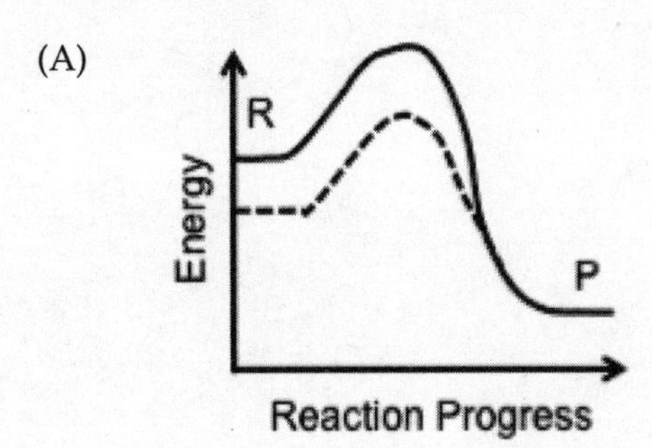

(B)

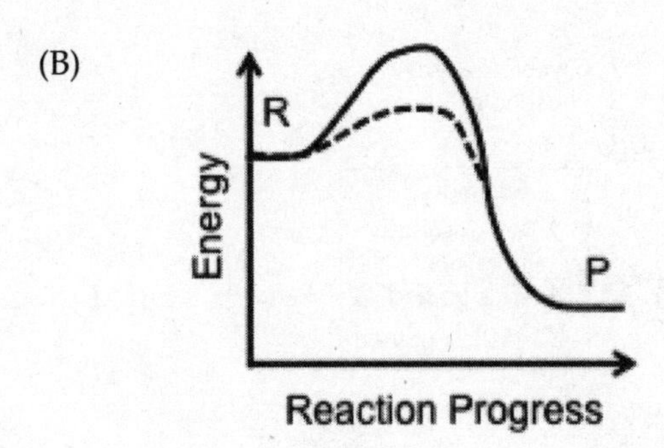

(C)

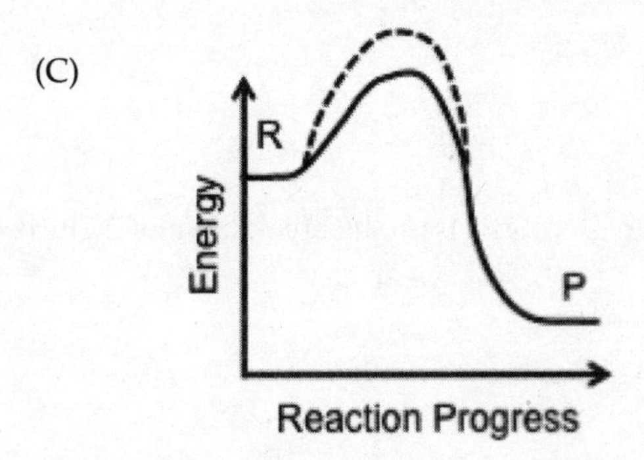

(D)

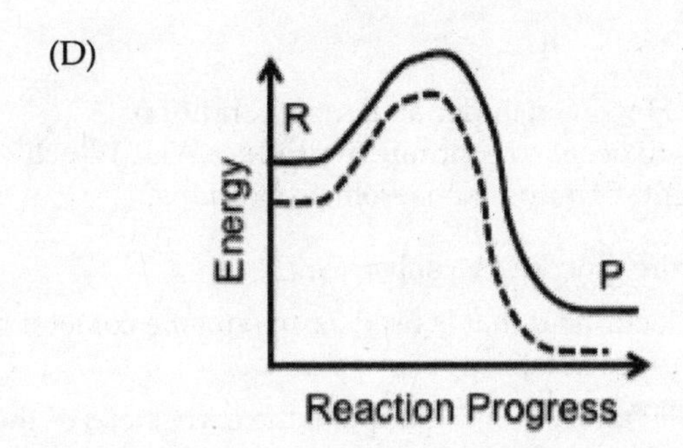

Questions 4 - 6 refer to the following diagram and paragraph

Glycolysis (shown below) is a critical metabolic pathway that is utilized by nearly all forms of life. The process of glycolysis occurs in the cytoplasm of the cell and converts 1 molecule of glucose into 2 molecules of pyruvic acid.

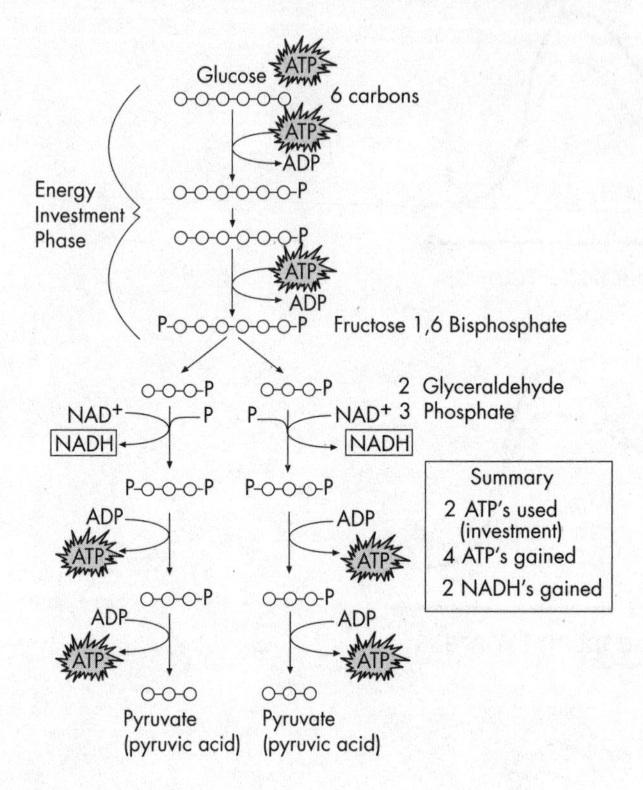

4. How many net ATP would be generated directly from glycolysis from the breakdown of 2 glucose molecules?

   (A) 2
   (B) 4
   (C) 8
   (D) 12

5. Glycolysis does not require oxygen to occur in cells. However, under anaerobic conditions, glycolysis normally requires fermentation pathways to occur to continue to produce ATP. Which best describes why glycolysis is dependent on fermentation under anaerobic conditions?

   (A) Glycolysis requires fermentation to produce more glucose as a substrate.
   (B) Glycolysis requires fermentation to synthesize lactic acid that is used as an enzyme cofactor of hexokinase (which catalyzes the first step of glycolysis).
   (C) Glycolysis requires fermentation to generate ATP molecules to complete the early steps of the pathway.
   (D) Glycolysis requires fermentation to regenerate NAD^+ from NADH to complete a later step in the pathway.

6. Which of the following most accurately describes the net reaction of glycolysis?

(A) It is an endergonic process because it results in a net increase in energy.
(B) It is an exergonic process because it results in a net increase in energy.
(C) It is an endergonic process because it results in a net decrease in energy.
(D) It is an exergonic process because it results in a net decrease in energy.

6

Photosynthesis

Plants and algae are producers. All they do is bask in the sun, churning out the glucose necessary for life.

As we discussed earlier, photosynthesis is the process by which light energy is converted to chemical energy. Here's an overview of photosynthesis:

$$6CO_2 + 6H_2O \xrightarrow{\text{sunlight}} C_6H_{12}O_6 + 6O_2$$

You'll notice from this equation that carbon dioxide and water are the raw materials plants use in manufacturing simple sugars. But remember, there's much more to photosynthesis than what's shown in the simple reaction above. We'll soon see that this beautifully orchestrated process occurs thanks to a whole host of special enzymes and pigments. But before we turn to the stages in photosynthesis, let's talk about where photosynthesis occurs.

THE ANATOMY OF A LEAF

Photosynthesis occurs in the leaves of plants. Here's a cross-sectional view of a typical leaf:

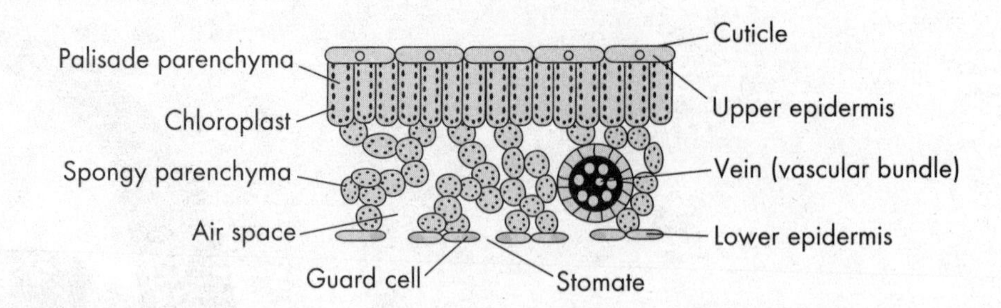

If you look closely at the leaf of any plant, the first thing you'll notice is the waxy covering called the cuticle. The **cuticle** is produced by the **upper epidermis** to protect the leaf from water loss through evaporation. Just below the upper epidermis is the **palisade parenchyma**. These cells contain lots of chloroplasts, which are the primary sites of photosynthesis.

Now let's look at an individual chloroplast. If you split the membrane of a chloroplast, you'll find a fluid-filled region called the **stroma**. Inside the stroma are structures that look like stacks of coins. These structures are the **grana**.

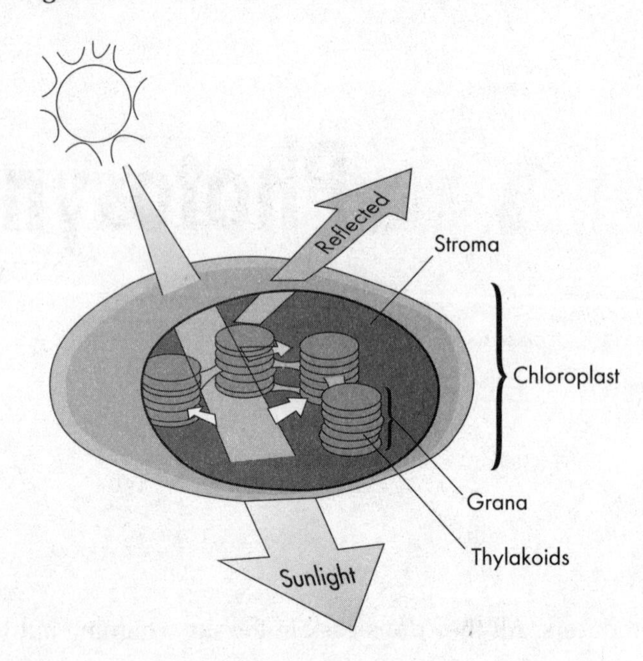

The many disk-like structures that make up the grana are called the **thylakoids.** They contain chlorophyll, the light-absorbing pigment that drives photosynthesis, as well as enzymes involved in the process.

Now we'll discuss some structures that are not involved in photosynthesis but are important for the AP test. Just below the palisade parenchyma, you'll find irregular-shaped cells called **spongy parenchyma**. It allows for diffusion of gases, especially CO_2, within the leaf. The **vascular bundles** are found in this layer of the leaf. The vascular bundles include **xylem** and **phloem**, "vessels" that transport materials throughout the plant. At the **lower epidermis** are tiny openings called **stomates,** which allow for gas exchange and transpiration. Surrounding each stomate are **guard cells**, which control the opening and closing of the stomates.

A CLOSER LOOK AT PHOTOSYNTHESIS

There are two stages in photosynthesis: the **light reaction** (also called the light-dependent reaction) and the **dark reaction** (also called the light-independent reaction). The whole process begins when the **photons** (or "energy units") of sunlight strike a leaf, activating chlorophyll and exciting electrons. The activated chlorophyll molecule then passes these excited electrons down to a series of electron carriers, ultimately producing ATP and NADPH. The whole point of the light reaction is to produce two things: (1) energy in the form of ATP and (2) electron carriers, specifically NADPH.

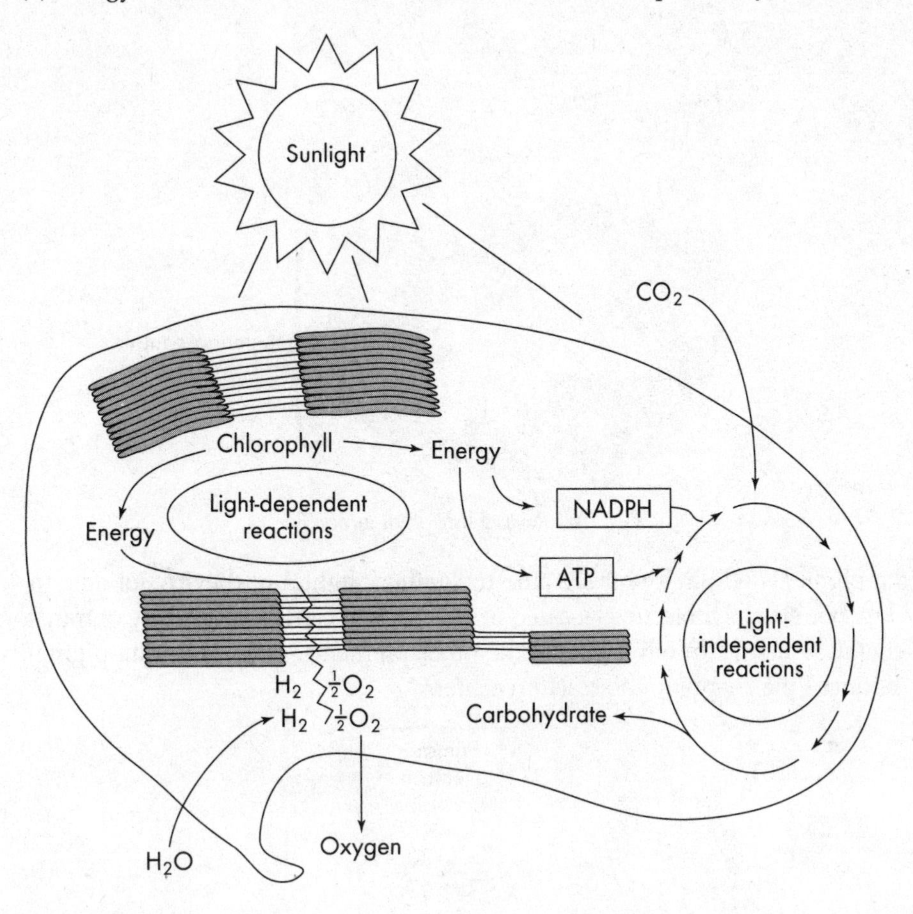

Both of these products, along with carbon dioxide, are then used in the dark reaction (light-independent) to make carbohydrates.

The Light Reaction

Many light-absorbing pigments participate in photosynthesis. Some of the more important ones are **chlorophyll *a*, chlorophyll *b***, and **carotenoids**. These pigments are clustered in the thylakoid membrane into units called antenna complexes.

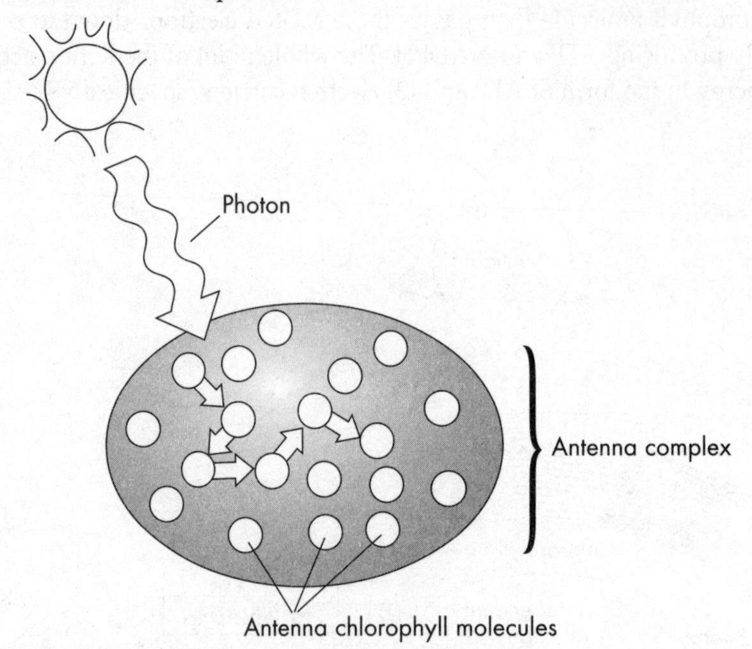

All of the pigments within a unit are able to "gather" light, but they're not able to "excite" the electrons. Only one special molecule—located in the **reaction center**—is capable of transforming light energy to chemical energy. In other words, the other pigments, called **antenna pigments**, "gather" light and "bounce" the energy to the reaction center.

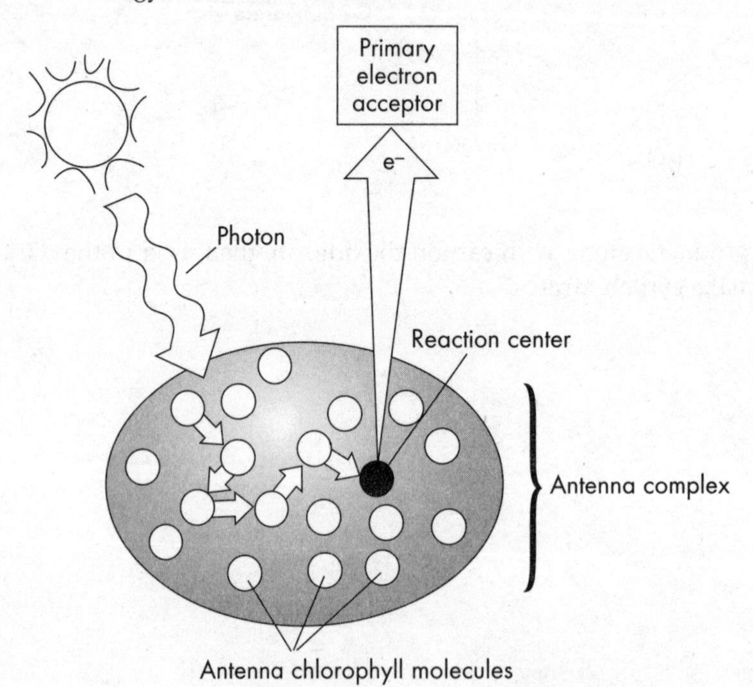

There are two types of reaction centers: **photosystem I (PS I)** and **photosystem II (PS II)**. The principal difference between the two is that each reaction center has a type of chlorophyll—chlorophyll a—that absorbs a particular wavelength of light. For example, **P680**, the reaction center of photosystem II, has a maximum absorption at a wavelength of 680 nanometers. The reaction center for photosystem I—**P700**—best absorbs light at a wavelength of 700 nanometers.

When light energy is used to make ATP, it is called **photophosphorylation**. We're using light (that's *photo*) and ADP and phosphates (that's *phosphorylation*) to produce ATP.

Noncyclic Photophosphorylation

The noncyclic method of photophosphorylation produces ATP using both photosystem I and photosystem II. When a leaf captures sunlight, the energy is sent to P680, the reaction center for photosystem II. The activated electrons are trapped by P680 and passed to a molecule called the primary acceptor. They are then passed down to carriers in the electron transport chain and eventually enter photosystem I. Some of the energy that dissipates as electrons move along the chain of acceptors will be used to "pump" protons across the membrane into the thylakoid lumen. When the reaction center P680 absorbs light, it also splits water into oxygen, hydrogen ions, and electrons. That process is called **photolysis**. The electrons from photolysis replace the missing electrons in photosystem II.

Remember the chemiosmotic theory mentioned in aerobic respiration? Well, the same mechanism applies to photosynthesis. Here's how it works: Hydrogen ions accumulate inside the thylakoids when photolysis occurs. A proton gradient is established. As the hydrogen ions move through ATP synthase, ADP and P_i produce ATP.

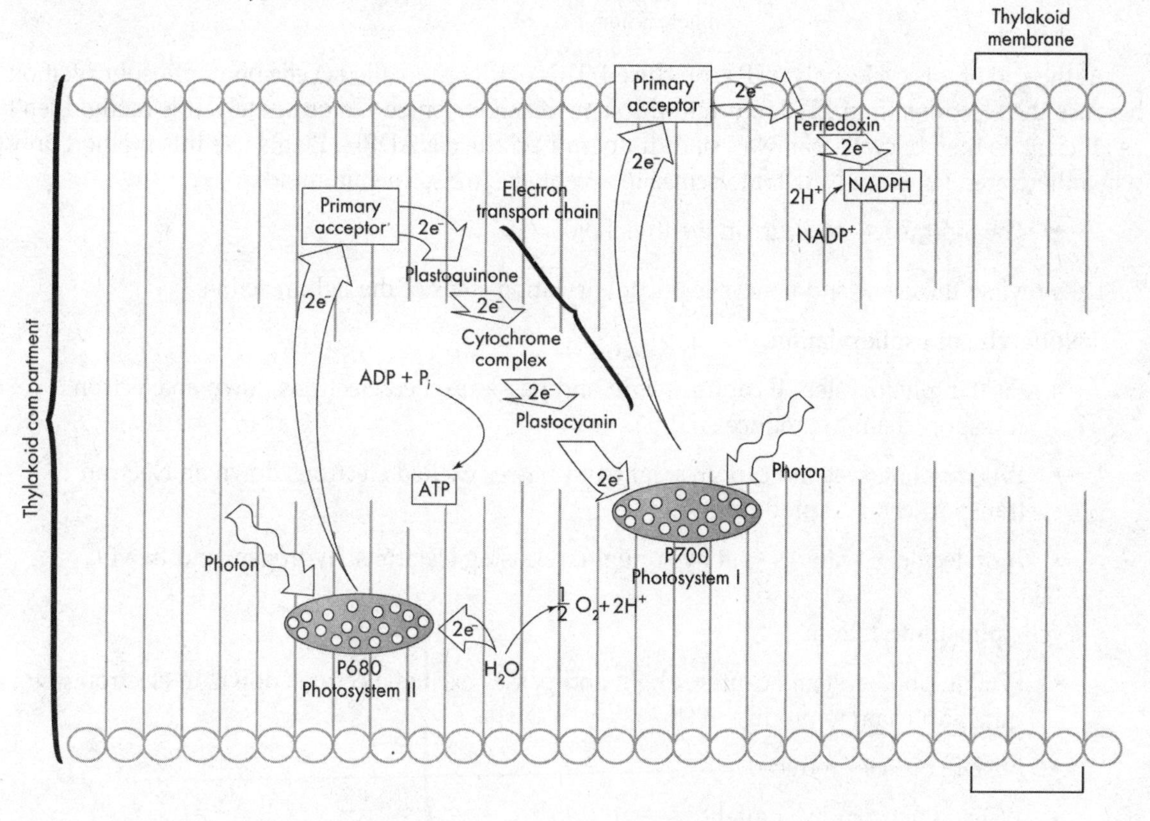

When these electrons in photosystem I receive a second boost, they're activated again. The electrons are passed through a second electron transport chain until they reach the final electron acceptor $NADP^+$ to make **NADPH**.

Cyclic Photophosphorylation

The cyclic method uses a much simpler pathway to generate ATP. The electrons in photosystem I are excited and leave the reaction center, P700. They are passed from carrier to carrier in the electron transport system and eventually return to P700:

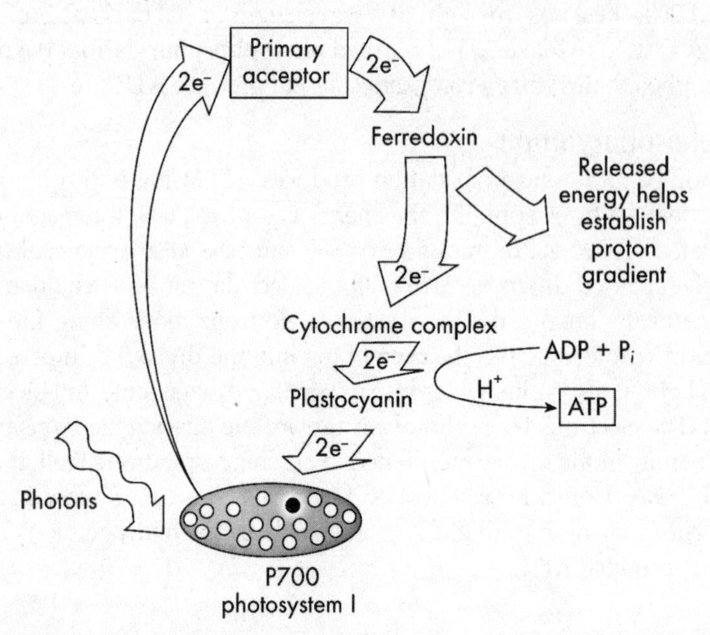

At the end of this cycle, only ATP is produced. This pathway is called cyclic photophosphorylation because the electrons from P700 return to the same reaction center. Unfortunately, this method isn't as efficient as the noncyclic pathway since it doesn't produce NADPH. Plants use this method only when there aren't enough NADP molecules to accept electrons. Keep in mind:

- The light reaction occurs in the thylakoids.

Let's review the cyclic and noncyclic phosphorylation steps of the light reaction:

Noncyclic phosphorylation:

- P680 in photosystem II captures light and passes excited electrons down an electron transport chain to produce ATP.
- P700 in photosystem I captures light and passes excited electrons down an electron transport chain to produce NADPH.
- A molecule of water is split by sunlight, releasing electrons, hydrogen, and free O_2.

Cyclic phosphorylation:

- P700 in photosystem I captures light and passes excited electrons down an electron transport chain to produce ATP.
- NADPH is *not* produced.
- Water is *not* split by sunlight.

Both reactions occur in the grana of chloroplasts, where the thylakoids are found. Remember: The light-absorbing pigments and enzymes for the light-dependent reactions are found within the thylakoids.

The Light-Independent Reaction

Now let's turn to the dark reaction. The dark reaction uses the products of the light reaction—ATP and NADPH—to make sugar. We now have energy to make glucose, but what do plants use as their carbon source? CO_2, of course. You've probably heard of the term **carbon fixation**. All this means is that CO_2 from the air is converted into carbohydrates. This step occurs in the stroma of the leaf.

The Calvin Cycle: The C_3 Pathway

We're finally ready to make glucose. CO_2 enters the **Calvin cycle** and combines with a 5-carbon molecule called **ribulose bisphosphate (RuBP)** to make an unstable six-carbon compound. The enzyme RuBP carboxylase, or *rubisco*, catalyzes this reaction.

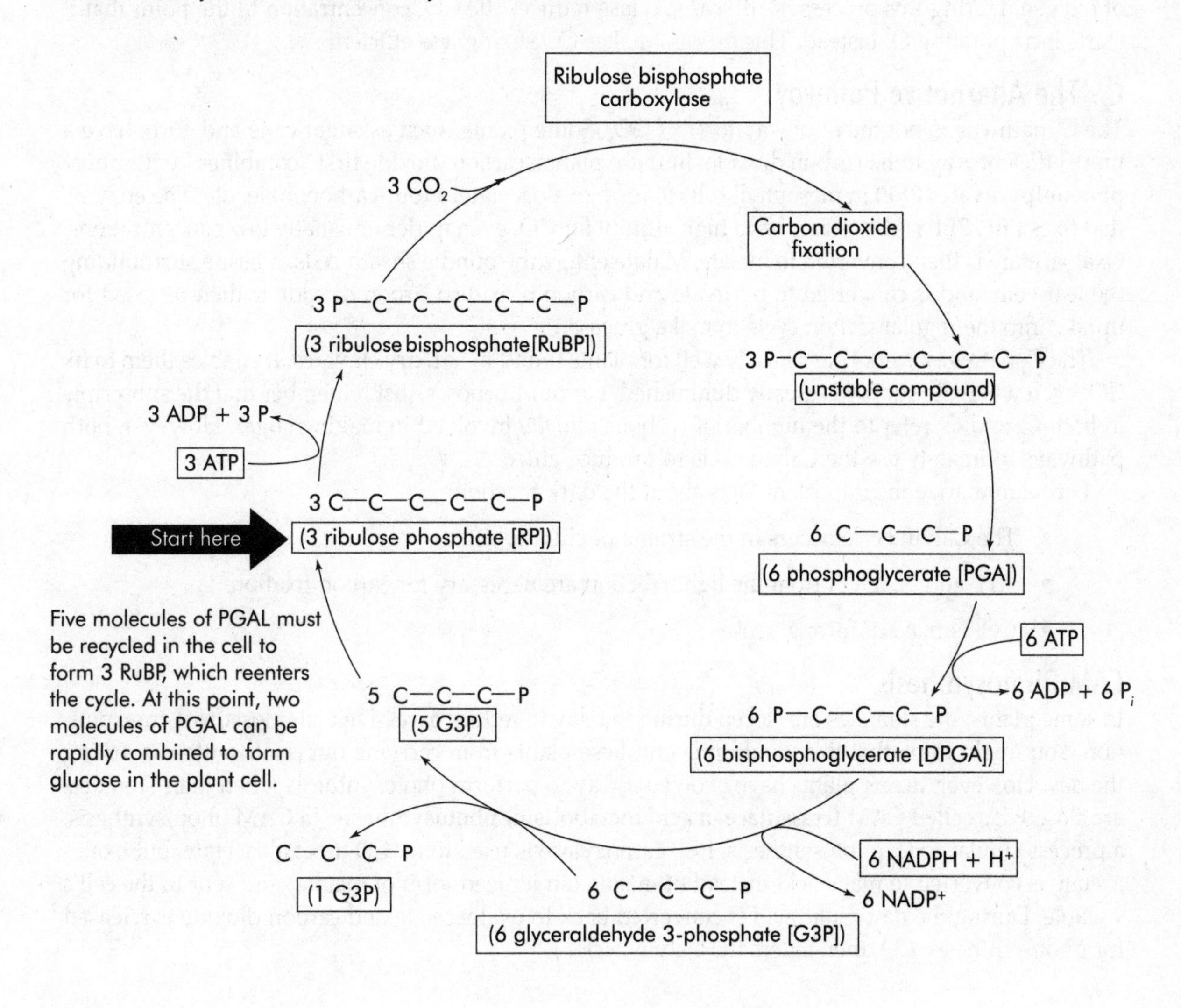

The easiest way to view the Calvin cycle is to consider 6 RuBP and 6 CO_2 at the start. Next, 12 ATP and 12 NADPH are used to convert 12 PGA to 12 G3P, an energy-rich molecule. ADP and $NADP^+$ are released and then recycled into the thylakoid where they will again be available for the light-dependent reactions. Two of the G3P are used to make glucose while the remaining 10 are rearranged into 6 RuBPs ready for the next round of the cycle. Since G3P, a three-carbon molecule, is the first stable product, this method of producing glucose is called the **C_3 pathway**.

Photorespiration

Sometimes intensely bright light tends to stunt the growth of C_3 plants. Why? Lighted conditions can trigger a process called **photorespiration**. Photorespiration is the pathway that leads to the fixation of oxygen. During this process, RuBP carboxylase reduces the CO_2 concentration to the point that it starts incorporating O_2 instead. This process makes CO_2-fixing less efficient.

C_4: The Alternative Pathway

The C_3 pathway is not the only way to "fix" CO_2. Some plants, such as sugar cane and corn, have a more efficient way to fix carbon dioxide. In these plants, carbon dioxide first "combines" with **phosphoenolpyruvate (PEP)** in mesophyll cells to form oxaloacetate, a four-carbon molecule. The enzyme that fixes PEP, **PEP carboxylase**, has a high affinity for CO_2 even under unusually low concentrations. Oxaloacetate is then converted to malate. Malate enters the bundle sheath cells, a tissue surrounding the leaf vein, and is converted to pyruvate and carbon dioxide. Carbon dioxide is then released for uptake into the regular Calvin cycle to make glucose.

The **C_4 pathway** works particularly well for plants found in hot, dry climates. It enables them to fix CO_2 even when the supply is greatly diminished. For our purposes, just remember that the subscripts in both C_3 and C_4 refer to the number of carbons *initially* involved in making sugar. However, both pathways ultimately use the Calvin cycle to produce glucose.

Let's summarize the important facts about the dark reaction:

- The Calvin cycle occurs in the stroma of chloroplasts.
- ATP and NADPH from the light reaction are necessary for carbon fixation.
- CO_2 is fixed to form glucose.

CAM Photosynthesis

In some plants, the stomates are closed during the day to reduce excessive water loss from transpiration. You might think that this would prevent these plants from carrying out photosynthesis during the day. However, desert plants have evolved a way to perform photosynthesis when their stomates are closed; it's called **CAM (crassulacean acid metabolism) photosynthesis**. In CAM photosynthesis, a process similar to C_4 photosynthesis, PEP carboxylase is used to fix CO_2 to oxaloacetate, but oxaloacetate is converted to malic acid instead of malate (an ionized form of malate) and sent to the cell's vacuole. During the day, malic acid is converted back to oxaloacetate and carbon dioxide is released for photosynthesis. CO_2 then enters the Calvin cycle.

KEY WORDS

cuticle
upper epidermis
palisade parenchyma
stroma
grana
thylakoids
spongy parenchyma
vascular bundles
xylem
phloem
lower epidermis
stomates
guard cells
light reaction (light-dependent reaction)
dark reaction (light-independent reaction)
photons
chlorophyll *a*
chlorophyll *b*
carotenoids
reaction center
antenna pigments
photosystem I (PS I)
photosystem II (PS II)
P680
P700
photophosphorylation
photolysis
NADPH
noncyclic photophosphorylation
cyclic photophosphorylation
carbon fixation
Calvin cycle
ribulose bisphosphate (RuBP)
C_3 pathway
photorespiration
phosphoenolpyruvate (PEP)
PEP carboxylase
C_4 pathway
CAM (crassulacean acid metabolism) photosynthesis

CHAPTER 6 REVIEW QUESTIONS

Answers can be found in Chapter 15.

1. Antenna complexes are located on the surface of thylakoid membranes and play an important role in photosynthesis. Surrounding the reaction center of an antenna complex are many antenna pigments. What is the functional role of these antenna pigments?

   (A) They are photosystems, which harness light and directly generate ATP by photophosphorylation.

   (B) They absorb and harness green light to ensure that only the correct wavelengths of light reach the reaction center.

   (C) They absorb light and transfer its energy to the reaction center.

   (D) They produce chlorophyll a and chlorophyll b to be used by the reaction centers for photophosphorylation.

2. The light-dependent and light-independent stages of photosynthesis are both linked together. Which of the following is generated by the light-independent reactions and is used in the light-dependent reactions?

   (A) ATP

   (B) NAD^+

   (C) NADPH

   (D) $NADP^+$

3. Carbon dioxide generated as a by-product of cellular respiration is recycled during photosynthesis. During which process of photosynthesis is carbon dioxide consumed?

   (A) Cyclic photophosphorylation

   (B) Noncyclic photophosphorylation

   (C) The C_3 pathway

   (D) Krebs cycle

Questions 4 – 6 refer to the following diagram and paragraph.

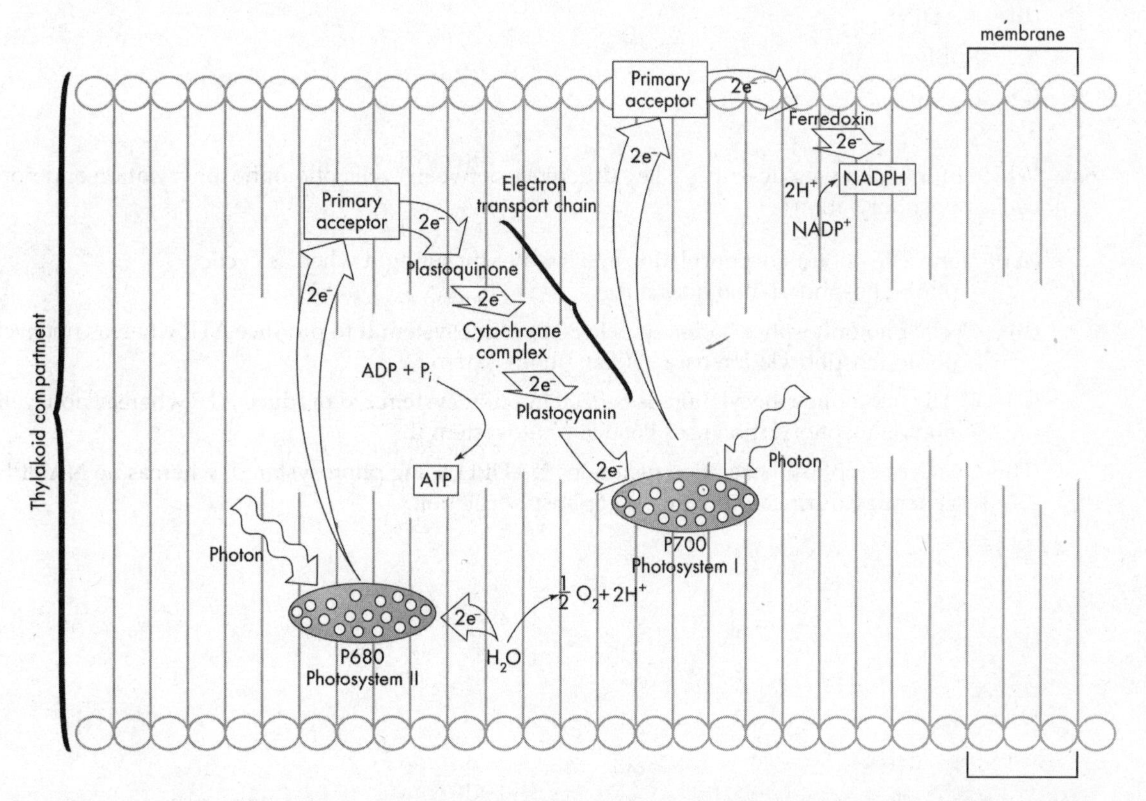

Noncyclic photophosphorylation (shown above) is a critical process by which photophosphorylation of ATP occur through the use of both photosystems I and II. During noncyclic photophosphorylation, light activates photosystem II that donates electrons to the primary acceptor molecule, beginning an electron transport chain cascade that ultimately generates ATP and activates photosystem I. Photosystem I donates electrons to another primary acceptor that can be used to generate NADPH. Although this pathway is more complex than cyclic photophosphorylation, it is the preferred mechanism of photophosphorylation in the chloroplast.

4. Why is noncyclic photophosphorylation preferred over cyclic photophosphorylation despite being more complex?

   (A) Cyclic photophosphorylation fails to produce ATP, instead producing lower energy ADP molecules.

   (B) Cyclic photophosphorylation fails to produce NADPH, which is necessary for the light-independent pathway.

   (C) Cyclic photophosphorylation fails to recycle CO_2 gas generated as a byproduct of cellular respiration in the mitochondria.

   (D) Cyclic photophosphorylation requires more energy input than noncyclic photophosphorylation.

5. Suppose a scientist treats a plant with a ferredoxin inhibitor. Which of the following molecules may still be generated under these conditions?

(A) Glucose

(B) NADPH

(C) RuBP

(D) O_2

6. Which of the following describes a key difference between cyclic photophosphorylation and noncyclic phosphorylation?

(A) Noncyclic photophosphorylation splits water by sunlight whereas cyclic photophosphorylation does not.

(B) Cyclic photophosphorylation uses P680 in photosystem II to produce ATP whereas noncyclic photophosphorylation uses P700 in photosystem I.

(C) Cyclic photophosphorylation uses P700 in photosystem I to produce ATP whereas noncyclic photophosphorylation uses P680 in photosystem II.

(D) Cyclic photophosphorylation generates NADPH using photosystem I, whereas no NADPH is generated during noncyclic photophosphorylation.

Molecular Genetics

DNA: THE BLUEPRINT OF LIFE

All living things possess an astonishing degree of organization. From the simplest single-celled organism to the largest mammal, millions of reactions and events must be coordinated precisely for life to exist. This coordination is directed from the nucleus of the cell, by deoxyribonucleic acid, or DNA. DNA is the hereditary blueprint of the cell.

The DNA of a cell is contained in structures called chromosomes. The chromosomes consist of DNA wrapped around proteins called histones. When the genetic material is in a loose form in the nucleus it is called **euchromatin,** and its genes are active, or available for transcription. When the genetic material is fully condensed into coils it is called **heterochromatin,** and its genes are generally inactive. Situated in the nucleus, chromosomes direct and control all the processes necessary for life, including passing themselves and their information on to future generations. In this chapter, we'll look at precisely how they accomplish this.

THE MOLECULAR STRUCTURE OF DNA

The DNA molecule consists of two strands that wrap around each other to form a long, twisted ladder called a **double helix**. The structure of DNA was brilliantly deduced in 1956 by two scientists named Watson and Crick.

DNA is made up of repeated subunits of **nucleotides**. Each nucleotide has a **five-carbon sugar,** a **phosphate**, and a **nitrogenous base**. Take a look at the nucleotide below. This particular nucleotide contains a nitrogenous base called **adenine**:

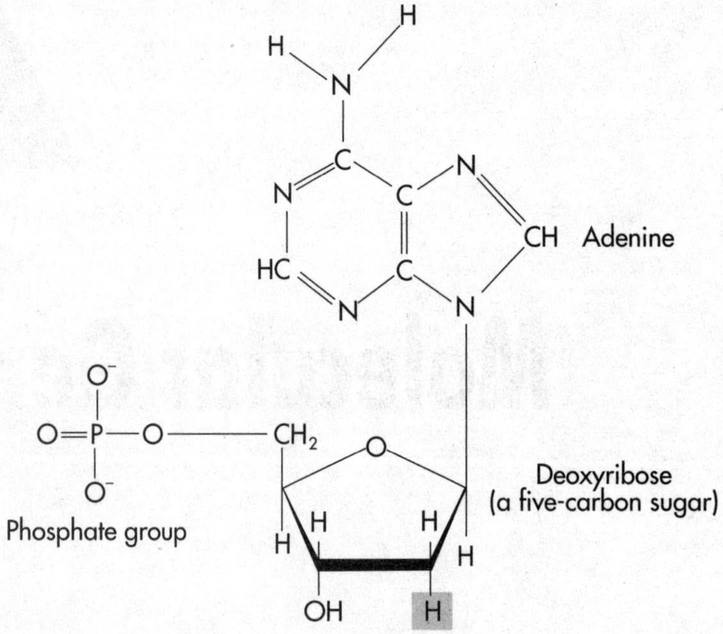

Adenine

Deoxyribose
(a five-carbon sugar)

Phosphate group

The name of the pentagon-shaped sugar in DNA is **deoxyribose**. Hence, the name *deoxyribo*nucleic acid. Notice that the sugar is linked to two things: a phosphate and a nitrogenous base. A nucleotide can have one of four different nitrogenous bases:

- **Adenine**—a purine (double-ringed nitrogenous base)
- **Guanine**—a purine (double-ringed nitrogenous base)
- **Cytosine**—a pyrimidine (single-ringed nitrogenous base)
- **Thymine**—a pyrimidine (single-ringed nitrogenous base)

Any of these four bases can attach to the sugar. As we'll soon see, this is extremely important when it comes to the "sense" of the genetic code in DNA.

The nucleotides can link up in a long chain to form a single strand of DNA. Here's a small section of a DNA strand:

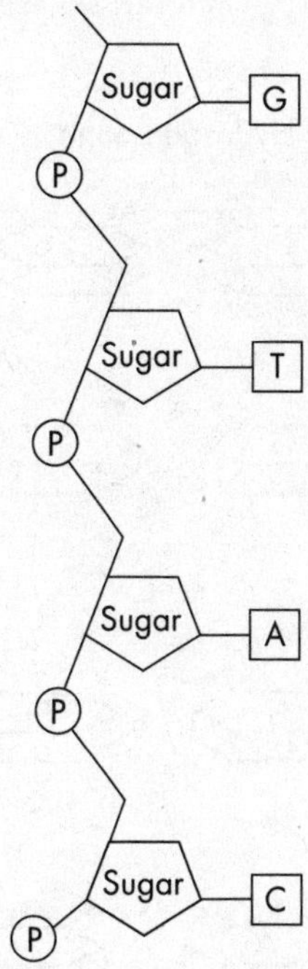

Sugar G
P
Sugar T
P
Sugar A
P
Sugar C
P

The nucleotides themselves are linked together by **phosphodiester bonds**.

Two DNA Strands

Now let's look at the way two DNA strands get together. Again, think of DNA as a ladder. The sides of the ladder consist of alternating sugar and phosphate groups, while the rungs of the ladder consist of a pair of nitrogenous bases:

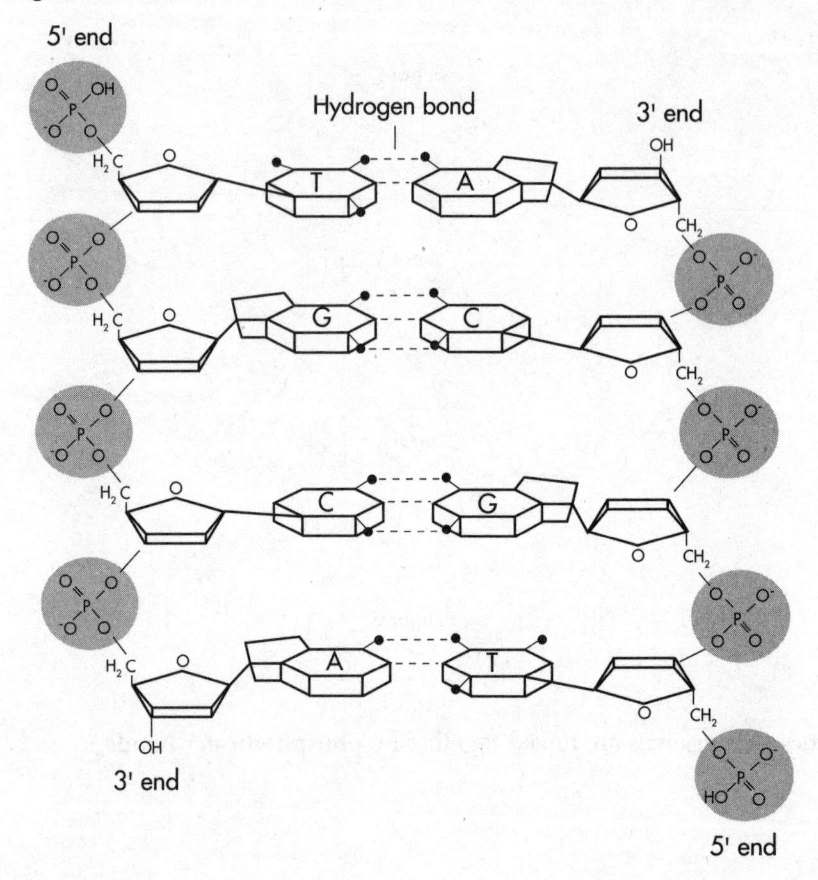

The nitrogenous bases pair up in a particular way. Adenine in one strand always binds to thymine (A–T or T–A) in the other strand. Similarly, guanine always binds to cytosine (G–C or C–G). This predictable matching of the bases is known as **base pairing**.

The two strands are said to be **complementary**. This means that if you know the sequence of bases in one strand, you'll know the sequence of bases in the other strand. For example, if the base sequence in one DNA strand is A–T–C, the base sequence in the complementary strand will be T–A–G.

The two DNA strands run in opposite directions. You'll notice from the figure above that each DNA strand has a 5' end and a 3' end, so-called for the carbon that ends the strand (i.e., the fifth carbon in the sugar ring is at the 5' end, while the third carbon is at the 3' end). The 5' end has a phosphate group and the 3' end has an OH, or "hydroxyl," group. The 5' end of one strand is always opposite to the 3' end of the other strand. The strands are therefore said to be **antiparallel**.

The DNA strands are linked by **hydrogen bonds**. Two hydrogen bonds hold adenine and thymine together and three hydrogen bonds hold cytosine and guanine together.

Before we go any further, let's review the base pairing in DNA.

- Adenine pairs up with thymine (A–T or T–A) by forming *two* hydrogen bonds.
- Cytosine pairs up with guanine (G–C or C–G) by forming *three* hydrogen bonds.

WHY DNA IS IMPORTANT

You'll recall from our discussion of bioenergetics in Chapter 5 that enzymes are proteins that are essential for life. This is true not only because they help liberate energy stored in chemical bonds, but also because they direct the construction of the cell. This is where DNA comes into the picture.

DNA's main role is directing the manufacture of proteins. These proteins, in turn, regulate everything that occurs in the cell. But DNA does not *directly* manufacture proteins. Instead, DNA passes its information to an intermediate molecule known as ribonucleic acid (RNA). These RNA molecules carry out the instructions in DNA, producing the proteins that determine the course of life.

The flow of genetic information is therefore:

$$\text{DNA} \rightarrow \text{RNA} \rightarrow \text{proteins}$$

This is the central dogma of molecular biology.

We said in the beginning of this chapter that DNA is the hereditary blueprint of the cell. By directing the manufacture of proteins, DNA serves as the cell's blueprint. But how is DNA inherited? For the information in DNA to be passed on, it must first be copied. This copying of DNA is known as **DNA replication**.

DNA REPLICATION

Because the DNA molecule is twisted over on itself, the first step in replication is to unwind the double helix by breaking the hydrogen bonds. This is accomplished by an enzyme called **helicase**. The exposed DNA strands now form a *y*-shaped **replication fork**:

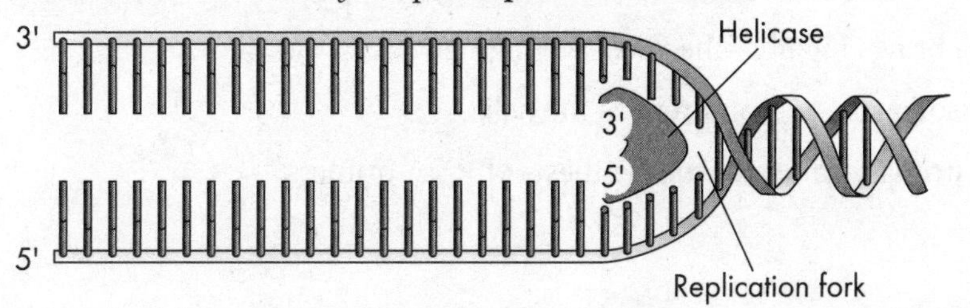

Now each strand can serve as a template for the synthesis of another strand. DNA replication begins at specific sites called **origins of replication**. Because the DNA helix twists and rotates during DNA replication, another class of enzymes, called DNA **topoisomerases**, cuts and rejoins the helix to prevent tangling. The enzyme that performs the actual addition of nucleotides alongside the naked strand is **DNA polymerase**. But DNA polymerase, oddly enough, can only add nucleotides to the 3' end of an existing strand. Therefore, to start off replication at the 5' end, DNA polymerase must add nucleotides to an **RNA primase**—a short strand of RNA nucleotides. The primer is later degraded by enzymes and the space is filled with DNA.

One strand is called the **leading strand**, and it is made continuously. That is, the nucleotides are steadily added one after the other by DNA polymerase. The other strand—the **lagging strand**—is made discontinuously. Unlike the leading strand, the lagging strand is made in pieces of nucleotides known as **Okazaki fragments**. Why is the lagging strand made in small pieces?

Normally, nucleotides are added only in the 5' to 3' direction. However, when the double-helix is "unzipped," one of the two strands is oriented in the opposite direction—3' to 5'. Because DNA polymerase doesn't work in this direction, nucleotides—the Okazaki fragments—need to be added in pieces. These fragments are eventually linked together by the enzyme **DNA ligase** to produce a continuous strand. Finally, hydrogen bonds form between the new base pairs, leaving two identical copies of the original DNA molecule.

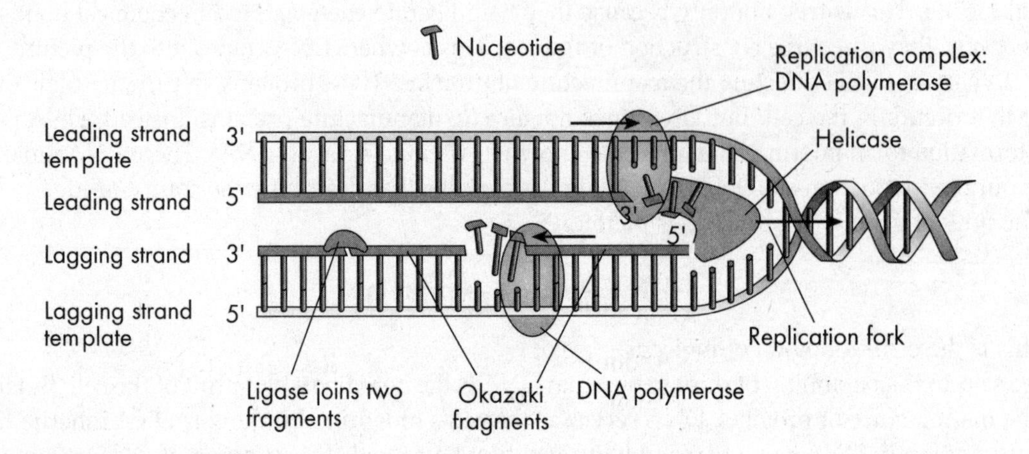

When DNA is replicated, we don't end up with two entirely new molecules. Each new molecule has half of the original molecule. Because DNA replicates in this way, by conserving half of the original molecule in each of the two new ones, it is said to be **semiconservative**.

Many enzymes and proteins are involved in DNA replication. The ones you'll need to know for the AP Biology Exam are DNA helicase, DNA polymerase, DNA ligase, topoisomerase, and RNA primase.

- **Helicase** unwinds our double helix into two strands.
- **Polymerase** adds nucleotides to an existing strand.
- **Ligase** brings together the Okazaki fragments.
- **Topoisomerase** cuts and rejoins the helix.
- **RNA primase** catalyzes the synthesis of RNA primers.

RNA

Now that we've seen how DNA is replicated, let's take a look at how the genetic code is expressed as proteins. As we mentioned earlier, genetic information is first passed to an intermediate molecule called RNA. Proteins called **transcription factors** control the transfer of genetic information from DNA to RNA by binding to specific DNA sequences. Here's a "roadmap" of how information is transferred from DNA to proteins:

$$\text{DNA} \xrightarrow[\text{in the nucleus}]{\text{transcription}} \text{RNA} \xrightarrow[\text{in the cytoplasm}]{\text{translation}} \text{Proteins}$$

Before we discuss what RNA does, let's talk about its structure. Although RNA is also made up of nucleotides, it differs from DNA in three ways:

1. RNA is single-stranded, not double-stranded.
2. The five-carbon sugar in RNA is **ribose** instead of deoxyribose.
3. The RNA nitrogenous bases are adenine, guanine, cytosine, and a different base called **uracil**. Uracil replaces thymine.

Here's a table to compare DNA and RNA. Keep these differences in mind—ETS loves to test you on them.

DIFFERENCES BETWEEN DNA AND RNA		
	DNA (double-stranded)	**RNA (single-stranded)**
Sugar:	deoxyribose	ribose
Bases:	adenine guanine cytosine thymine	adenine guanine cytosine uracil

There are three types of RNA: messenger RNA (mRNA), ribosomal RNA (rRNA), and transfer RNA (tRNA). All three types of RNA are key players in the synthesis of proteins.

- **Messenger RNA (mRNA)** copies the information stored in the strand of DNA and carries it to the ribosome.
- **Ribosomal RNA (rRNA)**, which is produced in the nucleolus, makes up part of the ribosomes. You'll recall from our discussion of the cell in Chapter 3 that the ribosomes are the sites of protein synthesis. We'll see how they function a little later on.
- **Transfer RNA (tRNA)** shuttles amino acids to the ribosomes. It is responsible for bringing the appropriate amino acids into place at the appropriate time. It does this by reading the message carried by the mRNA.

Now that we know about the different types of RNA, let's see how they direct the synthesis of proteins.

PROTEIN SYNTHESIS

Protein synthesis involves three basic steps: **transcription**, **RNA processing**, and **translation**.

TRANSCRIPTION

Transcription involves copying the genetic code directly from DNA. The initial steps in transcription are similar to the initial steps in DNA replication. The obvious difference is that whereas in replication we end up with a complete copy of the cell's DNA, in transcription we end up with only a partial copy in the form of mRNA.

Transcription involves three phases: initiation, elongation, and termination. As in DNA replication, the first initiation step in transcription is to unwind and unzip the DNA strands using helicase. Transcription begins at special sequences of the DNA strand called **promoters**. Because RNA is single-stranded, we have to copy only *one* of the two DNA strands. The strand that serves as the template is the **sense strand**. The other strand that lies dormant is the **antisense strand.**

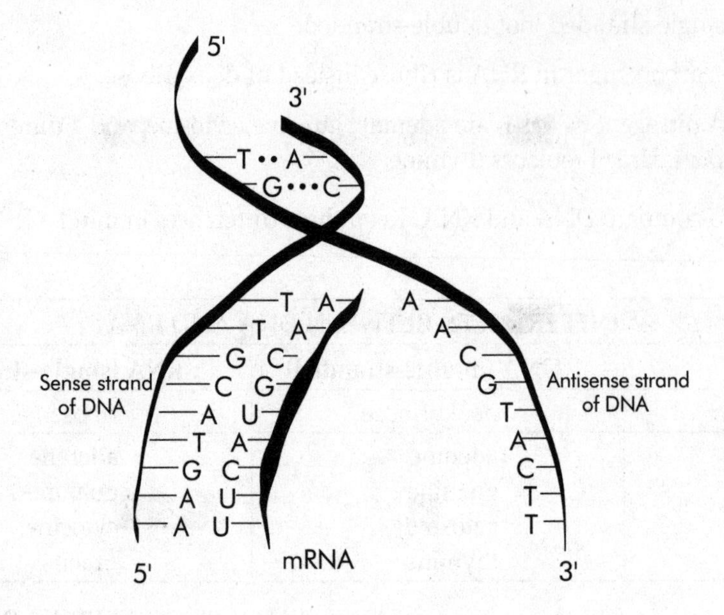

This time, RNA nucleotides line up alongside one DNA strand to form an mRNA strand. Another enzyme, **RNA polymerase,** brings free-floating RNA nucleotides to the DNA strand. As we saw earlier, guanines and cytosines pair up. However, the RNA molecule replaces the DNA's thymine molecules with uracils. In other words, the exposed adenines are now paired up with uracil instead of thymine.

Once the mRNA finishes adding on nucleotides and reaches the termination sequence, it separates from the DNA template, completing the process of transcription. The new RNA has now transcribed, or "copied," the sequence of nucleotide bases directly from the exposed DNA strand.

RNA Processing

Now the mRNA strand is ready to move out of the nucleus. But before the mRNA molecule can leave the nucleus, it must be processed. That means the mRNA has to be modified before it exits the nucleus.

A newly made mRNA molecule (called the heterogeneous nuclear RNA, or hnRNA) contains more nucleotides than it needs to code for a protein. The mRNA consists of both coding regions and noncoding regions. The regions that express the code for the polypeptide are **exons**. The noncoding regions in the mRNA are **introns**.

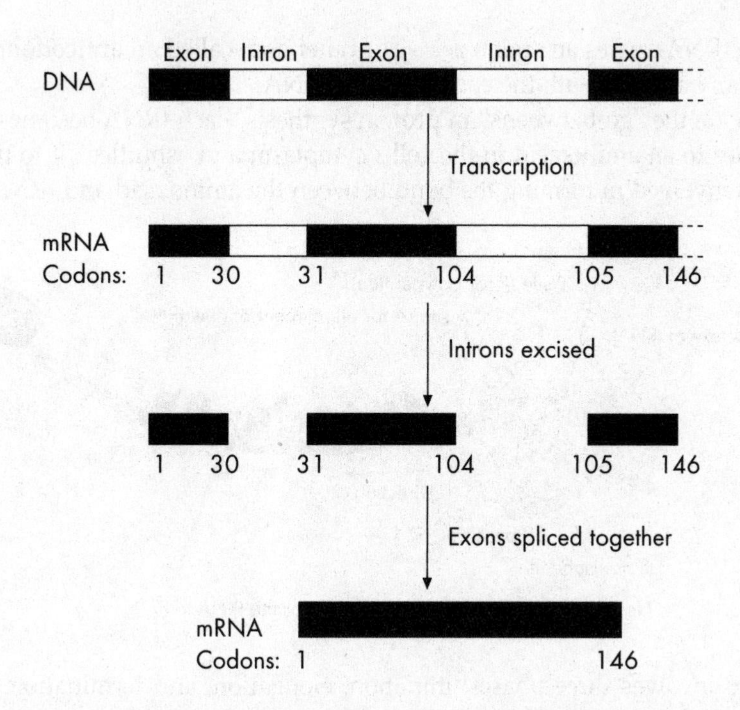

The introns—the intervening sequences—must be removed before the mRNA leaves the nucleus. The removal of introns is a complex process accomplished by an RNA-protein complex called a **spliceosome**. In addition, a **poly(A) tail** is added to the 3' end and a **5' GTP cap** is added to the 5' end. This process produces a final mRNA that is shorter than the transcribed mRNA.

Translation

Now, the mRNA leaves the nucleus and searches for a ribosome. The mRNA molecule carries the message from DNA in the form of **codons**, a group of three bases, or "letters," that corresponds to one of 20 amino acids. The genetic code is redundant, meaning that certain amino acids are specified by more than one codon.

The mRNA attaches to the ribosome to initiate translation and "waits" for the appropriate amino acids to come to the ribosome. That's where tRNA comes in. A tRNA molecule has a unique three-dimensional structure that resembles a four-leaf clover:

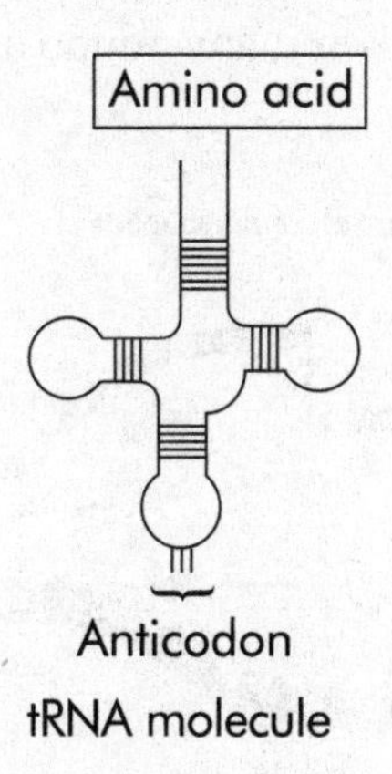

tRNA molecule

One end of the tRNA carries an amino acid. The other end, called an **anticodon**, has three nitrogenous bases that can base pair with the codon in the mRNA.

Transfer RNAs are the "go-betweens" in protein synthesis. Each tRNA becomes charged and enzymatically attaches to an amino acid in the cell's cytoplasm and "shuttles" it to the ribosome. The charging enzymes involved in forming the bond between the amino acid and tRNA require ATP.

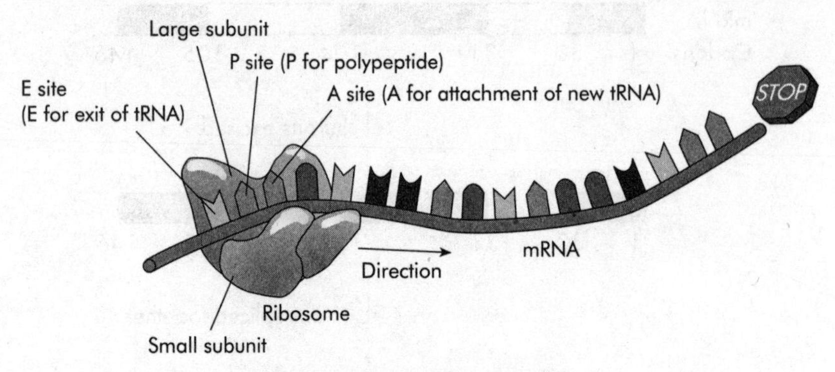

Note: Each site E, P, or A are grooves for the tRNA to fit.

Translation also involves three phases: initiation, elongation, and termination. Initiation begins when a ribosome attaches to the mRNA.

What does the ribosome do? It helps the process along by holding everything in place while the RNAs assist in assembling polypeptides.

Initiation

Ribosomes contain three binding sites: an **A site**, a **P site**, and an **E site**. An initiator tRNA serves to activate translation and occupies the P site. In all organisms the codon for the initiation of protein synthesis is A–U–G, which codes for the amino acid methionine. The tRNA with the complementary anticodon, U–A–C, is methionine's personal shuttle. Once the methionine tRNA is attached to the P site, the A site can be filled by the appropriate tRNA that corresponds to the next codon. The E site binds a free tRNA before it exits the ribosome.

Elongation

The addition of amino acids is called elongation. Remember that the mRNA contains many codons, or "triplets," of nucleotide bases. As each amino acid is brought to the mRNA, it is linked to its neighboring amino acid by a peptide bond. When many amino acids link up, a polypeptide is formed.

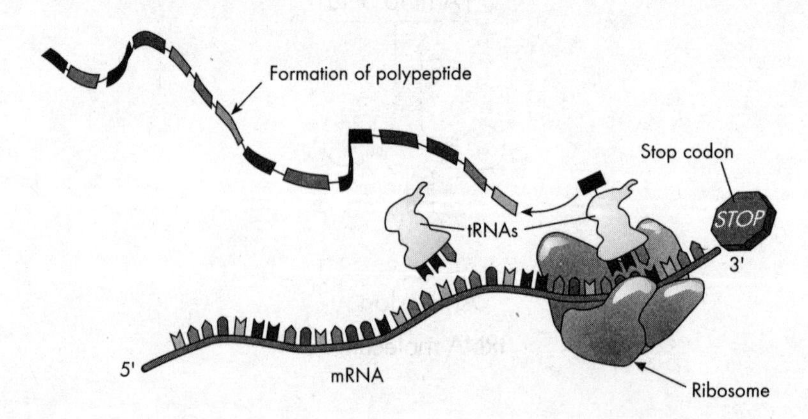

Termination

How does this process know when to stop? The synthesis of a polypeptide is ended by stop codons. A codon doesn't always code for an amino acid; there are three that serve as a stop codon. Termination occurs when the ribosome runs into one of these three stop codons.

Higher Protein Structure

The polypeptide has to go through several changes before it can officially be called a protein. Proteins can have four levels of structure. The linear sequence of the amino acids is called the **primary structure** of a protein. Now the polypeptide begins to twist, forming either a coil (known as an alpha helix), or zigzagging patterns (known as beta-pleated sheets). These are examples of proteins' **secondary structures**. Next, the polypeptide folds in a three dimensional pattern. This is called the **tertiary structure**. Finally, when two or more polypeptides get together, we call it a **quaternary structure**. Once that's complete the protein is ready to perform its task.

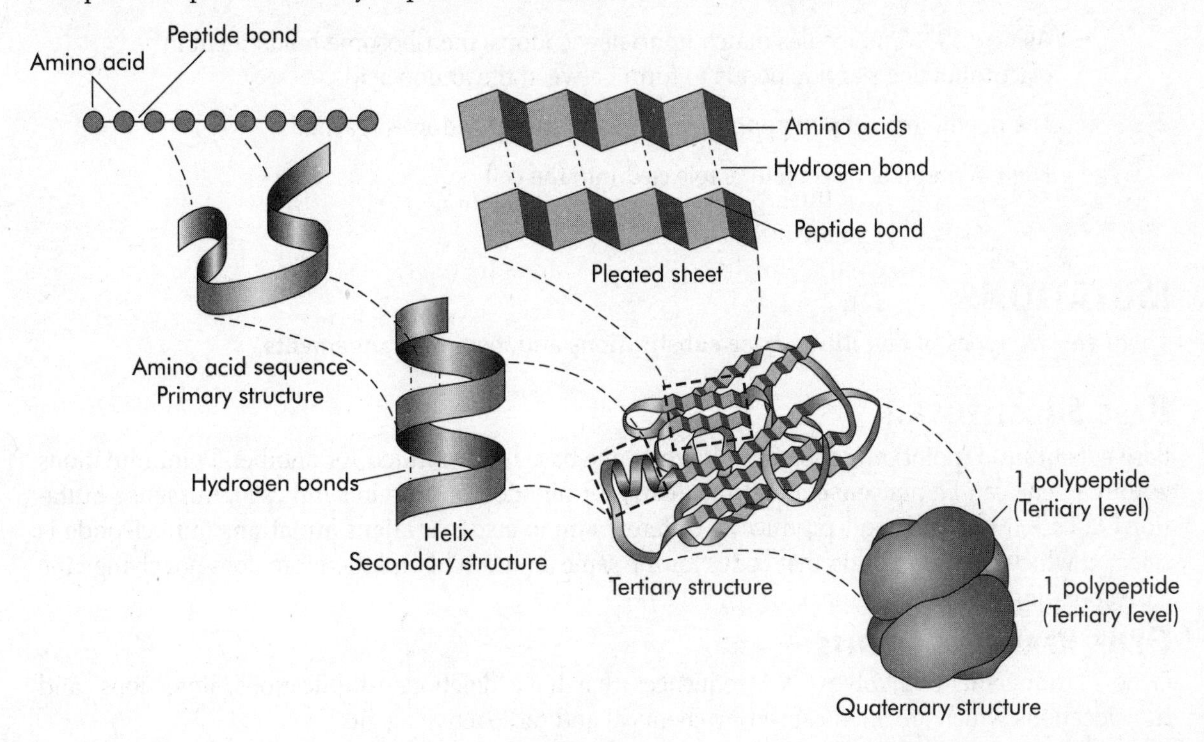

One more thing: In some cases, the folding of proteins involves other proteins known as **chaperone proteins** (or **chaperonins**). They help the protein fold properly and make the process more efficient.

How about a little review?

- In transcription, mRNA copies, or "transcribes," the code from an exposed strand of DNA in the nucleus.
- The mRNA is "processed" by having its introns, or noncoding sequences, removed.
- Now, ready to be translated, mRNA proceeds to the ribosome.
- Free-floating amino acids are picked up by tRNA and shuttled over to the ribosome, where mRNA awaits.
- In translation, the anticodon of a tRNA molecule carrying the appropriate amino acid base pairs with the codon on the mRNA.
- As new tRNA molecules match up to new codons, the ribosome holds them in place, allowing peptide bonds to form between the amino acids.
- The newly formed polypeptide grows until a stop codon is reached.
- The polypeptide or protein is released into the cell.

MUTATIONS

There are two types of **mutations**: **base substitutions** and **gene rearrangements.**

BASE SUBSTITUTION

Base substitution (point) mutations result when one base is substituted for another. Point mutations within a gene can be **nonsense mutations** (early termination of protein synthesis), **missense mutations** (a codon is altered and produces a different amino acid), or **silent mutations** (a nucleotide is selected which creates a codon that codes for the same amino acid and therefore does not change the corresponding protein sequence).

GENE REARRANGEMENTS

Gene rearrangements involve DNA sequences that have deletions, duplications, inversions, and translocations which are often caused by chemical and radioactive agents.

1. **Insertions** and **deletions** result in the gain or loss, respectively, of DNA or a gene. They can either involve the addition (insertion) or removal (deletion) of as little as a single base or much larger sequences of DNA. Insertions and deletions may have devastating consequences during protein translation because introduction or deletion of bases often results in a change in the sequence of codons used by the ribosome (called a frameshift) to synthesize a polyprotein.

   Example:

Original DNA Sequence:	ATG	TAT	AAA	CAT	TGA	
Original Polyprotein:	Met	Tyr	Lys	His	Stop	
DNA Sequence after Insertion:	ATG	CTA	TAA	ACA	TTG	A
Polyprotein Sequence after Insertion:	Met	Leu	Stop	-	-	

   In this example, the insertion of an additional cytosine nucleotide resulted in a frameshift and a premature stop codon.

2. **Duplications** can result in an extra copy of genes, and are usually caused by unequal crossing over during meiosis or chromosome rearrangements. This may result in new traits, since one copy can maintain the gene's original function and one copy man evolve a new function.
3. **Inversions** can result when changes occur in the orientation of chromosomal regions. This may cause harmful effects if the inversion involves a gene or an important sequence involved in the regulation of **gene expression.**
4. **Translocations** occur when a portion of two different chromosomes (or a single chromosome in two different places) breaks and rejoins in a way that the DNA sequence or gene is lost, repeated, or interrupted.

Either type of mutation, if not repaired, will be kept in subsequent rounds of replication. Mutations in somatic cells may damage the cell, make it cancerous or kill it. Mutations in a germ cell will be passed down to the next generation.

RECOMBINANT DNA

Recombinant DNA is generated by combining DNA from multiple sources to create a unique DNA that is not found in nature. A common application of recombinant DNA technology is the introduction of a eukaryotic gene of interest (such as insulin) into a bacterium for production. This branch of technology that produces new organisms or products by transferring genes between cells is called **genetic engineering.**

How does this process work? Unique sequences, which flank the gene of interest in eukaryotic DNA, are recognized and cut by special enzymes called restriction enzymes. Using the same **restriction enzymes,** a cut is made in a circular non-essential bacterial DNA called a **plasmid** to create a site for insertion of the eukaryotic DNA. Once the plasmid has been cut, the eukaryotic DNA fragment containing the gene of interest is inserted into the plasmid by the enzyme DNA ligase.

In order to introduce the recombinant DNA plasmid containing the eukaryotic gene of interest into a bacterial cell, the plasmid DNA is combined with bacteria and placed under conditions that favor uptake of the DNA. This process is called **transformation**. Once transformed, these bacteria may be expanded and the gene or protein of interest purified in large quantity.

The DNA fragments can be separated according to their molecular weight using **gel electrophoresis.** Because DNA and RNA are negatively charged, they migrate through the gel toward the positive pole of the electrical field. The smaller the fragments, the faster they move through the gel. Restriction enzymes are also used to create a molecular fingerprint. When restriction fragments between individuals of the same species are compared, the fragments differ in length because of polymorphisms, which are slight differences in DNA sequences. These fragments are called **restriction fragment length polymorphisms,** or **RFLPs.** In **DNA fingerprinting,** RFLPs produced from DNA left at a crime scene are compared to RFLPs from the DNA of suspects.

BIOTECHNOLOGY

POLYMERASE CHAIN REACTION (PCR)

A few years ago, it would take weeks of tedious experiments to identify and study specific genes. Today, thanks to **polymerase chain reaction (PCR)**, we are able to make billions of identical copies of genes within a few hours. To do PCR, the process of DNA replication is slightly modified. In a small PCR tube, DNA, primers, Taq Polymerase, and lots of DNA nucleotides (A's, C's, G's, and T's) are mixed together.

In a PCR machine, or thermocycler, the tube is heated, cooled, and warmed many times. Each time the machine is heated, the hydrogen bonds break, separating the double-stranded DNA. As it is cooled, the primers bind to the sequence flanking the region of the DNA we want to copy. When it is warmed, Taq Polymerase binds to the primers on both strands and adds nucleotides on each template strand. After this first cycle is finished, there are two identical double-stranded DNA molecules. When the second cycle is completed, these two double-stranded DNA segments will have been copied into four. The process repeats itself over and over, creating as much DNA as needed. Today, a thermocycler is commonplace in science labs. It is regularly used to study small amounts of DNA from crime scenes, determine the origin of our foods, detect diseases in animals and humans, and to better understand the inner workings of our cells.

TRANSFORMATION

Insulin, the protein hormone that lowers blood sugar levels, can now be made for medical purposes by bacteria. Yes, bacteria can be induced to use the universal DNA code to transcribe and translate a human gene! This can be done by transformation.

Genes of interest are first placed into a transformation vector, such as a plasmid. Plasmid vectors are small, circular pieces of DNA that contain genes for antibiotic resistance and restriction sites. Plasmids and the gene of interest are cut with the same restriction enzyme, creating compatible sticky ends. When placed together, the gene is inserted into the plasmid creating recombinant DNA.

The bacteria are then transformed using the recombinant plasmid. In most AP biology classes, this is done by the heat shock method. Because the plasmid contains a gene for antibiotic resistance, transformed bacteria will be able to grow on a medium that contains antibiotics whereas bacteria without the plasmid will die. This allows scientists to cleverly identify transformed bacteria.

This laboratory technique has not only been used to safely mass-produce important proteins used for medicine, like insulin, it has an important role in the study of gene expression.

HUMAN GENOME PROJECT

In 1990, an international, publicly funded consortium of scientists was determined to sequence every chromosome, base by base, in the human genome. The latest DNA sequencing machines and innovative computer programming skills were used in this tedious process. A draft of the sequence was first published in 2001. Today, in addition to the human genome, the genomes of many different species are available online for anyone to study at www.ornl.gov/sci/techresources/Human_Genome/home.shtml.

These genomes can be used to study our evolutionary history, genetic diseases, and can be used to make pharmaceuticals.

VIRUSES

Viruses are nonliving agents capable of infecting cells. Why are viruses considered nonliving? They require a host cell's machinery in order to replicate. A virus consists of two main components: a protein capsid and genetic material made of DNA or RNA, depending on the virus.

Viruses vary according to their replication needs as dictated by their host (the cell in which they are replicating) and their type of genome (e.g. DNA, RNA). A commonly studied virus is a bacteriophage (a virus that infects bacteria). Bacteriophages undergo two different types of replication cycles, the lytic cycle and the lysogenic cycle. In the lytic cycle, the lagging strand virus immediately starts using the host cell's machinery to replicate the genetic material and create more protein capsids. These spontaneously assemble into mature viruses and cause the cell to lyse, or break open, releasing new viruses into the environment. In the lysogenic cycle, the virus incorporates itself into the host genome and remains dormant until it is triggered to switch into the lytic cycle.

Animal viruses must overcome additional challenges not present in bacterial cells including immune pathways, organelle structures, and different cell machinery. Retroviruses like the HIV virus are RNA viruses use an enzyme called reverse transcriptase to convert their RNA genomes into DNA so that they can be inserted into a host genome. RNA viruses have extremely high rates of mutation because they lack error-proofing mechanisms when they replicate their genomes of mutation.

KEY WORDS

euchromatin
heterochromatin
double helix
nucleotides
five-carbon sugar
phosphate
nitrogenous base
adenine
deoxyribose
guanine
cytosine
thymine
phosphodiester bonds
base pairing
complementary
antiparallel
hydrogen bonds
DNA replication
helicase
replication fork
origins of replication
topoisomerase
DNA polymerase
RNA primase
leading strand
lagging strand
Okazaki fragments
DNA ligase
transcription factors
ribose
uracil
messenger RNA (mRNA)
ribosomal RNA (rRNA)
transfer RNA (tRNA)
protein synthesis
transcription
RNA processing
translation
sense strand
antisense strand
RNA polymerase
exons
introns
spliceosome
poly(A) tail
5′ GTP cap
codons
anticodon
A site
P site
E site
stop codons
primary structure
secondary structure
tertiary structure
quaternary structure
chaperon proteins (or chaperonins)
mutation
base substitution
gene rearrangement
nonsense mutation
missense mutation
silent mutation
deletions
duplications
inversions
gene expression
translocations
genetic engineering
restriction enzyme
sticky end
plasmid
cloning vector
transformation
gel electrophoresis
restriction fragment length polymorphism (RFLPs)
DNA fingerprinting
polymerase chain reaction (PCR)
lytic cycle
lysogenic cycle
lyse
reverse transcriptase
retrovirus

CHAPTER 7 REVIEW QUESTIONS

Answers can be found in Chapter 15.

1. A geneticist has discovered a yeast cell, which encodes a DNA polymerase that may add nucleotides in both the 5′ to 3′ and 3′ to 5′ directions. Which of the following structures would this cell not likely generate during DNA replication?

   (A) RNA primers
   (B) Okazaki fragments
   (C) Replication fork
   (D) Nicked DNA by topoisomerases

2. A eukaryotic gene, which does not normally undergo splicing, was exposed to benzpyrene, a known carcinogen and mutagen. Following exposure, the protein encoded by the gene was shorter than before exposure. Which of the following types of genetic rearrangements or mutations was likely introduced by the mutagen?

   (A) Silent mutation
   (B) Missense mutation
   (C) Nonsense mutation
   (D) Duplication

3. DNA Replication occurs through a complex series of steps involving several enzymes. Which of the following represents the correct order beginning with the earliest activity of enzymes involved in DNA replication?

   (A) Helicase, ligase, RNA primase, DNA polymerase
   (B) DNA polymerase, RNA primase, helicase, ligase
   (C) RNA primase, DNA polymerase, ligase, helicase
   (D) Helicase, RNA primase, DNA polymerase, ligase

4. If a messenger RNA codon is UAC, which of the following would be the complementary anticodon triplet in the transfer RNA?

   (A) ATG
   (B) AUC
   (C) AUG
   (D) ATT

5. During post-translational modification, the polypeptide from a eukaryotic cell typically undergoes substantial alteration that results in

   (A) excision of introns
   (B) addition of a poly(A) tail
   (C) formation of peptide bonds
   (D) a change in the overall conformation of a polypeptide

6. Which of the following represents the maximum number of amino acids that could be incorporated into a polypeptide encoded by 21 nucleotides of messenger RNA?

(A) 3

(B) 7

(C) 21

(D) 42

8

Cell Reproduction

CELL DIVISION

Every second, thousands of cells are dying throughout our bodies. Fortunately, the body replaces them at an amazing rate. In fact, **epidermal**, or skin, cells die off and are replaced so quickly that the average 18-year-old grows an entirely new skin every few weeks. The body keeps up this unbelievable rate thanks to the mechanisms of **cell division**.

This chapter takes a closer look at how cells divide. But remember, cell division is only a small part of the life cycle of a cell. Most of the time, cells are busy carrying out their regular activities. Since we covered DNA replication in Chapter 7, let's now look at how cells pass their genetic material to their offspring.

THE CELL CYCLE

Every cell has a life cycle—the period from the beginning of one division to the beginning of the next. The cell's life cycle is known as the **cell cycle**. The cell cycle is divided into two periods: **interphase** and **mitosis**. Take a look at the cell cycle of a typical cell on the next page.

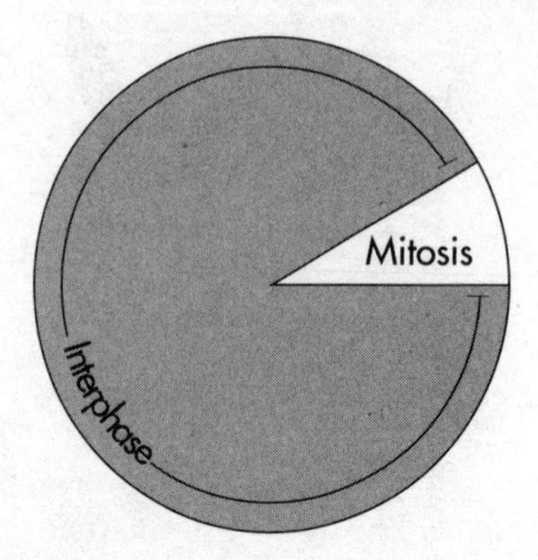

Notice that most of the life of a cell is spent in interphase.

INTERPHASE: THE GROWING PHASE

Interphase is the time span from one cell division to another. We call this stage interphase (*inter-* means between) because the cell has not yet started to divide. Although biologists sometimes refer to interphase as the "resting stage," the cell is definitely not inactive. This phase is when the cell carries out its regular activities. All the proteins and enzymes it needs to grow are produced during interphase.

The Three Stages of Interphase

Interphase can be divided into three stages: **G1**, **G2**, and **S phase**.

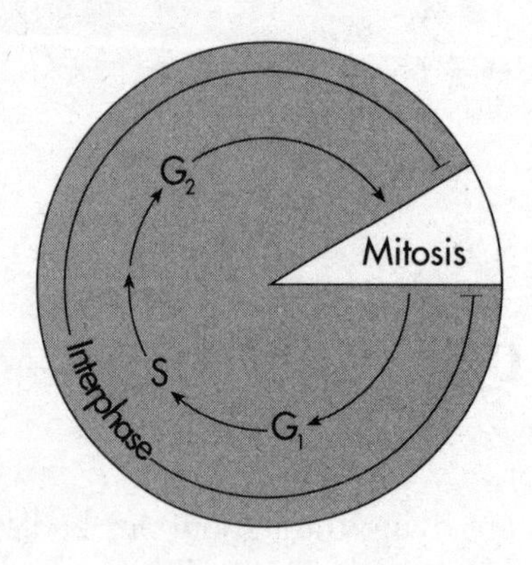

The most important phase is the S phase. That's when the cell replicates its genetic material. The first thing a cell has to do before undergoing mitosis is to duplicate all of its chromosomes, which contain the organism's DNA "blueprint." During interphase, every single chromosome in the nucleus is duplicated.

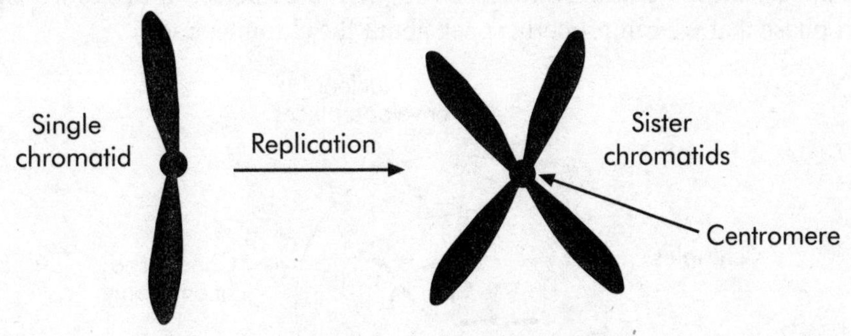

You'll notice that the original chromosome and its duplicate are still linked, like Siamese twins. These identical chromosomes are now called **sister chromatids** (each individual structure is called a chromatid). The chromatids are held together by a structure called the **centromere**. Although the chromosomes have been duplicated, they are still considered a single unit. Once duplication has been done, we're ready for the big breakup: mitosis.

We've already said that replication occurs during the S phase of interphase, so what happens during G1 and G2? During these stages, the cell produces proteins and enzymes. For example, during G1 the cell produces all of the enzymes required for DNA replication (as we saw in Chapter 7, that means DNA helicase, DNA polymerase, and DNA ligase). By the way, "G" stands for "gap," but we can also associate it with "growth." These three phases are highly regulated by checkpoints and special proteins called cyclins and cyclin-dependent kinases (CDKs).

Let's recap:

- The cell cycle consists of two things: interphase and mitosis.
- During the S phase of interphase, the chromosomes replicate.
- Growth and preparation for mitosis occur during the G1 and G2 stages of interphase.

MITOSIS: THE DANCE OF THE CHROMOSOMES

Once the chromosomes have replicated, the cell is ready to begin mitosis. Mitosis is the period when the cell divides. Mitosis consists of a sequence of four stages: **prophase**, **metaphase**, **anaphase**, and **telophase**.

Stage 1: Prophase

One of the first signs of prophase is the disappearance of the nucleolus. In prophase, the chromosomes thicken, forming coils upon coils, and become visible. (During interphase, the chromosomes are not visible. Rather, the genetic material is scattered throughout the nucleus and is called **chromatin**. It is only during prophase that we can properly speak about the chromosomes.)

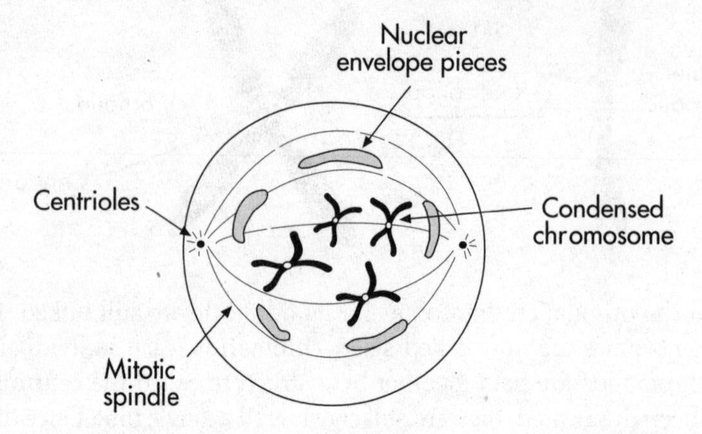

Now the cell has plenty of room to "sort out" the chromosomes. Remember centrioles? During prophase, these cylindrical bodies found within microtubule organizing centers (MTOCs) start to move away from each other, toward opposite ends of the cell. The centrioles will spin out a system of microtubules known as the **spindle fibers**. These spindle fibers will attach to a structure on each chromatid called a **kinetochore**. The kinetochores are part of the centromere.

Stage 2: Metaphase

The next stage is called metaphase. The chromosomes now begin to line up along the equatorial plane, or the **metaphase plate**, of the cell. That's because the spindle fibers are attached to the kinetochore of each chromatid.

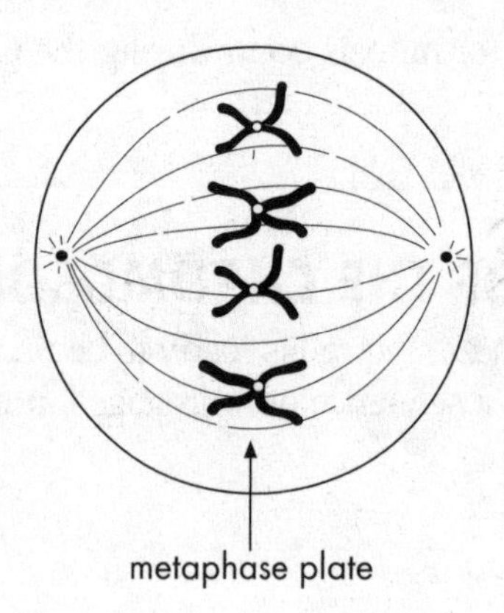

Stage 3: Anaphase

During anaphase, the sister chromatids of each chromosome separate at the centromere and migrate to opposite poles. The chromatids are pulled apart by the microtubules, which begin to shorten. Each half of a pair of sister chromatids now moves to opposite poles of the cell. Non-kinetochore microtubules elongate the cell.

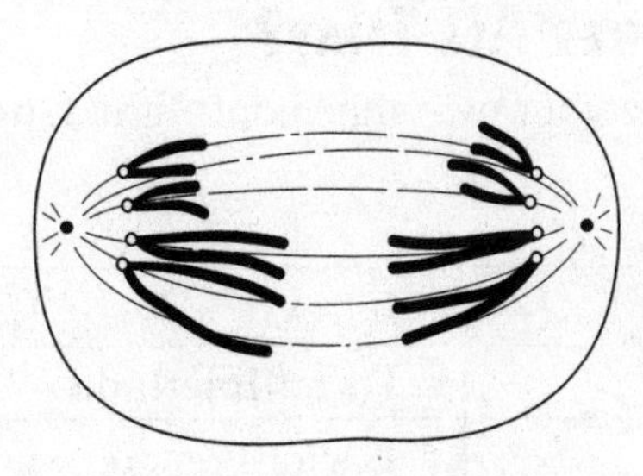

Stage 4: Telophase

The final phase of mitosis is telophase. A nuclear membrane forms around each set of chromosomes and the nucleoli reappear.

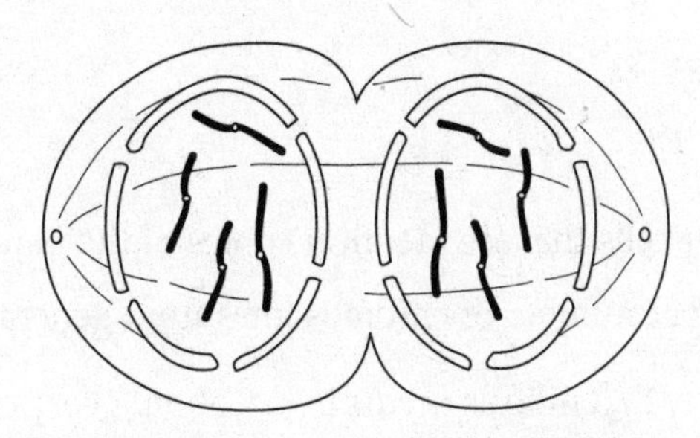

The nuclear membrane is ready to divide. Now it's time to split the cytoplasm in a process known as **cytokinesis**. Look at the figure below and you'll notice that the cell has begun to split along a **cleavage furrow** (which is produced by actin microfilaments):

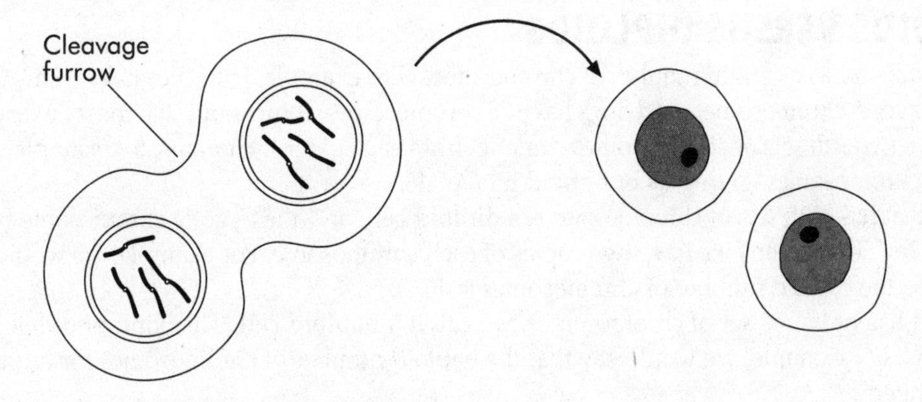

A cell membrane forms about each cell and they split into two distinct daughter cells. The division of the cytoplasm yields two daughter cells.

Here's one thing to remember: Cytokinesis occurs differently in plant cells. The cell doesn't form a cleavage furrow. Instead, a partition called a **cell plate** forms down the middle region.

STAGE 5: INTERPHASE

Once the daughter cells are produced, they reenter the initial phase—interphase—and the whole process starts over. The cell goes back to its original state. Once again, the chromosomes become invisible, and the genetic material is called chromatin.

BUT HOW WILL I REMEMBER ALL THAT?

For mitosis, you may already have your own mnemonic. If not, here's a table with a mnemonic we created for you.

IPMAT	
Interphase	I is for **Interlude**
Prophase	P is for **Prepare**
Metaphase	M is for **Meet**
Anaphase	A is for **Apart**
Telophase	T is for **Tear**

PURPOSE OF MITOSIS

Mitosis has two purposes:

- to produce daughter cells that are identical copies of the parent cell
- to maintain the proper number of chromosomes from generation to generation

For our purposes, we can say that mitosis occurs in just about every cell except sex cells. When you think of mitosis, remember: "Like begets like." Hair cells "beget" other hair cells; skin cells "beget" other skin cells, etc. Mitosis is involved in growth, repair, and asexual reproduction.

HAPLOIDS VERSUS DIPLOIDS

Every organism has a certain number of chromosomes. For example, fruit flies have 8 chromosomes, humans have 46 chromosomes, and dogs have 78 chromosomes. It turns out that most eukaryotic cells in fact have two full sets of chromosomes—one set from each parent. Humans, for example, have two sets of 23 chromosomes, giving us our grand total of 46.

A cell that has both sets of chromosomes is a **diploid cell**, and the zygotic chromosome number is given as "2n." That means we have two copies of each chromosome. For example, we would say that for humans the diploid number of chromosomes is 46.

If a cell has only one set of chromosomes, we call it a **haploid cell**. This kind of cell is given the symbol "n." For example, we would say that the haploid number of chromosomes for humans is 23.

Remember:

- *Diploid* refers to any cell that has two sets of chromosomes.
- *Haploid* refers to any cell that has one set of chromosomes.

Why do we need to know the terms *haploid* and *diploid*? Because they are extremely important when it comes to sexual reproduction. As we've seen, 46 is the normal diploid number for human beings. We can say, therefore, that human cells have 46 chromosomes. However, this isn't entirely correct.

Human chromosomes come in pairs called **homologues**. So while there are 46 of them altogether, there are actually only 23 *distinct* chromosomes. The **homologous chromosomes** which make up each pair are similar in size and shape and express similar traits. This is the case in all sexually reproducing organisms. In fact, this is the essence of sexual reproduction: Each parent donates half its chromosomes to its offspring.

GAMETES

Although most cells in the human body are diploid (i.e., filled with pairs of chromosomes), there are special cells that are haploid (i.e., unpaired). These haploid cells are called **sex cells**, or **gametes**. Why do we have haploid cells?

As we've said, an offspring has one set of chromosomes from each of its parents. A parent, therefore, contributes a gamete with one set that will be paired with the set from the other parent to produce a new diploid cell, or zygote. This produces offspring that are genetically different from either parent.

AN OVERVIEW OF MEIOSIS

To preserve the diploid number of chromosomes in an organism, each parent must contribute only half of its chromosomes. This is the point of **meiosis**. Meiosis is the production of gametes. Since sexually reproducing organisms need only haploid cells for reproduction, meiosis is limited to sex cells in special sex organs called **gonads**. In males, the gonads are the **testes**, while in females they are the **ovaries**. The special cells in these organs—also known as **germ cells**—produce haploid cells (n), which then combine to restore the diploid (2n) number during fertilization:

female gamete (n) + male gamete (n) = zygote (2n)

When it comes to genetic variation, meiosis is a big plus. Variation, in fact, is the driving force of evolution. The more variation there is in a population, the more likely it is that some members of the population will survive extreme changes in the environment. Meiosis is far more likely to produce these sorts of variations than is mitosis, and therefore confers selective advantage on sexually reproducing organisms. We'll come back to this theme in Chapter 11.

A CLOSER LOOK AT MEIOSIS

Meiosis actually involves two rounds of cell division called **meiosis I** and **meiosis II**.

Before meiosis begins, the diploid cell goes through interphase. Just as in mitosis, double-stranded chromosomes are formed during this phase.

Meiosis I

Meiosis I consists of four stages: prophase I, metaphase I, anaphase I, and telophase I.

Prophase I

Prophase I is a little more complicated than regular prophase. As in mitosis, the nuclear membrane disappears, the chromosomes become visible, and the centrioles move to opposite poles of the nucleus. But that's where the similarity ends.

The major difference involves the movement of the chromosomes. In meiosis, the chromosomes line up side-by-side with their counterparts (homologues). This event is known as **synapsis**.

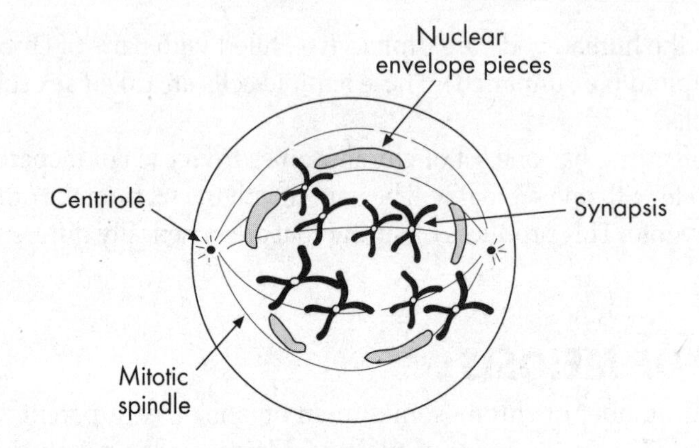

Synapsis involves two sets of chromosomes that come together to form a **tetrad** (or a **bivalent**). A tetrad consists of four chromatids. Synapsis is followed by **crossing-over**, the exchange of segments between homologous chromosomes.

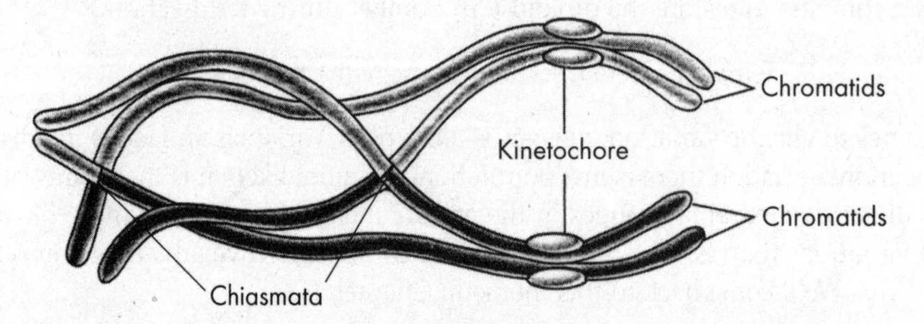

What's unique in prophase I is that "pieces" of chromosomes are exchanged between the homologous partners. This is one of the ways organisms produce genetic variation. By the end of prophase I, the chromosomes will have exchanged regions containing several **alleles**, or different forms of the same gene. By the end of prophase, the homologous chromosomes are held together only at specialized regions called **chiasmata**.

Metaphase I

As in mitosis, the chromosome pairs—now called tetrads—line up at the metaphase plate. By contrast, you'll recall that in regular metaphase the chromosomes lined up individually. One important concept to note is that the alignment during metaphase is random, so which copy of each chromosome that ends up in a daughter cell is random.

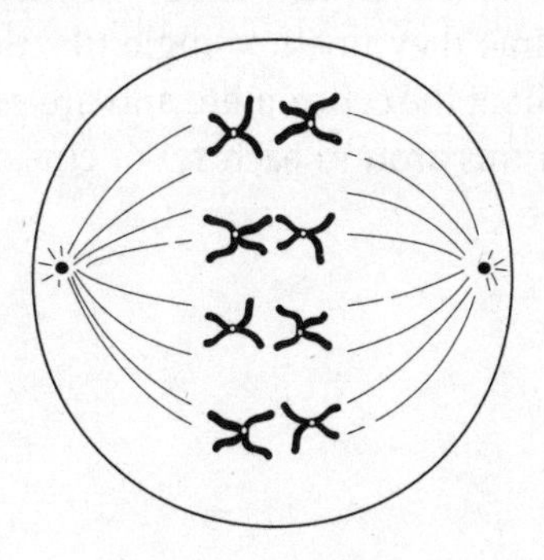

Anaphase I

During anaphase I, one of each pair of chromosomes within a tetrad separates and moves to opposite poles. Notice that the chromosomes do not separate at the centromere. They separate with their centromeres intact.

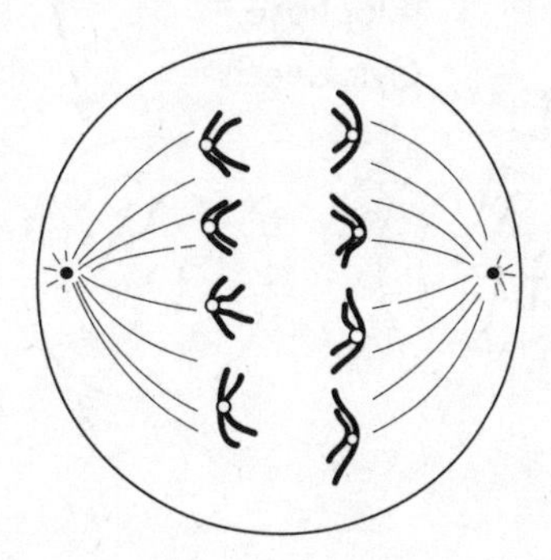

The chromosomes now go on to their respective poles.

Telophase I

During telophase I, the nuclear membrane forms around each set of chromosomes.

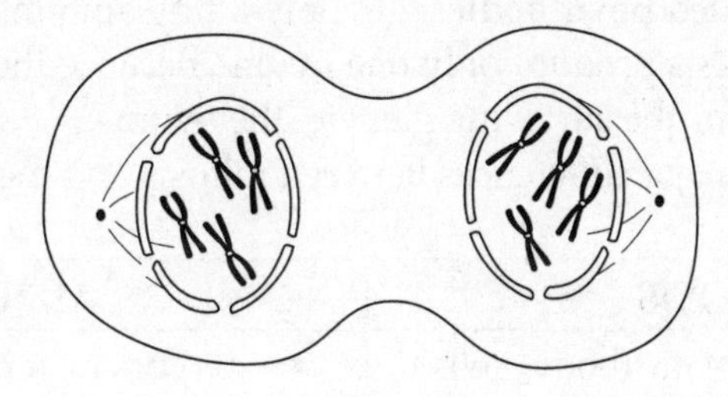

Finally, the cells undergo cytokinesis, leaving us with two daughter cells. Notice that at this point the nucleus contains the haploid number of chromosomes, but each chromosome is a duplicated chromosome.

MEIOSIS II

The purpose of the second meiotic division is to separate the duplicated chromosomes, and is virtually identical to mitosis. Let's run through the steps in meiosis II.

After a brief period, the cell undergoes a second round of cell division. During prophase II, the chromosomes once again condense and become visible. In metaphase II, the chromosomes move toward the metaphase plate. This time they line up single file, not as pairs. During anaphase II, the chromatids of each chromosome split at the centromere and are pulled to opposite ends of the cell. At telophase II, a nuclear membrane forms around each set of chromosomes and a total of *four* haploid cells are produced:

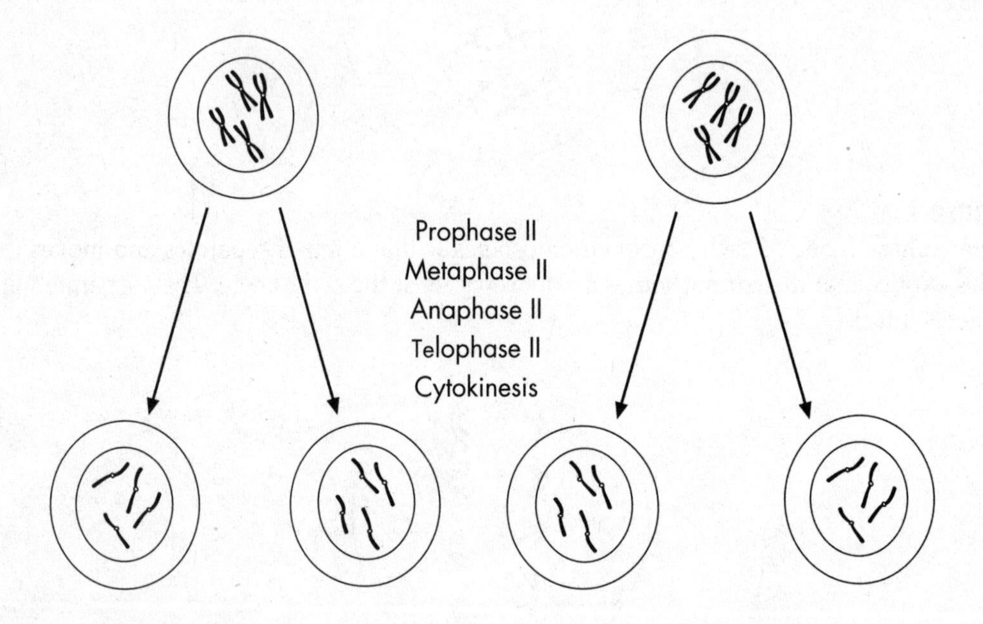

GAMETOGENESIS

Meiosis is also known as **gametogenesis.** If sperm cells are produced then meiosis is called **spermatogenesis**. During spermatogenesis, four sperm cells are produced for each diploid cell. If an egg cell or an ovum is produced, this process is called **oogenesis**.

Oogenesis is a little different from spermatogenesis. Oogenesis produces only one ovum, not four. The other three cells, called **polar bodies**, get only a tiny amount of cytoplasm and eventually degenerate. Why does oogenesis produce only one ovum? Because the female wants to conserve as much cytoplasm as possible for the surviving gamete, the **ovum**.

Here's a summary of the major differences between mitosis and meiosis:

MITOSIS	MEIOSIS
• Occurs in somatic (body) cells	• Occurs in germ (sex) cells
• Produces identical cells	• Produces gametes
• Diploid cell → diploid cells	• Diploid cell → haploid cells
• 1 cell becomes 2 cells	• 1 cell becomes 4 cells
• Number of divisions: 1	• Number of divisions: 2

MUTATIONS

Sometimes, a set of chromosomes has an extra or a missing chromosome. This occurs because of **non-disjunction**—the chromosomes failed to separate properly during meiosis. This error, which produces the wrong number of chromosomes in a cell, results in severe genetic defects. For example, humans typically have 23 pairs of chromosomes, but individuals with **Down syndrome** have three—instead of two—copies of the 21st chromosome.

Chromosomal abnormalities also occur if one or more segments of a chromosome break. The most common example is translocation (a segment of a chromosome moves to another chromosome).

Here's an example of a translocation:

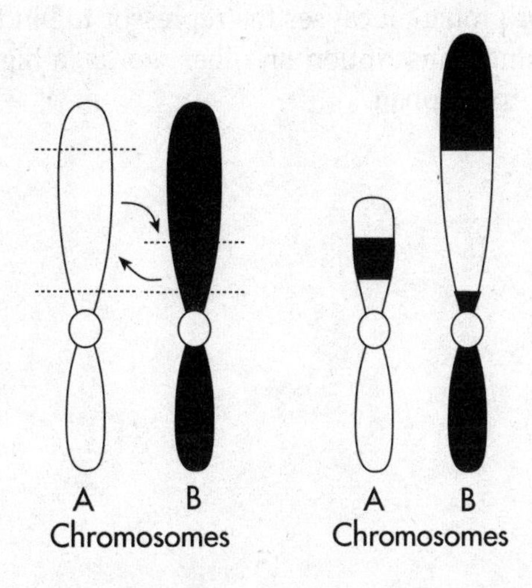

Translocation is a mutation that occurs during meiosis when a chromosome breaks and the fragment attaches to a nonhomologous chromosome.

Fortunately, in most cases, damaged DNA can usually be repaired with special repair enzymes.

GENE REGULATION

What controls gene transcription, and how does an organism express only certain genes? Most of what we know about gene regulation comes from our studies of *E. coli.* In bacteria, the region of bacterial DNA that regulates gene expression is called an **operon**. One of the best-understood operons is the *lac* operon, which controls expression of the enzymes that break down lactose.

The operon consists of four major parts: structural genes, the regulatory gene, the promoter gene, and the operator.

- **Structural genes** are genes that code for enzymes needed in a chemical reaction. These genes will be transcribed at the same time to produce particular enzymes. In the *lac* operon, three enzymes (beta galactosidase, galactose permease, and thiogalactoside transacetylase) involved in digesting lactose are coded for.
- The **promoter gene** is the region where the RNA polymerase binds to begin transcription.
- The **operator** is a region that controls whether transcription will occur.

- The **regulatory gene** codes for a specific regulatory protein called the *repressor*. The repressor is capable of attaching to the operator and blocking transcription. If the repressor binds to the operator, transcription will not occur. On the other hand, if the repressor does not bind to the operator, RNA polymerase moves right along the operator and transcription occurs. In the *lac* operon, the **inducer**, lactose, binds to the repressor, causing it to fall off the operator, and "turns on" transcription.

Other operons, such as the *trp* operon, operate in a similar manner except that this mechanism is continually "turned on" and is only "turned off" in the presence of high levels of the amino acid, tryptophan. Tryptophan is a product of the pathway that codes for the *trp* operon. When tryptophan combines with the *trp* repressor protein, it causes the repressor to bind to the operator, which turns the operon "off" thereby blocking transcription. In other words, a high level of tryptophan acts to repress the further synthesis of tryptophan.

KEY WORDS

epidermal
cell division
cell cycle
interphase
mitosis
G1 phase
G2 phase
S phase
sister chromatids
centromere
prophase
metaphase
anaphase
telophase
chromatin
spindle fibers
kinetochores
metaphase plate
cytokinesis
cleavage furrow
cell plate
diploid cell
haploid cell
homologues
homologous chromosomes
sex cells (or gametes)
meiosis
gonads
testes
ovaries
germ cells
meiosis I
meiosis II
synapsis
tetrad (or bivalent)
crossing-over
alleles
chiasmata
gametogenesis
spermatogenesis
oogenesis
polar bodies
ovum
nondisjunction
Down syndrome
translocation
operon
structural genes
promoter gene
operator
regulatory gene
inducer

CHAPTER 8 REVIEW QUESTIONS

Answers can be found in Chapter 15.

1. A scientist is testing new chemicals designed to stop the cell cycle at various stages of mitosis. Upon applying one of the chemicals, she notices that all of the cells appear as shown below. Which of the following best explains how the chemical is likely acting on the cells?

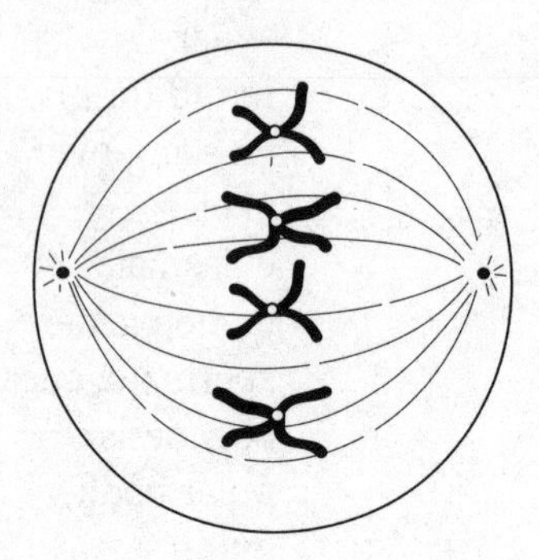

   (A) The chemical has arrested the cells in prophase and has prevented attachment of the spindle fibers to the kinetochore.
   (B) The chemical has arrested the cells in metaphase and has prevented dissociation of the spindle fibers from the kinetochore.
   (C) The chemical has arrested the cells in metaphase and is preventing the shortening of the spindle fibers.
   (D) The chemical has arrested the cells in anaphase and is preventing the formation of a cleavage furrow.

2. A unique lineage of cells was identified which undergoes one round of meiosis, but results in four daughter cells without exchange of alleles. It is suspected that there is a defect during the crossing-over event. During which phase of meiosis is the defect likely occurring?

   (A) Prophase I
   (B) Metaphase I
   (C) Telophase I
   (D) Prophase II

3. In the *lac* operon, lactose binds to a molecule, which blocks RNA polymerase activity, causing it to dissociate and permit transcription. Which of the following terms most accurately describes the role of lactose and its molecule target?

   (A) Lactose is a repressor and its target molecule is an inducer.
   (B) Lactose is a promoter and its target molecule is an inducer.
   (C) Lactose is an inducer and its target molecule is a repressor.
   (D) Lactose is a repressor and its target molecule is a promoter.

Questions 4 – 5 refer to the following graph and paragraph.

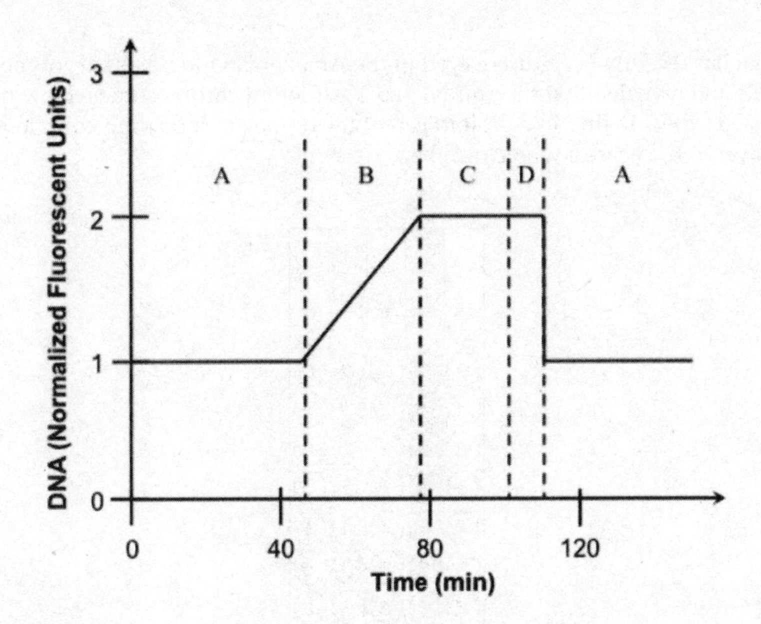

An experiment is performed to evaluate the amount of DNA present during a complete cell cycle. All of the cells were synced prior to the start of the experiment. During the experiment, a fluorescent chemical was applied to cells, which would only fluoresce when bound to DNA. The results of the experiment are shown above. Differences in cell appearance by microscopy or changes in detected DNA were determined to be phases of the cell cycle and are labeled with the letters A – D.

4. Approximately how long does S phase take to occur in these cells?

   (A) 15 min
   (B) 20 min
   (C) 30 min
   (D) 40 min

5. During which of the labeled phases of the experiment would the cell undergo anaphase?

   (A) Phase A
   (B) Phase B
   (C) Phase C
   (D) Phase D

Question 6 represents a question requiring a numeric answer. Calculate the correct answer for each question.

6. A new mammal has recently been discovered in the Amazonian jungle. A karyotype was performed on gametic cells and revealed that the animal had 13 different chromosomes. How many chromosomes would you expect to find in a diploid cell of the organism following completion of S phase? Give your answer to the nearest whole number.

9

Heredity

GREGOR MENDEL: THE FATHER OF GENETICS

What is genetics? In its simplest form, genetics is the study of heredity. It explains how certain characteristics are passed on from parents to children. Much of what we know about genetics was discovered by the monk **Gregor Mendel** in the 19th century. Since then, the field of genetics has vastly expanded. As scientists study the mechanisms of genetics, they've developed new ways of manipulating genes. For example, scientists have isolated the gene that makes insulin, a human hormone, and now use bacteria to make large quantities of it. But before we get ahead of ourselves, let's study the basic rules of genetics.

Let's begin then with some of the fundamental points of genetics:

- Every **trait**—or expressed characteristic—is produced by hereditary factors known as **genes**. A gene is a segment of a chromosome. Within a chromosome, there are many genes, each controlling the inheritance of a particular trait. For example, in pea plants, there's a gene on the chromosome that codes for seed coat. The position of a gene on a chromosome is called a **locus**.
- Diploid organisms (organisms that have two sets of chromosomes) usually have two copies of a gene, one on each homologous chromosome. These copies may be different from one another—that is, they may be **alleles**, or alternate forms of the same gene. For example, if we're talking about the height of a pea plant, there's an allele for tall and an allele for short. In other words, both alleles are alternate forms of the gene for height.
- An allele can be **dominant** or **recessive**. In simple cases, an organism can express contrasting conditions. For example, a plant can be tall or short. The convention is to assign one of two letters for the two different alleles. The dominant allele receives a capital letter and the recessive allele receives a lowercase of the same letter. For instance, we might give the dominant allele for height in pea plants a "T" for tall. This means that the recessive allele would be "t."
- When an organism has two identical alleles for a given trait, the organism is **homozygous**. For instance, TT and tt both represent homozygous organisms. TT is homozygous dominant and tt is homozygous recessive. If an organism has two *different* alleles for a given trait, Tt, the organism is **heterozygous**.
- When discussing the physical appearance of an organism, we refer to its **phenotype**. The phenotype tells us what the organism looks like. When talking about the genetic makeup of an organism, we refer to its **genotype**. The genotype tells us which alleles the organism possesses.

One of the major ways ETS likes to test genetic information is by having you do crosses. Crosses involve the mating of hypothetical organisms with specific phenotypes and genotypes. We'll look at some examples in a moment, but for now, keep these test-taking tips in mind:

- Label each generation in the cross. The first generation in an experiment is always called the **parent**, or **P1 generation**. The offspring of the P1 generation are called the **filial**, or **F1 generation**. The next generation, the grandchildren, is called the **F2 generation**.
- Always write down the symbols you're using for the alleles, along with a key to remind yourself what the symbols refer to. Use uppercase for dominant alleles and lowercase for recessive alleles.

Now let's look at some basic genetic principles.

MENDELIAN GENETICS

One of Mendel's hobbies was to study the effects of cross-breeding on different strains of pea plants. Mendel worked exclusively with true-breeder pea plants. This means the plants he used were genetically pure and consistently produced the same traits. For example, tall plants always produced tall plants; short plants always produced short plants. Through his work he came up with three principles of genetics: the **law of dominance**, the **law of segregation**, and the **law of independent assortment**.

The Law of Dominance

Mendel crossed two true-breeding plants with contrasting traits: tall pea plants and short pea plants. This type of cross is called a **monohybrid cross**, which means that only one trait is being studied. In this case, the trait was height.

To his surprise, when Mendel mated these plants, the characteristics didn't blend to produce plants of average height. Instead, all the offspring were tall.

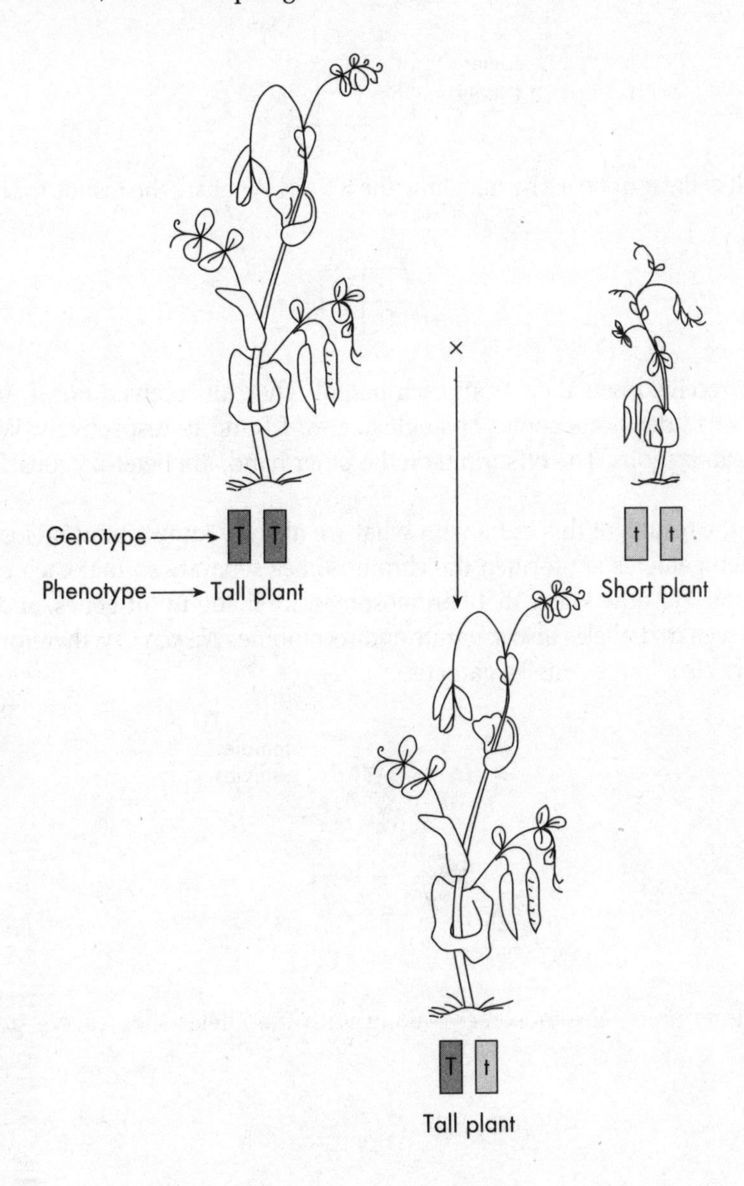

Mendel recognized that one trait must be masking the effect of the other trait. This is called the law of dominance. The dominant, tall allele, T, somehow masked the presence of the recessive, short allele, t. Consequently, all a plant needs is one tall allele to make it tall.

MONOHYBRID CROSS

A simple way to represent a cross is to set up a **Punnett square**. Punnett squares are used to predict the results of a cross. Let's construct a Punnett square for the cross between Mendel's tall and short pea plants. Let's first designate the alleles for each plant. As we saw earlier, we can use the letter "T" for the tall, dominant allele and "t" for the short, recessive allele.

Since one parent was a pure, tall pea plant, we'll give it two dominant alleles (TT homozygous dominant). The other parent was a pure, short pea plant, so we'll give it two recessive alleles (tt homozygous recessive). Let's put the alleles for one of the parents across the top of the box, and the alleles for the other parent along the side of the box.

	T	T ← One parent
Other parent → t		
t		

Now we can fill in the four boxes by matching the letters. What are the results for the F1 generation?

	T	T
t	Tt	Tt
t	Tt	Tt

Each offspring received one allele from each parent. They all received one T and one t. They're all Tt! Our parents had duplicate copies of single alleles, TT and tt, respectively. We could therefore refer to them as homozygous. The offspring, on the other hand, are heterozygous: They possess one copy of each allele.

Let's compare the results of this cross with what we already know about meiosis. From meiosis, we know that when gametes are formed the chromosomes separate so that each cell gets one copy of each chromosome. We now know that chromosomes are made up of genes, and genes consist of alleles. We've just seen that alleles also separate and recombine. We can say, therefore, that each allele in a Punnett square also "represents" a gamete:

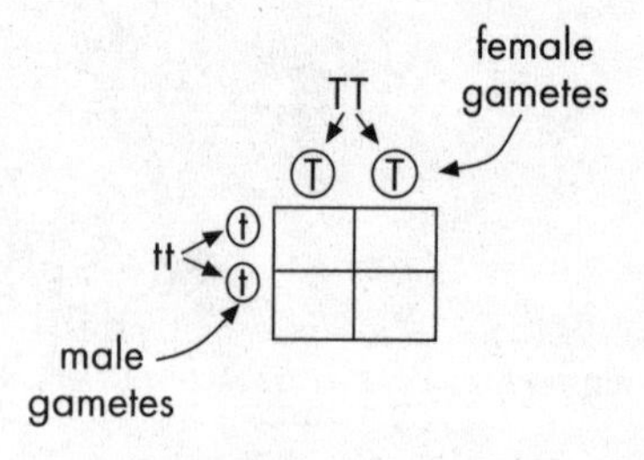

When fertilizaton occurs, chromosomes—along with the alleles they carry—get paired up in a new combination.

The Law of Segregation

Next, Mendel took the offspring and self-pollinated them. Let's use a Punnett square to spell out the results. This time we're starting with the offspring of the first generation—F1. Take a look at the results:

	T	t
T	TT	Tt
t	Tt	tt

F2 Generation

One of the offspring could be a short pea plant! The short-stemmed trait reappeared in the F2 generation. How could that happen? Once again, the alleles separated and recombined, producing a new combination for this offspring. The cross resulted in one offspring with a pair of recessive alleles, tt. Because there is no T (dominant) allele around to mask the expression of the short, recessive allele, our new plant could wind up short.

Although all of the F1 plants appear to be tall, the alleles separate and recombine during the cross. This is an example of the law of segregation.

What about the genotype and phenotype for this cross? Remember, genotype refers to the genetic makeup of an organism, whereas phenotype refers to the appearance of the organism. Using the results of our Punnett square, what is the ratio of phenotypes and genotypes in the offspring?

Let's sum up the results. We have four offspring with two different phenotypes: three of the offspring are tall, whereas one of them is short. On the other hand, we have three genotypes: 1 TT, 2 Tt, and 1 tt.

Here's a summary of the results:

- The ratio of phenotypes is 3 : 1 (three tall: one short).
- The ratio of genotypes is 1 : 2 : 1 (one TT: two Tt: one tt).

The Law of Independent Assortment

So far, we have looked at only one trait: tall versus short. What happens when we study two traits at the same time? The two traits also segregate randomly. This is an example of independent assortment. For example, let's look at two traits in pea plants: height and color. When it comes to height, a pea plant can be either tall or short. As for color, the plant can be either green or yellow, with green being dominant. This gives us four alleles. By the law of independent assortment, these four alleles can combine to give us four different gametes:

TG Tg tG tg

Dihybrid Cross

Keep in mind that the uppercase letter refers to the dominant allele. Therefore, "T" refers to tall and "G" refers to green, whereas "t" refers to short and "g" refers to yellow. Now let's set up a cross between plants differing in two characteristics—called a **dihybrid cross**—using these four gametes and see what happens.

Each trait will act independently, meaning that a plant that is tall can be either green or yellow. Similarly, a green plant can be either tall or short.

Here is the Punnett Square for a cross between two double heterozygotes (Tt Gg):

	TG	Tg	tG	tg
TG	TTGG	TTGg	TtGG	TtGg
Tg	TTGg	TTgg	TtGg	Ttgg
tG	TtGG	TtGg	ttGG	ttGg
tg	TtGg	Ttgg	ttGg	ttgg

This is an example of the law of independent assortment. Each of the traits segregated independently. Don't worry about the different combinations in the cross—you'll make yourself dizzy with all those letters. Simply memorize the phenotype ratio of the pea plants. For the 16 offspring there are:

- 9 tall and green
- 3 tall and yellow
- 3 short and green
- 1 short and yellow

That's 9 : 3 : 3 : 1. Since Mendel's laws hold true for most of the traits they'll ask you about on the AP test, simply learning the ratios of offspring for this type of cross will help you nail any questions that come up.

The Punnett square method works well for monohybrid crosses and helps us visualize the possible combinations. However, a better method for predicting the likelihood of certain results from a dihybrid cross is to apply the law of probability. For dihybrid ratios, the law states that the probability that two or more independent events will occur simultaneously is equal to the *product* of the probability that each will occur independently. To illustrate the product rule, let's consider again the cross between two identical dihybrid tall, green plants with the genotype TtGg. To find the probability of having a tall, yellow plant, simply *multiply* the probabilities of each event. If the probability of being tall is $\frac{3}{4}$ and the probability of being yellow is $\frac{1}{4}$, then the probability of being tall *and* yellow is $\frac{3}{4} \times \frac{1}{4} = \frac{3}{16}$.

One more thing: Probability can be expressed as a fraction, percentage, or a decimal. Remember that this rule works only if the results of one cross are not affected by the results of another cross.

Let's summarize Mendel's three laws.

SUMMARY OF MENDEL'S LAWS	
Laws	**Definition**
Law of Dominance	One trait masks the effects of another trait.
Law of Segregation	Alleles can segregate and recombine.
Law of Independent Assortment	Traits can segregate and recombine independently of other traits.

Test Cross

Suppose we want to know if a tall plant is homozygous (TT) or heterozygous (Tt). Its physical appearance doesn't necessarily tell us about its genetic makeup. The only way to determine its genotype is to cross the plant with a recessive, short plant, tt. This is known as a **test cross**, or back cross. Using the recessive plant, there are only two possibilities: (1) TT × tt or (2) Tt × tt. Let's take a look.

homozygous

Recessive (testcross) →

	T	T
t	Tt	Tt
t	Tt	Tt

heterozygous

Recessive (testcross) →

	T	t
t	Tt	tt
t	Tt	tt

If none of the offspring is short, our original plant must have been homozygous, TT. If, however, even one short plant appears in the bunch, we know that our original pea plant was heterozygous, Tt. In other words, it wasn't a pure-breeding plant. A test cross uses a recessive organism to determine the genotype of an organism of unknown genotype.

BEYOND MENDELIAN GENETICS

Not all patterns of inheritance obey the principles of Mendelian genetics. In fact, many traits we observe are due to a combined expression of alleles. Here are a couple of examples of non-Mendelian forms of inheritance:

- **Incomplete dominance (blending inheritance):** In some cases, the traits will blend. For example, if you cross a white snapdragon plant (genotype WW) with a red snapdragon plant (RR), the resulting progeny will be pink (RW).
- **Codominance:** Sometimes you'll see an equal expression of both alleles. For example, an individual can have an AB blood type. In this case, each allele is equally expressed. That is, both the A allele and the B allele are expressed (I^AI^B). That's why the person is said to have AB blood.
- **Polygenic inheritance:** In some cases, a trait results from the interaction of *many* genes. Each gene will have a small effect on a particular trait. Height, skin color, and weight are all examples of polygenic traits.
- **Multiple alleles**: Some traits are the product of many different alleles that occupy a specific gene locus. The best example is the ABO blood group system in which three alleles (I^A, I^B, and *i*) determine blood type.
- **Linked genes**: Sometimes genes on the *same* chromosome stay together during assortment and move as a group. The group of genes is considered linked and tends to be inherited together. For example, the genes for flower color and pollen shape are linked on the same chromosomes and show up together. This pattern has led to methods for mapping human chromosomes.

Since linked genes are found on the same chromosome, they cannot segregate independently. That means they do not follow the probability rule and the expected results from a dihybrid cross.

In addition, the frequency of crossing-over between any two linked alleles is proportional to the distance between them. That is, the farther apart two linked alleles are on a chromosome, the more often the chromosome will break between them. This finding led to recombination mapping—mapping of linkage groups with each map unit being equal to 1 percent recombination. For example, if two linked genes, A and B, recombine with a frequency of 15 percent, and B and C recombine with a frequency of 9 percent, and A and C recombine with a frequency of 24 percent, what is the sequence and the distance between them?

The sequence and the distance of A-B-C is:

15 units 9 units

A_________________B____________C

If the recombination frequency between A and C had been 6 percent instead of 24 percent, the sequence and distance of A-B-C would instead be:

6 units 9 units

A______C____________B

SEX-LINKED TRAITS

We already know that humans contain 23 pairs of chromosomes. Twenty-two of the pairs of chromosomes are called **autosomes**. They code for many different traits. The other pair contains the **sex chromosomes**. This pair determines the sex of an individual. A female has two X chromosomes. A male has one X and one Y chromosome—an X from his mother and a Y from his father. Some traits, such as **color blindness** and **hemophilia**, are carried on sex chromosomes. These are called **sex-linked traits**. Most sex-linked traits are found on the X chromosome and are more properly referred to as "X-linked."

Since males have one X and one Y chromosome, what happens if a male has a defective X chromosome? Unfortunately, he'll express the sex-linked trait. Why? Because his one and only X chromosome is defective. He doesn't have another X to mask the effect of the bad X. However, if a female has only one defective X chromosome, she won't express the sex-linked trait. For her to express the trait, she has to inherit two defective X chromosomes. A female with one defective X is called a **carrier**. Although she does not exhibit the trait, she can still pass it on to her children.

You can also use the Punnett square to figure out the results of sex-linked traits. Here's a classic example: A male who has normal color vision and a woman who is a carrier for color blindness have children. How many of the children will be color-blind? To figure out the answer, let's set up a Punnett square:

	X	$\bar{X}$ ← Mother
Father → X	XX	$\bar{X}$X
Y	XY	$\bar{X}$Y

$\bar{X}$ = defective X

Notice that we placed a bar above any defective X to indicate the presence of a defective allele. And now for the results. The couple would have one son who is color-blind, a normal son, a daughter who is a carrier, and a normal daughter. The color-blind child is a son.

Barr Bodies

A look at the cell nucleus of normal females will reveal a dark-staining body known as a **Barr body**. A Barr body is an X chromosome that is condensed and visible. In every female cell, one X chromosome is activated and the other X chromosome is deactivated during embryonic development. Surprisingly, the X chromosome destined to be inactivated is randomly chosen in each cell. Therefore, in every tissue in the adult female one X chromosome remains condensed and inactive. However, this X chromosome is replicated and passed on to a daughter cell.

Other Sex-linked Diseases

- Duchenne muscular dystrophy
- Vitamin D resistant rickets
- Hereditary nephritis
- Juvenile gout

KEY WORDS

Gregor Mendel
trait
genes
locus
alleles
dominant
recessive
homozygous
heterozygous
phenotype
genotype
parent, or P1 generation
filial, or F1 generation
F2 generation
law of dominance
law of segregation
law of independent assortment
monohybrid cross
Punnett square
dihybrid cross
test cross
incomplete dominance
codominance
polygenetic inheritance
multiple alleles
epistasis
pleiotrophy
linked genes
autosomes
sex chromosomes
color blindness
hemophilia
sex-linked traits
carrier
Barr body

CHAPTER 9 REVIEW QUESTIONS

Answers can be found in Chapter 15.

1. Two expecting parents wish to determine all possible blood types of their unborn child. If both parents have an AB blood type, which of the following blood types will their child NOT possess?

   (A) A
   (B) B
   (C) AB
   (D) O

2. A new species of tulip was recently discovered. A population of pure red tulips was crossed with a population of pure blue tulips. The resulting F1 generation was all purple. This result is an example of which of the following?

   (A) Complete dominance
   (B) Incomplete dominance
   (C) Codominance
   (D) Linkage

3. In pea plants, tall (T) is dominant over short (t) and green (G) is dominant over yellow (g). If a pea plant heterozygous for both traits is crossed with a plant that is recessive for both traits, approximately what percentage of the progeny plants will be tall and yellow?

   (A) 0%
   (B) 25%
   (C) 66%
   (D) 75%

Questions 4-6 refer to the following pedigree tree and paragraph.

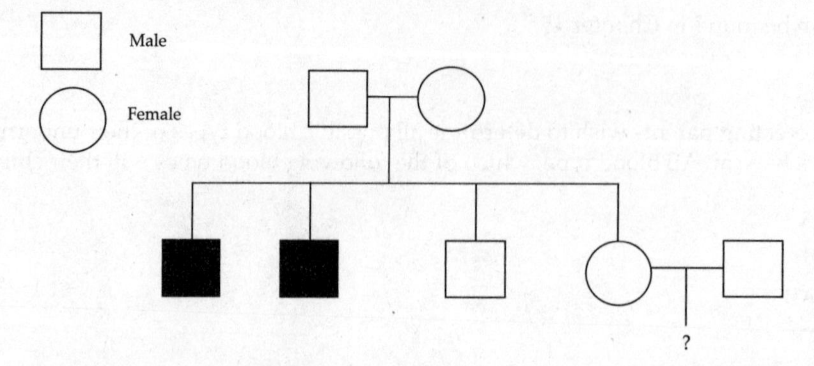

Hemophilia is an X-linked disease associated with inability to produce specific proteins in the blood-clotting pathway. Shown above is a family pedigree tree where family members afflicted with the disease are shown with filled-in squares (male) or circles (females). A couple is trying to determine the likelihood of passing on the disease to their future children (represented by the ? symbol above) since the hemophilia runs in the woman's family.

4. Assuming that the woman is a carrier, what is the probability that the couple's first son will have hemophilia?

(A) 0%
(B) 25%
(C) 50%
(D) 100%

5. Why were both women in the family tree above free of the disease?

(A) They were lucky since they didn't receive an X chromosome with the diseased allele.
(B) They cannot get hemophilia since it is only associated with a diseased Y chromosome.
(C) They are carriers and will only get the disease if they have children.
(D) They have at least one normal (unaffected) copy of the X chromosome.

6. Turner syndrome is a disease where an individual is born with only a single X chromosome. Suppose the woman in the couple is a carrier for hemophilia and has a child with Turner Syndrome, would this child have the disease?

(A) Yes, because the child would only have one copy of the X chromosome and it would be affected.
(B) No, because women cannot be affected by hemophilia.
(C) No, because the child would have to receive a normal copy of the X chromosome from its mother.
(D) Maybe, it depends on which X chromosome she receives from her mother.

10

Animal Structure and Function

To carry on with the business of life, higher organisms must all contend with the same basic challenges: obtaining nutrients, distributing them throughout their bodies, voiding wastes, responding to their environments, and reproducing. To accomplish these basic tasks, nature has come up with solutions. However, for all their differences, most animals have remarkably similar ways of dealing with these challenges.

This chapter looks at the basic structures of animals and the ways in which they function. Since the AP Biology Exam includes many questions on human anatomy and physiology, we'll focus primarily on how these systems have evolved in human beings.

The systems we'll look at include the following:

- The digestive system
- The respiratory system
- The circulatory system
- The immune system
- The excretory system
- The nervous system
- The musculoskeletal system
- The endocrine system
- The reproductive system
- Morphogenesis, or "development"

THE DIGESTIVE SYSTEM

All organisms need nutrients to survive. But where do the nutrients—the raw building blocks—come from? That depends on whether the organism is an autotroph or heterotroph. As you may recall, autotrophs make their own food through photosynthesis, and all of the building blocks—CO_2, water, and sunlight—come from their immediate environment. Heterotrophs, on the other hand, can't make their own food; they must acquire their energy from outside sources.

When we talk about digestion, we're talking about the breakdown of large food molecules into simpler compounds. These molecules are then absorbed by the body to carry out cell activities. In fact, everything we'll discuss in this section revolves around three simple questions.

1. What do organisms need from the outside world in order to survive?
2. How do they get those things?
3. What do they do with them once they get them?

Multicellular organisms have come up with a variety of ways of getting their nutrients. In simple animals, food is digested through **intracellular digestion**—that is, digestion occurs within food vacuoles. For example, a hydra encloses the food it captures in a food vacuole. Lysosomes containing digestive enzymes then fuse with the vacuole and break down the food. More complex animals have evolved a digestive tract and digest food through **extracellular digestion**. That is, the food is digested in a gastrovascular cavity. For example, in grasshoppers, food passes through specialized regions of the gut: the **mouth**, **esophagus**, **crop** (a storage organ), **stomach**, **intestine**, **rectum**, and **anus**.

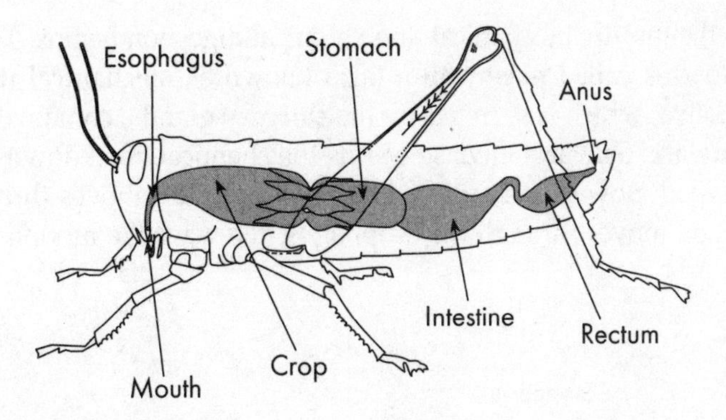

The Human Digestive System

The human digestive tract consists of the mouth, esophagus, stomach, **small intestine**, **large intestine**, and **accessory organs** (liver, pancreas, gall bladder, and salivary glands). Four groups of molecules must be broken down by the digestive tract: starch, proteins, fats, and nucleic acids.

The Mouth

The first stop in the digestive process is the mouth, or **oral cavity**.

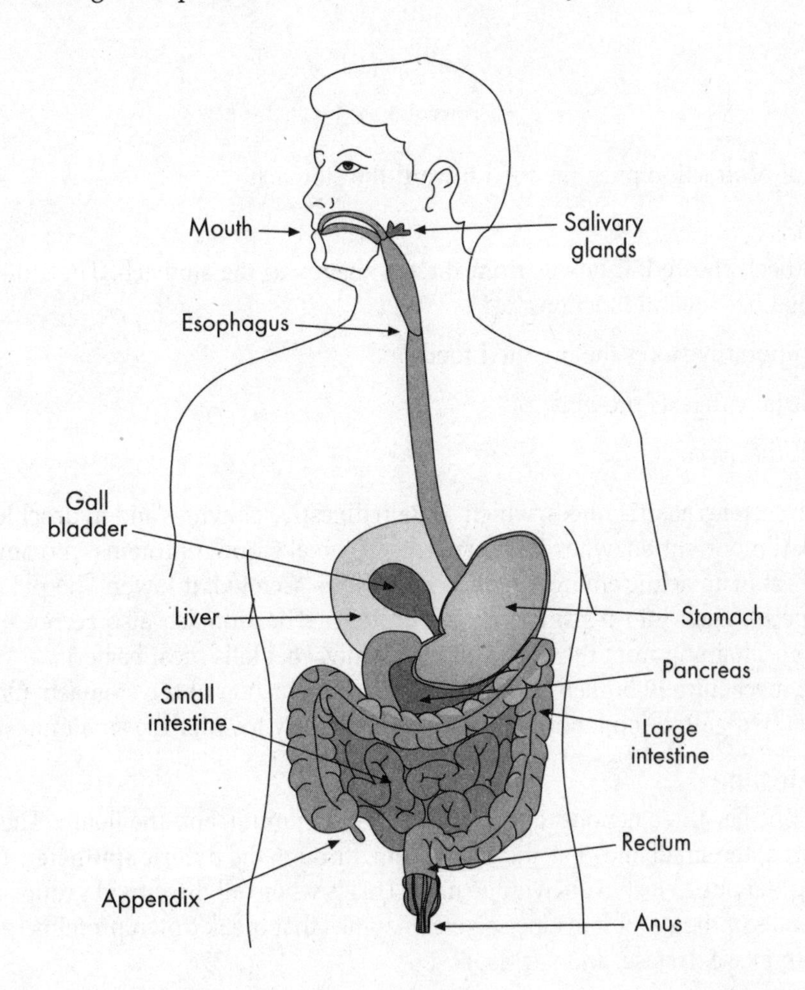

When food enters the mouth, mechanical and chemical digestion begins. The chewing, softening and breaking up of food is called **mastication** (also known as **mechanical digestion**). The mouth also has **saliva** in it. Saliva, which is secreted by the **salivary glands**, contains an important enzyme known as **salivary amylase**. Salivary amylase begins the chemical breakdown of *starch* into maltose.

Once chewed, the food, now shaped into a ball called a **bolus**, moves through the **pharynx** and into the esophagus. Food moves through the esophagus in a wavelike motion known as **peristalsis**.

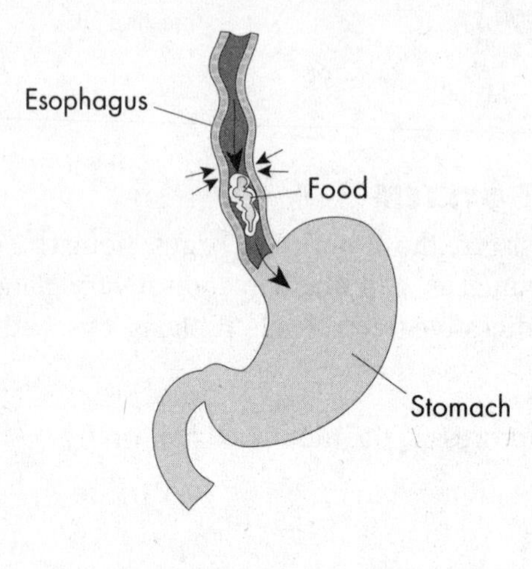

Peristalsis

The waves of contraction push the food toward the stomach.

The Stomach

Once food has been chewed, it moves from the esophagus to the stomach. The stomach is a thick, muscular sac that has several functions.

- It temporarily stores the ingested food.
- It partially digests proteins.
- It kills bacteria.

The stomach secretes **gastric juices**, which contain digestive enzymes and hydrochloric acid (HCl). One of the most important enzymes is **pepsin**, which breaks down proteins into smaller peptides. Pepsin works best in an acidic environment. When HCl is secreted, it lowers the pH of the stomach and activates pepsinogen into pepsin to digest proteins. The stomach also secretes mucus, which protects the stomach lining from the acidic juices. Finally, HCl kills most bacteria.

Food is also mechanically broken down by the churning action of the stomach. Once that's complete, this partially digested food, now called **chyme**, is ready to enter the small intestine.

The Small Intestine

The small intestine has three regions: the duodenum, the jejunum, and the ileum. The chyme moves into the first part of the small intestine, the duodenum, through the **pyloric sphincter**. The small intestine is very long—about 23 feet in an average man. This is where all three food groups are completely digested. The walls of the small intestine secrete enzymes that break down proteins (peptidases) and carbohydrates (maltase, lactase, and sucrase).

The Pancreas

The **pancreas** secretes a number of enzymes into the small intestine: **trypsin**, **chymotrypsin**, **pancreatic lipase**, and **pancreatic amylase**. Trypsin and chymotrypsin break down proteins into dipeptides. Pancreatic lipase breaks down lipids into fatty acids and glycerol. Pancreatic amylase, on the other hand, breaks down starch into disaccharides. Ribonuclease and deoxyribonuclease break down nucleic acids into nucleotides.

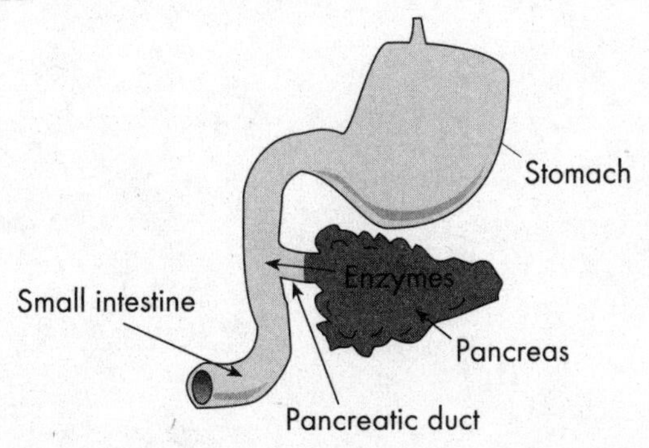

These enzymes are secreted into the small intestine via the **pancreatic duct**.

Another substance that works in the small intestine is called **bile**. Bile is not a digestive enzyme. It's an **emulsifier**, meaning that it mechanically breaks up fats into smaller fat droplets. This process increases the surface area of the fat to be digested, making it more accessible to pancreatic lipase. Bile enters the small intestine by the bile duct, which merges with the pancreatic duct.

Here's something you should memorize:

Bile is made in the **liver** and stored in the **gall bladder**.

Once food is broken down, it is absorbed by tiny, fingerlike projections of the intestine called **villi** and **microvilli**. Villi and microvilli are folds that increase the surface area of the small intestine for food absorption. Within each of the villi is a capillary that absorbs the digested food and carries it into the bloodstream. Within each villus there are also lymph vessels, called **lacteals**, which absorb fatty acids.

Don't forget that hormones are also involved in the digestive system: **gastrin** (which stimulates stomach cells to produce gastric juice), **secretin** (which stimulates the pancreas to produce bicarbonate and digestive enzymes), and **cholecystokinin** (which stimulates the secretion of pancreatic enzymes and the release of bile).

Here's a summary of the pancreatic enzymes.

THE PANCREATIC ENZYMES AND THE FOODS THEY DIGEST	
Pancreatic Enzymes	**Food Substance**
pancreatic amylase, pancreatic lipase, trypsin, chymotrypsin	starch fat protein
proteolytic enzymes, maltase, lactase	proteins carbohydrates

The Large Intestine

The large intestine is much shorter and thicker than the small intestine. The large intestine has an easy job: It reabsorbs water and salts. The large intestine also harbors harmless bacteria that are actually quite useful. These bacteria break down undigested food and in the process provide us with certain essential vitamins, like **Vitamin K**. The leftover undigested food, called **feces,** then moves out of the large intestine and into the rectum.

KEY WORDS

intracellular digestion
extracellular digestion
mouth
esophagus
crop
stomach
intestine
rectum
anus
small intestine
large intestine
accessory organs
oral cavity
mastication (or mechanical digestion)
saliva
salivary glands
salivary amylase
bolus
pharynx
peristalsis
gastric juices
pepsin
chyme
pyloric sphincter
pancreas
trypsin
chymotrypsin
pancreatic lipase
pancreatic amylase
pancreatic duct
bile
emulsifier
liver
gall bladder
villi
microvilli
lacteals
gastrin
secretin
cholecystokinin
Vitamin K
feces

DIGESTIVE SYSTEM REVIEW QUESTIONS

Answers can be found in Chapter 15.

1. Absorption of lipids presents a unique challenge for the digestive system. All of the following enzymes and structures are involved with the breakdown and absorption of lipids EXCEPT:

   (A) Lacteals
   (B) Bile
   (C) Trypsin
   (D) Pancreatic lipase

2. The human stomach contains harsh conditions including low pH juices and enzymes which breakdown proteins to facilitate the chemical breakdown of food. In the image below, food is approaching the stomach. What terms are used to describe food just before it reaches the stomach and immediately after it has been partially digested there?

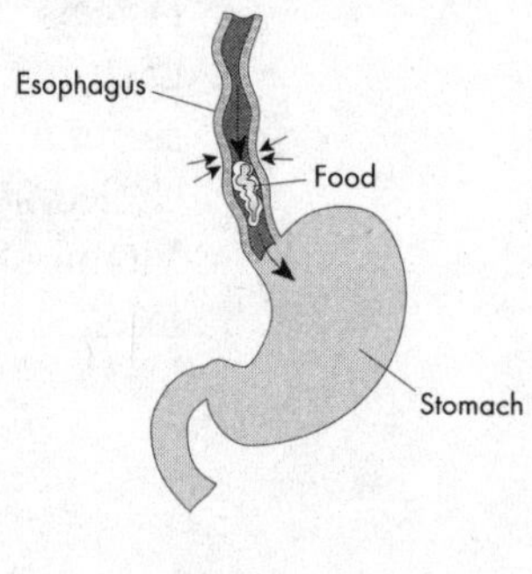

Peristalsis

   (A) Food is referred to as starch before the stomach and bolus after stomach digestion.
   (B) Food is referred to as a bolus before the stomach and feces after stomach digestion.
   (C) Food is referred to as a bolus before the stomach and chyme after stomach digestion.
   (D) Food is referred to as chyme before the stomach and a bolus after stomach digestion.

3. The colon or large intestine is inhabited by a wide variety of bacterial species which are beneficial to human health for all of the following reasons EXCEPT:

   (A) They provide the body with vitamins such as Vitamin K.
   (B) They help facilitate digestion of undigested food.
   (C) They prevent pathogenic bacteria from colonizing the colon by competing for resources.
   (D) They help the body absorb water and salts.

4. The digestive system must facilitate digestion of all four major groups of macromolecules: carbohydrates, proteins, lipids, and nucleic acids. Which of these groups of molecules is the primary target of digestion in the mouth?

   (A) Carbohydrates
   (B) Proteins
   (C) Lipids
   (D) Nucleic acids

THE RESPIRATORY SYSTEM

All cells need oxygen for aerobic respiration. For simple organisms, such as Platyhelminthes, no special structures are needed because the gases can easily diffuse across every cell membrane. In other multicellular organisms, however, the cells are not in direct contact with the environment. These organisms must find other ways of getting oxygen into their systems. For some animals, such as segmented worms, gas exchange occurs directly through their skin. Others, such as insects, have special tubes called **tracheae**. Air enters these tubes through tiny openings called **spiracles**. Among vertebrates, the respiratory structures you should be familiar with are **lungs** and **gills** (used by many aquatic creatures). Fish use counter current-exchange in order to transfer oxygen from the water to their blood.

THE HUMAN RESPIRATORY SYSTEM

Let's talk about how air gets into the body. Air enters through the nose or mouth:

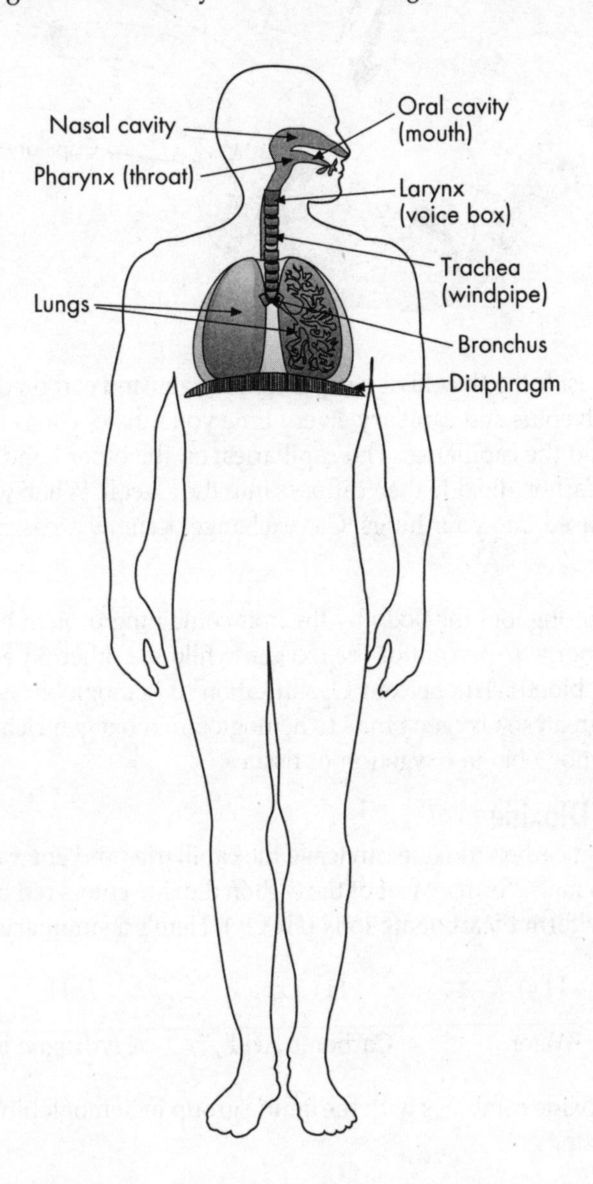

The nose cleans, warms, and moistens the incoming air and passes it through the **pharynx** (throat) and **larynx** (voice box). Next, air enters the trachea. A special flap called the **epiglottis** covers the trachea when you swallow, preventing food from going down the wrong pipe. The trachea also has cartilage rings to help keep the air passage open as air rushes in.

The trachea then branches into two bronchi: the **left bronchus** and the **right bronchus**. These two tubes service the lungs. In the lungs, the passageways break down into smaller tubes known as **bronchioles**. Each bronchiole ends in a tiny air sac known as an **alveolus**. These sacs enable the lungs to have an enormous surface area for maximum gas exchange: about 100 square meters. Let's take a look at one of these tiny air sacs.

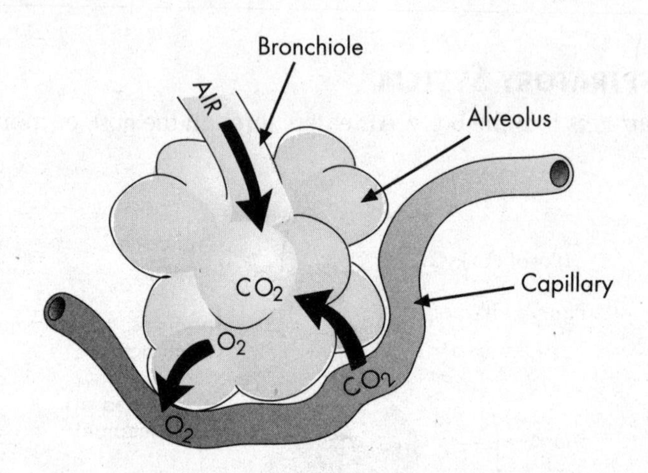

You'll notice that alongside the alveolus is a **capillary**. Oxygen and carbon dioxide diffuse across the membrane of both the alveolus and capillary. Every time you inhale, you send oxygen to the alveoli. Oxygen then diffuses into the capillaries. The capillaries, on the other hand, have a high concentration of carbon dioxide. Carbon dioxide then diffuses into the alveoli. When you exhale, you expel the carbon dioxide that diffused into your lungs. Gas exchange occurs via passive diffusion.

Transport of Oxygen

Oxygen is transported throughout the body by the iron-containing protein **hemoglobin** in red blood cells. Hemoglobin transports 97 percent of the oxygen while the other 3 percent is dissolved in the **plasma** (the fluid of the blood). The percent O_2 saturation of hemoglobin is highest where the concentration of oxygen is greatest. Oxygen binds to hemoglobin in oxygen-rich blood leaving the lungs and dissociates from hemoglobin in oxygen-poor tissues.

Transport of Carbon Dioxide

We've just mentioned that carbon dioxide can leave the capillaries and enter the lungs. However, carbon dioxide can travel in many forms. Most of the carbon dioxide enters red blood cells and combines with water to eventually form **bicarbonate ions** (HCO_3^-). Here's a summary of the reaction:

$$CO_2 + H_2O \leftrightarrows H_2CO_3 \leftrightarrows H^+ + HCO_3^-$$

Carbon dioxide, Water, Carbonic Acid, Hydrogen ion, Bicarbonate ion

Sometimes carbon dioxide combines with the amino group in hemoglobin and mixes with plasma, or is transported to the lungs.

Mechanics of Breathing

What happens to your body when you take a deep breath? Your diaphragm and intercostal muscles contract and your rib cage expands. This action increases the volume of the lungs, allowing air to rush in. This process of taking in oxygen is called **inspiration**. When you breathe out and let carbon dioxide out of your lungs, that's called **expiration**. Your respiratory rate is controlled by **chemoreceptors**. As your blood pH decreases, chemoreceptors send nerve impulses to the **diaphragm** and intercostal muscles to increase your respiratory rate.

KEY WORDS

trachea
spiracles
lungs
gills
pharynx
larynx
epiglottis
left bronchus
right bronchus
bronchioles
alveolus
capillary
hemoglobin
plasma
bicarbonate ions
inspiration
expiration
chemoreceptors
diaphragm

RESPIRATORY SYSTEM REVIEW QUESTIONS

Answers can be found in Chapter 15.

1. Which of the following accurately describes the flow of air through the respiratory tract beginning with the nose?

   (A) Pharynx, larynx, trachea, bronchi, bronchioles, alveoli
   (B) Trachea, pharynx, larynx, alveoli, bronchi, bronchioles
   (C) Pharynx, larynx, trachea, alveoli, bronchi, bronchioles
   (D) Larynx, pharynx, trachea, bronchi, bronchioles, alveoli

Questions 2–4 refer to the following diagram and paragraph.

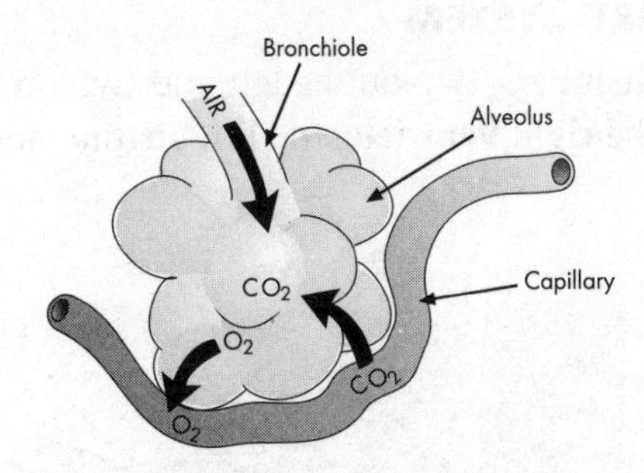

Gas exchange occurs at the interface of pulmonary capillaries and alveoli as shown above. Each time an individual breathes, oxygen and carbon dioxide gases are being exchanged across the membranes of these structures by passive diffusion. The mechanisms by which oxygen and carbon dioxide are carried in the blood and delivered to tissues differ.

2. Oxygen is predominately carried in the blood by which of the following mechanisms?

   (A) Oxygen attaches to hemoglobin in erythrocytes.
   (B) Oxygen is dissolved directly in the blood.
   (C) Oxygen combines with water to create bicarbonate ions, which are carried in the blood.
   (D) Oxygen combines with water to create hydrogen peroxide, which is carried in the blood.

3. Carbon dioxide is predominately carried in the blood by which of the following mechanisms?

   (A) Carbon dioxide attaches to hemoglobin in erythrocytes (red blood cells).
   (B) Carbon dioxide is dissolved directly in the blood.
   (C) Carbon dioxide combines with water to create bicarbonate ions, which are carried in the blood.
   (D) Carbon dioxide combines with water to create hydrogen peroxide, which is carried in the blood.

4. Immediately prior to reaching the lungs, which of the following most accurately describes the concentrations of oxygen and carbon dioxide in the blood?

   (A) Oxygen rich and carbon dioxide poor
   (B) Oxygen rich and carbon dioxide rich
   (C) Oxygen poor and carbon dioxide poor
   (D) Oxygen poor and carbon dioxide rich

THE CIRCULATORY SYSTEM

Most organisms need to carry out two tasks: (1) supply their bodies with nutrients and oxygen, and (2) get rid of wastes. Many simple aquatic organisms have no trouble moving materials across their membranes since their metabolic needs are met by diffusion. Larger organisms, on the other hand, particularly terrestrial organisms, can't depend on diffusion. They therefore need special circulatory systems to accomplish internal transport.

There are two types of circulatory systems: an **open circulatory system** and a **closed circulatory system**. In an open circulatory system, blood is carried by open-ended blood vessels that spill blood into the body cavity. In arthropods, for example, blood vessels from the heart open into internal cavities known as **sinuses**. Other organisms have closed circulatory systems. That is, blood flows continuously through a network of blood vessels. Earthworms and some mollusks have a closed circulatory system, as do vertebrates.

THE HUMAN CIRCULATORY SYSTEM

The heart is divided into four chambers, two on the left and two on the right. The four chambers of the heart are the **right atrium**, the **right ventricle**, the **left atrium**, and the **left ventricle**. Let's take a look at a picture of the heart:

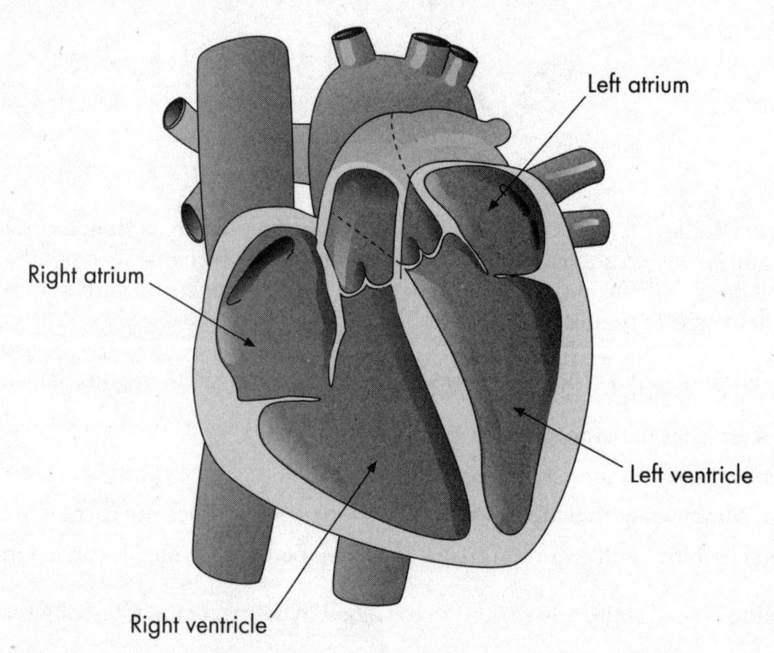

The heart pumps blood in a continuous circuit. Since blood makes a circuit in the body, it doesn't matter where we begin to trace the flow of blood. For our purposes, we'll begin at the point in the circulatory system where the blood leaves the heart and enters the body: the left ventricle. When blood leaves the left ventricle it will make a tour of the body. We call this **systemic circulation**.

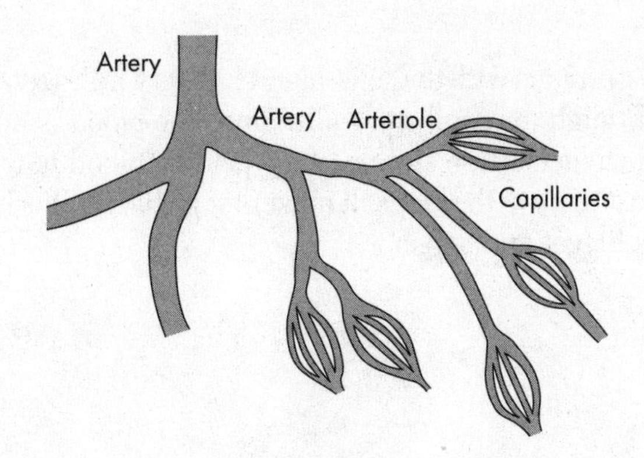

Systemic Circulation

Blood leaves the heart through the **aortic semilunar valve** and enters a large blood vessel called the **aorta**. The aorta is the largest artery in the body. The aorta then branches out into smaller vessels called **arteries**.

Arteries always carry blood away from the heart. Just remember "A" stands for "away" from the heart. They're able to carry the blood because arteries are thick-walled, elastic vessels. The arteries become even smaller vessels called **arterioles**, and then the smallest vessels called **capillaries**.

There are thousands of capillaries. In fact, some estimate that the capillary routes in your bloodstream are as long as 100 kilometers! These vessels are so tiny that red blood cells must "squeeze" through them in single file. Capillaries intermingle with the tissues and exchange nutrients, gases, and wastes. Oxygen and nutrients leave the capillaries and enter the tissues; carbon dioxide and wastes leave the tissues and enter the capillaries.

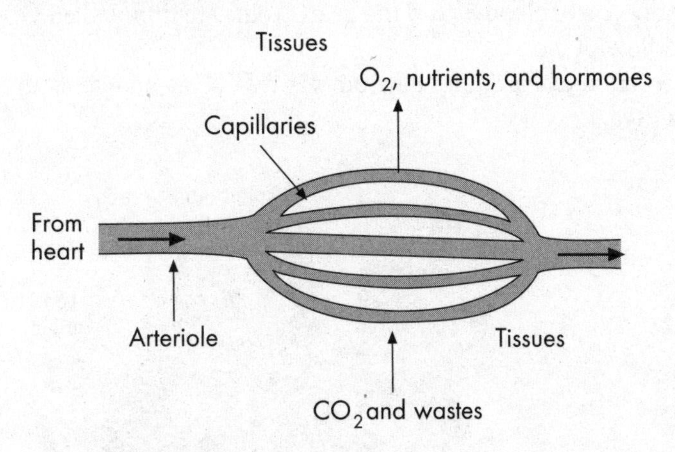

Before we take a look at the next stage of circulation, let's recap the pathway of blood through the body.

1. Blood leaves the heart's left ventricle via the aorta.
2. It travels through the arteries to the arterioles, and eventually to the capillaries.
3. Gas and nutrient/waste exchange occurs between the blood and the tissues through the capillary walls.

Back to the Heart

After exchanging gases and nutrients with the cells, blood has very little oxygen left. Most of its oxygen was donated to the cells through the capillary walls. Since the blood is now depleted of oxygen, it is said to be **deoxygenated**. To get a fresh supply of oxygen the blood now needs to go to the lungs.

But the blood doesn't go *directly* to the lungs. It must first go back to the heart. As the blood returns to the heart, the vessels get bigger and bigger.

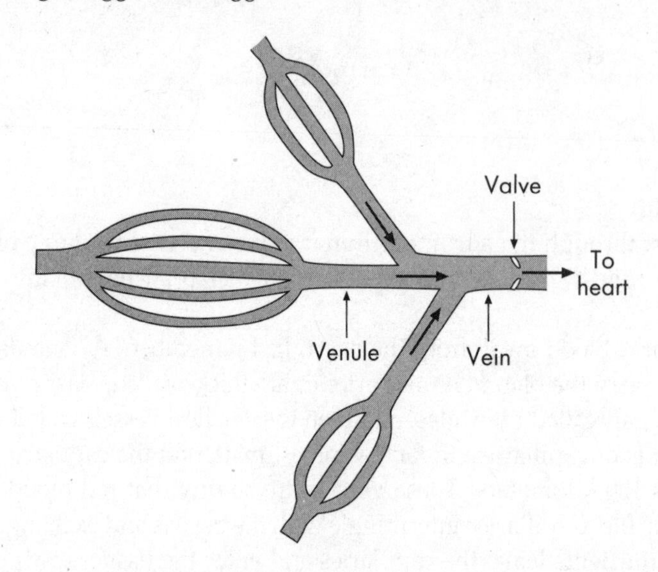

From the capillaries, blood travels through vessels called **venules** and then through larger vessels called **veins**. Veins always carry blood *toward* the heart. Veins are thin-walled vessels with valves that prevent the backward flow of blood.

Blood eventually enters the heart's right atrium via two veins known as the **superior vena cava** and the **inferior vena cava.**

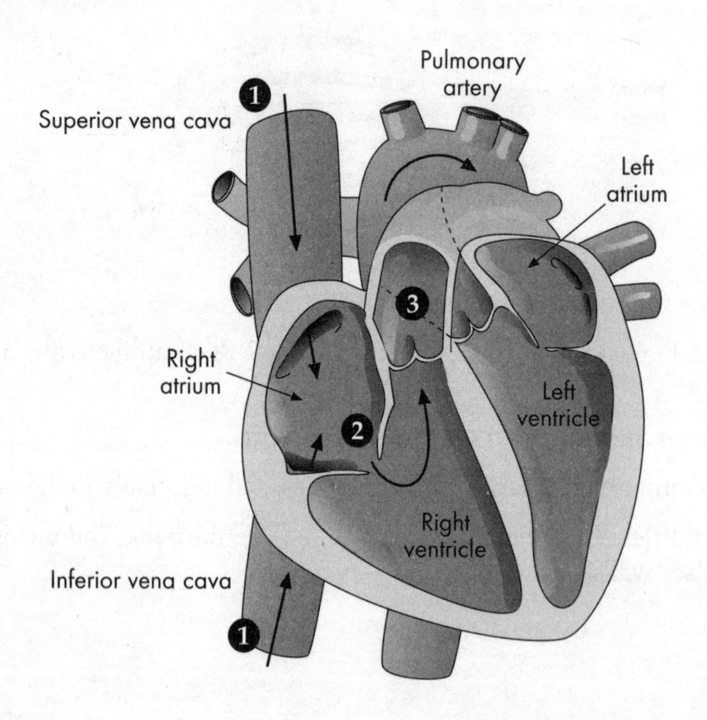

Blood now moves through the heart. Blood travels from the right atrium to the right ventricle through the **right atrioventricular valve** (or **tricuspid**). From the right ventricle, blood will go out again into the body, but this time toward the lungs. This is called **pulmonary circulation**.

The Pulmonary System

Blood leaves the right ventricle through the **pulmonary semilunar valve** and enters a large artery known as the **pulmonary artery**. Remember what we said about arteries? Blood vessels that leave the heart are always called *arteries*.

There's one major feature you must remember about the blood in the pulmonary system. Whereas in systemic circulation the blood was rich with oxygen, the pulmonary artery is carrying *deoxygenated* blood. The pulmonary artery branches into the right and left pulmonary arteries which lead, respectively, to the right and left lungs. These arteries become smaller arterioles and then once again capillaries.

We just said that these vessels carry deoxygenated blood. In the lungs, the blood will pick up oxygen and dump carbon dioxide. Sound familiar? It should. It's just like the gas exchange we discussed in the respiratory system. In the lungs, the blood fills with oxygen, or becomes **oxygenated**. The blood returns to the heart via the **pulmonary veins** and enters the left atrium.

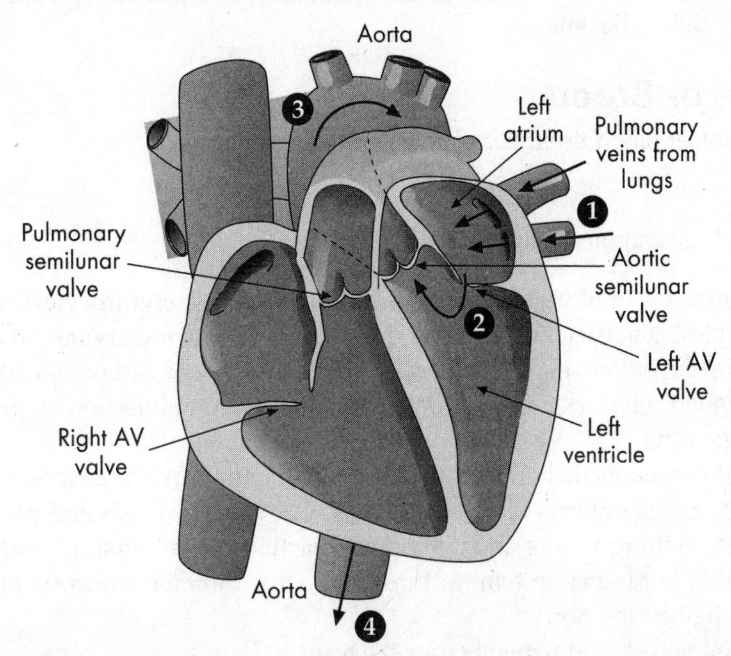

Blood then moves to the left ventricle through the **left atrioventricular valve** (or **bicuspid** or **mitral valve**). Now we've completed our tour of the heart. Let's recap the events in pulmonary circulation.

1. Deoxygenated blood leaves the right ventricle via the pulmonary artery.
2. The pulmonary artery branches into the right and left pulmonary arteries, carrying the blood to the lungs.
3. Blood travels from the arteries to the arterioles, and eventually to the capillaries.
4. Gas exchange occurs between the capillaries and alveoli in the lungs.
5. Once the blood is oxygenated, it returns to the heart through the pulmonary veins.

Thermoregulation

Human homeostasis is characterized by **thermoregulation**, or the maintenance of a fairly stable body temperature regardless of external conditions. Animals that regulate their internal temperature are known as **endotherms**. Frequently referred to as "warm-blooded," endotherms have arteries and veins arranged in a way that enables them to conserve heat through a process known as **counter current exchange**. The arteries carrying warm blood from the core to the outside are right next to the veins carrying cold blood in the opposite direction. Heat from the arterial blood warms the cold venous blood returning to the heart. **Ectotherms**, or "cold-blooded" animals, on the other hand, gain and lose heat by way of the environment.

HEART CYCLE

Your heart contracts and relaxes automatically, about 72 times a minute. A special conduction system makes sure that your heart beats rhythmically. The beat begins in tissues in the right atrium called the **sinoatrial (SA) node** ("the pacemaker"). The impulse then spreads through both atria and conducts directly to the **atrioventricular (AV) node**. From the AV node, the action potential spreads to the **bundle of His** and then to the **Purkinje fibers** in the walls of both ventricles. This generates a strong contraction. The part of the cycle in which contraction occurs is called **systole**, and the part in which relaxation occurs is called **diastole**.

THE CONTENTS OF BLOOD

Now let's take a look at blood itself. Blood consists of two things:

- Plasma
- Cells and cell fragments suspended in the fluid

Blood carries three types of cells: **red blood cells** (also called **erythrocytes**), **white blood cells** (also called **leukocytes**), and **platelets**. Red blood cells are the oxygen-carrying cells in the body. They contain hemoglobin, the protein that actually carries the oxygen (and carbon dioxide) throughout the body. Mature red blood cells lack a nucleus. White blood cells fight infection by protecting the body against foreign organisms.

Platelets are cell fragments that are involved in blood clotting. When a blood vessel is damaged, platelets stick to the collagen fibers of the vessel wall. The damaged cells and platelets release substances that activate clotting factors and a series of reactions occur. First, a prothrombin activator converts prothrombin (a plasma protein) to thrombin. Then thrombin converts fibrinogen to fibrin threads, which strengthen the clot.

Here's something you should remember for the test:

> All of the blood cells are made in the **bone marrow**. The bone marrow is located in the center of the bones.

Blood Types

There are four blood groups: **A**, **B**, **AB**, and **O**. Blood types are pretty important and are based on the type of antigen(s) found on red blood cells. If a patient is given the wrong type of blood in a transfusion, it could be fatal! Why? Because red blood cells in the blood will clump if they are exposed to the wrong blood type. For example, if you've got blood type A (i.e., your red cells carry the A antigen) and you receive a blood transfusion of blood type B, your blood will clump. That's because your blood contains **antibodies**, an immune substance that will bind and destroy the foreign blood.

What is important to remember about the different blood types is that type O blood is the universal donor and that type AB is the universal recipient. This means that anyone can receive a blood transfusion of type O blood, while those with type AB blood (which is very rare among Americans—only about 4 percent of the population) can receive any kind of blood without risk. **Rh factors** are also antigens found on red blood cells. Persons with these antigens are Rh^+ and those without them are Rh^-.
platelets

KEY WORDS

open circulatory system
closed circulatory system
sinuses
right atrium
right ventricle
left atrium
left ventricle
systemic circulation
aortic semilunar valve
aorta
arteries
arterioles
capillaries
deoxygenated blood
venules
veins
superior vena cava
inferior vena cava
right AV (tricuspid) valve
pulmonary circulation
pulmonary semilunar valve
pulmonary artery
oxygenated blood
pulmonary vein
left atrioventricular valve (or bicuspid or mitral valve)
thermoregulation
endotherm
counter current exchange
ectotherm
sinoatrial (SA) node
atrioventricular (AV) node
bundle of His
Purkinje fibers
systole
diastole
red blood cells (or erythrocytes)
white blood cells (or leukocytes)
platelets
bone marrow
blood type A
blood type B
blood type AB
blood type O
antibodies
Rh factor

CIRCULATORY SYSTEM REVIEW QUESTIONS

Answers can be found in Chapter 15.

1. A patient enters the emergency room and requires immediate invasive surgery. One of the first tests performed is to determine the blood type of the patient in case they require additional units of blood. If the patient has AB blood, which of the following blood types may be used?

   (A) Only AB blood
   (B) A, B, or AB blood
   (C) Only O blood
   (D) Any blood type (A, B, AB, or O)

2. An individual has recently suffered a myocardial infarction (heart attack) resulting in partial cell death in the muscle surrounding the right ventricle. This individual would have difficulty directly pumping blood to which of the following?

   (A) right atrium
   (B) left atrium
   (C) aorta and rest of the body
   (D) pulmonary artery and the lungs

3. Erythrocytes and leukocytes are two well-studied and important blood types. Which of the following is a common feature of mature cells of both blood types?

   (A) They are both derived from bone marrow.
   (B) They have a nucleus.
   (C) They carry oxygen bound to hemoglobin.
   (D) They protect the body from foreign antigens.

4. The electrical activity of the heart is carefully timed. Which of the following accurately describes the pathway that the electrical impulse takes during a beat?

   (A) AV node, SA node, Purkinje fibers, bundle of His
   (B) SA node, AV node, Purkinje fibers, bundle of His
   (C) Bundle of His, Purkinje fibers, SA node, AV node
   (D) Purkinje fibers, SA node, AV node, bundle of His

THE LYMPHATIC AND IMMUNE SYSTEMS

In addition to the circulatory system, vertebrates have another system called the lymphatic system. The **lymphatic system** is made up of a network of vessels that conduct **lymph**, a clear, watery fluid formed from interstitial fluid. Lymph vessels are found throughout the body along the routes of blood vessels and plays an important role in fluid homeostasis.

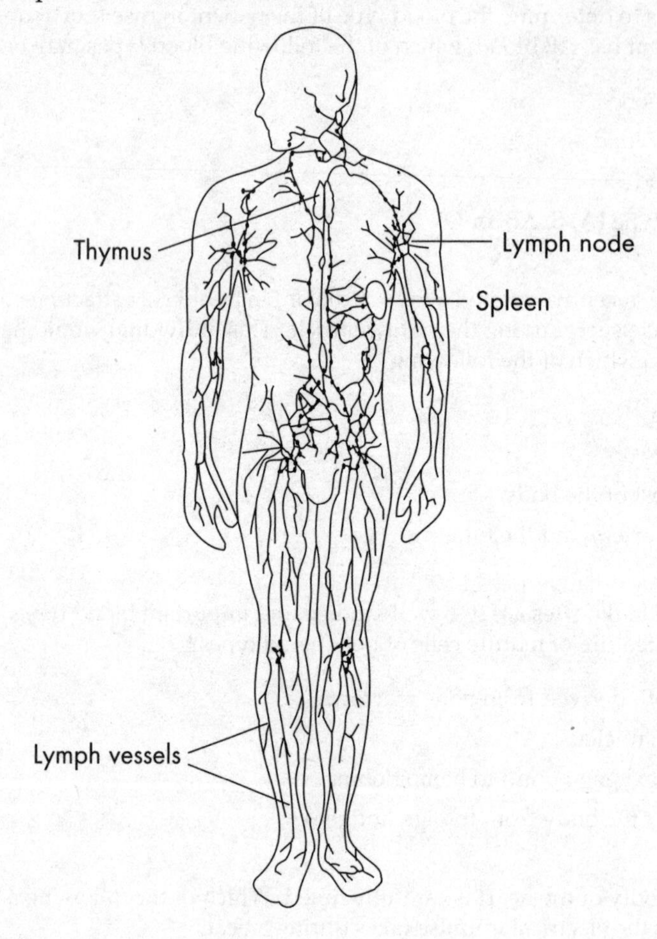

The lymphatic system has three functions.

- It collects, filters, and returns fluid to the blood by the contraction of adjacent muscles.
- It fights infection using lymphocytes, cells found in lymph nodes.
- It removes excess fluid from body tissue.

Sometimes a lymph vessel will form a **lymph node**, a mass of tissue found along the course of a lymph vessel. A lymph node contains a large number of lymphocytes:

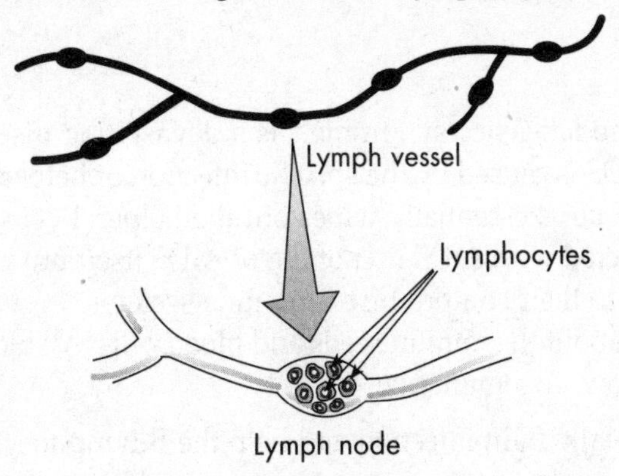

Lymphocytes are important in fighting infection. They multiply rapidly when they come in contact with an **antigen**, or foreign substance recognized by the **immune system**. (We'll talk about this in a second.) The lymph nodes swell when they're fighting an infection. That's why when you have a sore throat, one of the first things a doctor does is touch the sides of your throat to see if your lymph nodes are swollen, a probable sign of infection.

THE IMMUNE SYSTEM

The immune system, generally speaking, is one of the body's defense systems. It is a carefully and closely coordinated system of specialized cells, each of which plays a specific role in the war against bodily invaders. As we mentioned above, foreign molecules—be they viral, bacterial, or simply chemical—that can trigger an immune response are called antigens. The appearance of antigens in the body stimulates a defense mechanism that produces antibodies.

The body's first line of defense against foreign substances is the skin and mucous lining of the respiratory and digestive tracts. If these defenses are not sufficient, other nonspecific defense mechanisms are activated. These include **phagocytes** (which engulf antigens), **complement proteins** (which lyse the cell wall of the antigen), **interferons** (which inhibit viral replication and activate surrounding cells that have antiviral actions), and **inflammatory response** (a series of events in response to antigen invasion or physical injury).

Types of Immune Cells

The primary cells of the immune system are lymphocytes: B-cells and T-cells. When an individual becomes infected by a **pathogen**, a disease-causing agent, B and T lymphocytes get activated. B-lymphocytes mature in bone marrow and are involved in the humoral response, which defends the body against pathogens present in extracellular fluids, like lymphatic fluid or blood. When B-lymphocytes encounter pathogens, they are activated and produce clones. Some B-cells become memory cells that remain in circulation, allowing the body to mount a quicker response if a second exposure to the pathogen should occur. Other B-cells become plasma cells that produce antibodies, which are specific proteins that bind to the antigens on the surface of pathogens that originally activated them.

T-lymphocytes, maturing in the thymus, are involved in **cell-mediated immunity.** The plasma membrane of cells has **major histocompatibility complex** (MHC) markers that distinguish between self and foreign cells. When T-cells encounter cells infected by pathogens, they recognize the foreign antigen-MHC markers on the cell surface. Once activated, T-cells multiply and give rise to clones. Some T-cells become **memory T-cells,** whereas others become **helper T-cells.** Helper T-cells activate

B-lymphocytes and other T-cells in responding to the infected cells. Memory T-cells recognize pathogens they have encountered before. Another type of T-cells, the **cytotoxic T-cell,** recognizes and kill infected cells.

AIDS

AIDS, or "acquired immunodeficiency syndrome," is a devastating disease that interferes with the body's immune system. AIDS is caused by the specific infection of helper T cells by human immunodeficiency virus (HIV). The virus essentially wipes out the helper T-cells, preventing the body from defending itself. Those afflicted with AIDS do not die of AIDS itself but rather of infections that they can no longer fight off due to their compromised immune systems.

One thing to remember about the immune cells and blood cells: All blood cells, white and red, are produced in the bone marrow. To summarize:

- *T-lymphocytes* actually fight infection and help the B-lymphocytes proliferate.
- *B-lymphocytes* produce *antibodies.*

KEY WORDS

lymphatic system
lymph
lymph node
lymphocytes
antigen
immune system
phagocytes
complement proteins
interferons
inflammatory response
major histocompatibility complex markers (MHC markers)
T-lymphocytes
memory T-cells
helper T-cells
B-lymphocytes
cytotoxic T-cells
cell-mediated response
humoral immunity
macrophages
AIDS

LYMPHATIC AND IMMUNE SYSTEMS REVIEW QUESTIONS

Answers can be found in Chapter 15.

1. Lymph nodes are often evaluated when a patient is believed to have an active infection or cancer. Why?

   (A) Active infections or cancer often generate large quantities of fluids as cells rupture, the lymph nodes often swell as a result of the buildup of these fluids.

   (B) During an active infection, lymphocytes expand rapidly in lymph nodes causing swelling of the structures.

   (C) Lymph nodes are the most common tissue site of active infections or cancer.

   (D) Cancer often metastasizes to lymph nodes causing them to expand as new cancerous cells multiply.

Questions 2 – 4 refer to the following chart and paragraph.

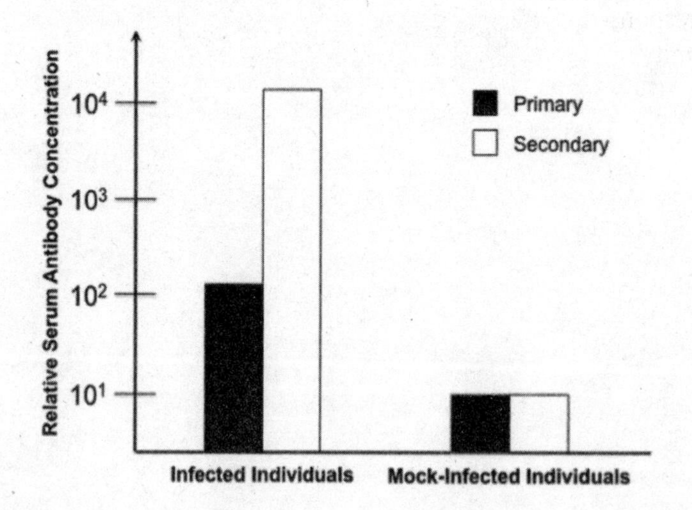

A scientist wishes to test the efficacy and immunogenicity of a newly developed live-attenuated virus vaccine strain. To test for immunogenicity, the scientist acquires 100 healthy volunteers and challenges them with either the vaccine strain or a placebo strain (consisting of isotonic saline solution). 12 days following the initial primary challenge, the serum antibody concentration is determined by an ELISA analysis. A follow-up challenge to test for existing memory was performed on day 60 using the same 100 individuals. Serum antibody titers were determined for the secondary challenge on day 72. These data are shown in the chart above.

2. The placebo strain was included in the experimental design in order to

   (A) evaluate the immune response to isotonic saline solution.

   (B) determine primary and secondary immune responses to the virus.

   (C) compare antibody titers of infected and mock-infected individuals as a control.

   (D) practice before challenging the same individuals with the vaccine strain.

3. What type of immune cells is responsible for generating the antibodies generated in this experiment?

(A) Erythrocytes
(B) B Cells
(C) T Cells
(D) Phagocytes

4. How can the scientist increase the statistical significance of these data?

(A) They could repeat the experiment with a greater sample size.
(B) They could use more control variables.
(C) They could perform a third challenge on the same 100 individuals.
(D) They could use a different statistical test, which provides the necessary statistical significance.

THE EXCRETORY SYSTEM

As you already know, all organisms must get rid of wastes. In this chapter, we'll focus primarily on how organisms get rid of **nitrogenous wastes** (products containing nitrogen) and regulate water. When cells break down proteins, one of the byproducts is **ammonia** (NH_3), a substance that is toxic to the body. Consequently, organisms had to develop ways of converting ammonia to a less poisonous substance. Some animals convert ammonia to **uric acid**, while others convert ammonia to **urea**. Some examples of excretory organs among invertebrates are **nephridia** (found in earthworms) and **Malpighian tubules** (found in arthropods).

Ammonia	Uric acid	Urea
Fish	Birds and reptiles	Most mammals

Excretory System of an Earthworm

Excretory System of a Grasshopper

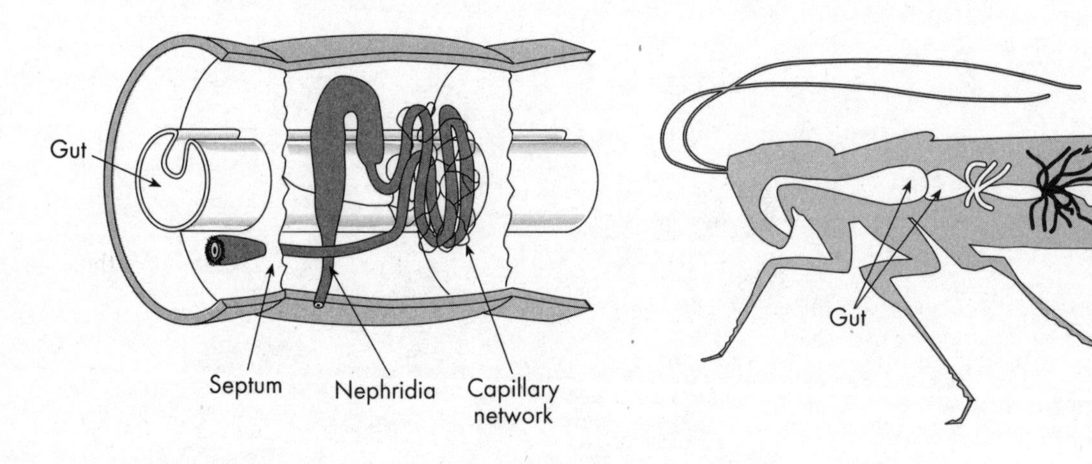

The Human Excretory System

In humans, the major organ that regulates excretion is the **kidney**. Each kidney is made up of a million tiny structures called **nephrons**. Nephrons are the functional units of the kidney. A nephron consists of several regions: the **Bowman's capsule**, the **proximal convoluted tubule**, the **loop of Henle**, the **distal convoluted tubule**, and the **collecting duct**. The **renal cortex**, or outer-most section, contains the Bowman's capsule and the proximal convoluted tubules. The **renal medulla**, the inner section, contains the loop of Henle and the distal convoluted tubules.

Here's an illustration of a nephron:

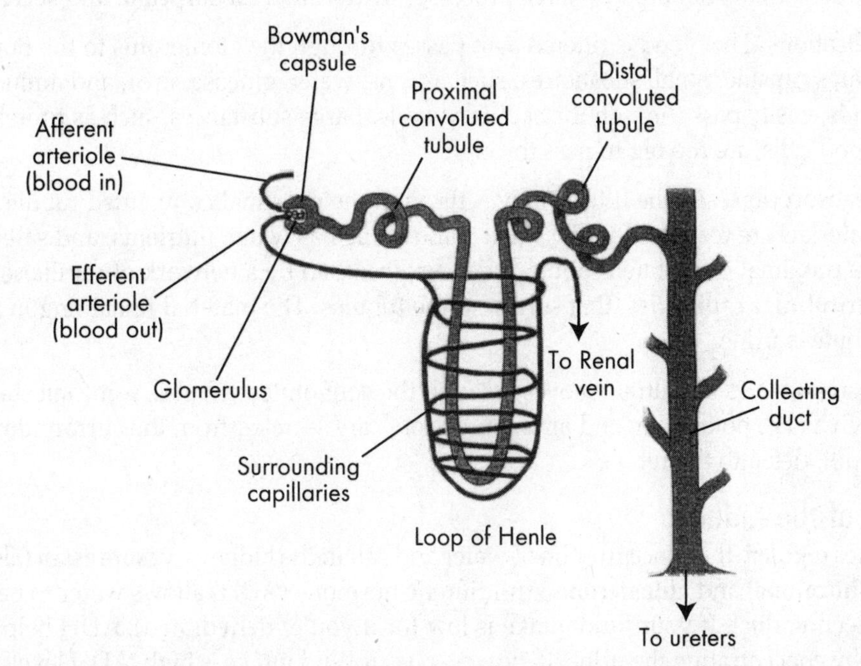

How does a nephron work? Let's trace the flow of blood in a nephron. Blood enters the nephron at the Bowman's capsule. A blood vessel called the **renal artery** leads to the kidney and branches into arterioles, then tiny capillaries. A ball of capillaries that "sits" within a Bowman's capsule is called a **glomerulus**. Blood is filtered as it passes through the glomerulus and the plasma is forced out of the capillaries into the Bowman's capsule. This plasma is now called a **filtrate**.

The filtrate travels along the entire nephron. From the Bowman's capsule, the filtrate passes through the proximal convoluted tubule, then the loop of Henle, then the distal convoluted tubule, and finally the collecting duct. As it travels along the tube, the filtrate is modified to form **urine**.

What happens next? The concentrated urine moves from the collecting ducts into the **ureters**, then into the **bladder**, and finally out through the **urethra**.

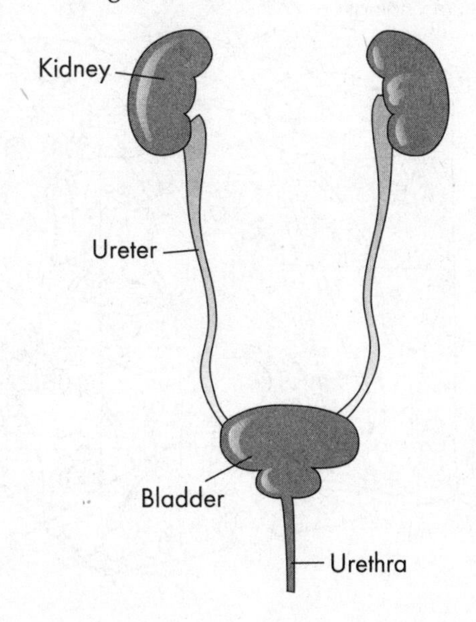

How Urine Is Made

Urine is produced in the nephron by three processes: **filtration**, **reabsorption**, and **secretion**.

- Filtration—The blood is filtered as it passes through the glomerulus to the Bowman's capsule. Small substances, such as ions, water, glucose, urea, and amino acids, easily pass through the capillary walls. Large substances, such as proteins and blood cells, are too big to pass through.
- Reabsorption—As the filtrate moves through the proximal convoluted tubule, some materials are reabsorbed. The small solutes, such as water, nutrients, and salts, leave the proximal convoluted tubule and are reabsorbed by a network of capillaries, the **peritubular capillaries**, that surround the tubules. The material remaining in the tubule is urine.
- Secretion—As the filtrate moves through the convoluted tubules, some substances, such as H^+, potassium, and ammonium ions, are secreted from the surrounding capillaries into the tubule.

Hormones of the Kidney

Two hormones regulate the concentration of water and salt in the kidneys: **vasopressin** (also known as **antidiuretic hormone**) and **aldosterone**. Antidiuretic hormone (ADH) allows water to be reabsorbed from the collecting duct. If your fluid intake is low (or if you're dehydrated), ADH helps your body retain water by concentrating the urine. If, however, your fluid intake is high, ADH levels will be low, and the body won't reabsorb most of the water. Your urine will contain lots of water and therefore be dilute. For now, just remember that ADH controls the volume of urine. The other hormone, aldosterone, is responsible for regulating sodium reabsorption at the distal convoluted tubule.

Skin

The **skin** is also an excretory organ that gets rid of excess water and salts from the body. Believe it or not, your skin is the largest organ in the body! It contains 2.5 million sweat glands that secrete water and ions through pores. The skin's primary function is to regulate body temperature.

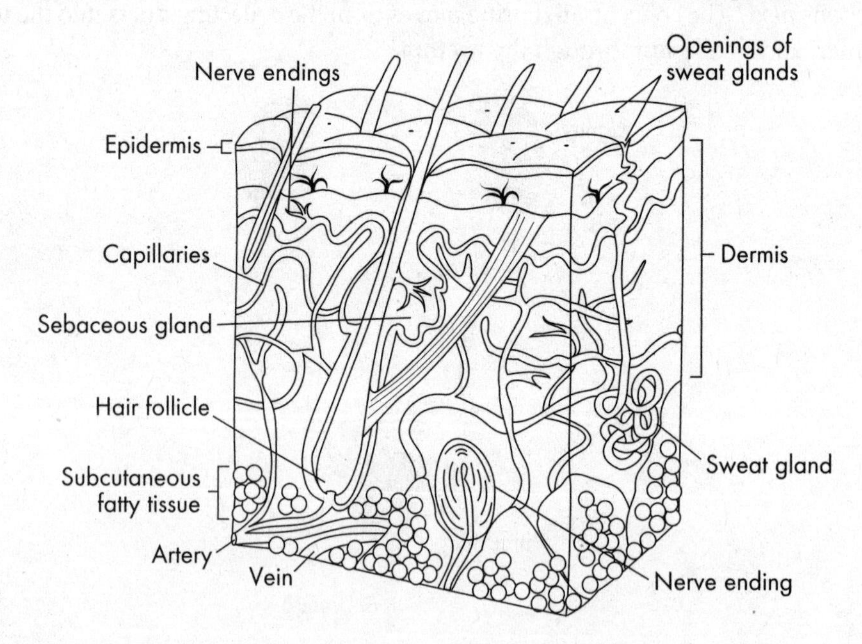

The skin has three layers: the **epidermis**, **dermis**, and **subcutaneous tissue** (or hyperdermis). Sweat glands are found in the dermis layer along with blood vessels, nerves, and sebaceous—or oil—glands.

The epidermis is covered by a layer of dead cells called the **stratum corneum**. These cells form a barrier against invading microorganisms. The bottom layer of skin, the subcutaneous tissue, is mostly fats.

To summarize, in humans, two organs control fluid balance and dispose of metabolic wastes:

- The kidney—gets rid of nitrogenous wastes and reabsorbs water and salt.
- The skin—gets rid of excess salt and water.

KEY WORDS

nitrogenous wastes
ammonia
uric acid
urea
nephridia
Malpighian tubules
kidney
nephrons
Bowman's capsule
proximal convoluted tubule
loop of Henle
distal convoluted tubule
collecting duct
renal cortex
renal medulla
renal arteries
glomerulus
filtrate
urine
ureters
bladder
urethra
filtration
reabsorption
secretion
peritubular capillaries
vasopressin (antidiuretic hormone)
aldosterone
skin
epidermis
dermis
subcutaneous tissue
stratum corneum

EXCRETORY SYSTEM REVIEW QUESTIONS

Answers can be found in Chapter 15.

1. The kidney performs a vital function of filtering out contaminants and substances that are in excess or unneeded from the blood. Which of the following should NOT be present in urine?

   (A) Urea
   (B) Water
   (C) Proteins
   (D) Salts

2. A patient has had an abnormal build up of fluids in the body as a result of a bacterial infection. The doctor prescribes a diuretic medication. Which of the following best describes how this medication works on the kidney?

   (A) The diuretic will result in increased sodium reabsorption in the distal convoluted tubule, which will trigger production of ADH.
   (B) The diuretic will result in increased potassium reabsorption in the distal convoluted tubule, which will trigger production of aldosterone.
   (C) The diuretic will result in increased water reabsorption by triggering production of ADH.
   (D) The diuretic will result in decreased water reabsorption by inhibiting production of ADH.

3. Approximately 2% of urine is comprised of urea. Why must urea be generated and excreted by the kidney?

   (A) Urea is a base, which neutralizes the acidity of urine to allow for its safe disposal through the ureter and urethra.
   (B) Urea is a waste product created by the breakdown of lipids and must be excreted to prevent cholesterol toxicity in the blood.
   (C) Urea is a waste product created by the breakdown of proteins and must be excreted to prevent ammonia toxicity in the blood.
   (D) Urea is a salt, which is critically important for maintaining the concentration gradient used in the nephrons of the kidney for filtration.

4. A urinary tract infection has resulted in destruction of the capillary beds within the Bowman's capsule of several nephrons. Which of the following types of structures will be directly affected by this injury?

   (A) Collecting duct
   (B) Ureter
   (C) Loop of Henle
   (D) Glomerulus

THE NERVOUS SYSTEM

All organisms must be able to react to changes in their environment. As a result, organisms have evolved systems that pick up and process information from the outside world. The task of coordinating this information falls to the nervous system. The simplest nervous system is found in the hydra. It has a **nerve net** made up of a network of nerve cells, the impulse of which travels in both directions. As animals became more complex, they developed clumps of nerve cells called **ganglia**. These cells are like primitive brains. More complex organisms have a brain with specialized cells called **neurons**.

NEURONS

The functional unit in the nervous system is a neuron. That's because neurons receive and send the neural impulses that trigger organisms' responses to their environments. Let's talk about the parts of a neuron. A neuron consists of a **cell body**, **dendrites**, and an **axon**.

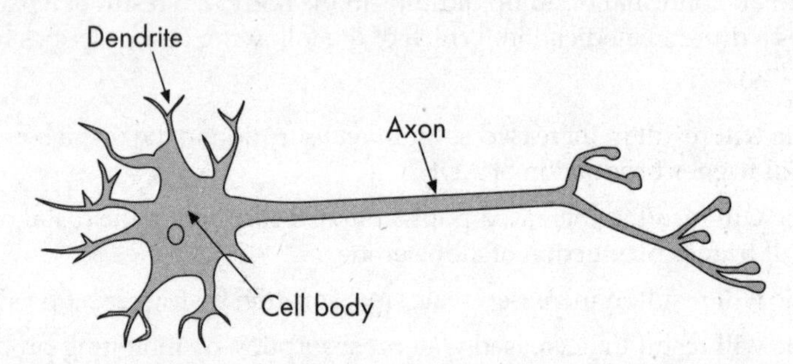

The cell body contains the nucleus and all the usual organelles found in the cytoplasm. Dendrites are short extensions of the cell body that receive stimuli. The axon is a long, slender extension that transmits an impulse from the cell body to another neuron or to an organ. A nerve impulse begins at the top of the dendrites, passes through the dendrites to the cell body, and moves down the axon.

Types of Neurons

Neurons can be classified into three groups: **sensory neurons**, **motor (effector) neurons**, and **interneurons**. Sensory neurons receive impulses from the environment and bring them to the body. For example, sensory neurons in your hand are stimulated by touch. A motor neuron transmits the impulse to muscles or glands to produce a response. The muscle will respond by contracting or the gland will respond by secreting a substance (e.g., a hormone). Interneurons are the links between sensory neurons and motor neurons. They're found in the brain or spinal cord:

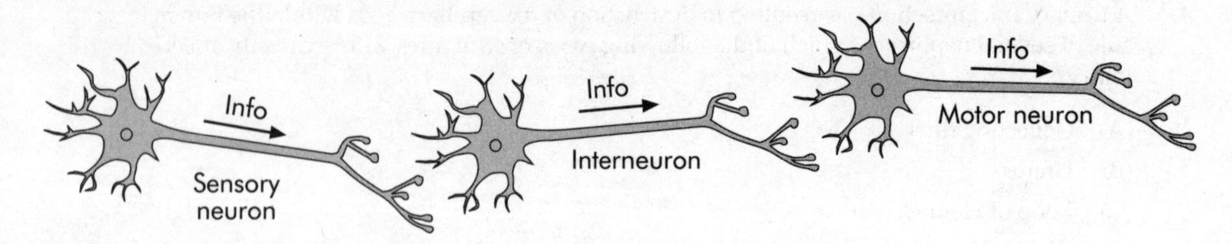

How Neurons Communicate

Before we talk about the events related to the transmission of a nerve impulse, let's review how neurons interact. There are billions of neurons running throughout the body, firing all the time. More often than not, one or more neurons are somewhat "connected." This means that one neuron has its dendrites next to another neuron's axon. In this way, the dendrites of one cell can pick up the impulse sent from the axon of another cell. The second neuron can then send the impulse to its cell body and down its axon, passing it on to yet another cell.

Resting Potential

Neurons are not always transmitting signals. The transmission of an impulse depends on the ionic gradients that exist across the axonal membrane. In humans, the concentration of sodium ions is higher in the extracellular fluid than the concentration of potassium ions. The reverse is true inside the axonal membrane. The concentration of potassium ions is higher in the cytosol than the concentration of sodium ion. Because there are many potassium channels that are open but only a relatively small number of sodium channels open the resting potential results from the diffusion of sodium and potassium ions through open ions channels.

The resting potential arises from two activities:

- The Na^+K^+ ATPase—This pump pushes two potassium ions (K^+) into the cell for every three sodium ions (Na^+) it pumps out of the cell which leads to a net loss of positive charges within the cell.
- Leaky protein channels—Some potassium channels in the plasma membrane are "leaky" allowing a slow diffusion of K^+ out of the cell.

Both the Na^+K^+ ATPase pump and the leaky channels cause a potential difference between the inside of the neuron and the surrounding interstitial fluid. The resting membrane potential is always negative inside the cell, and the neuronal membrane is said to be **polarized**. In humans, the negative charge is –70mV.

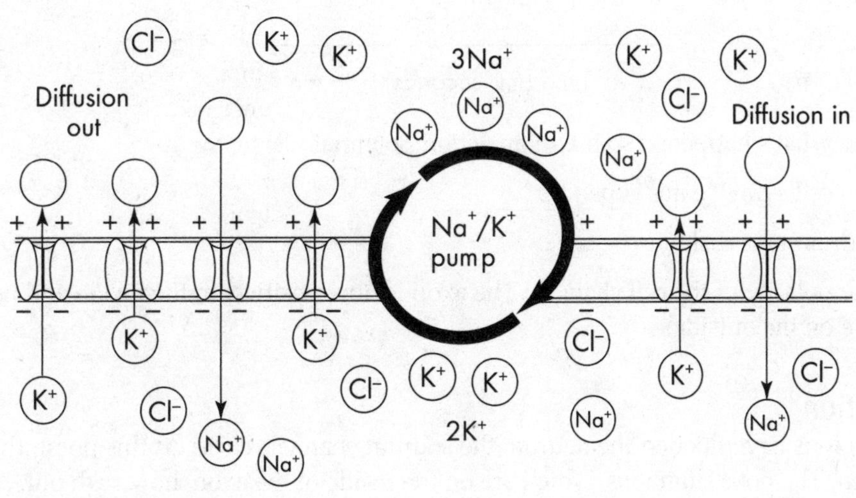

Action Potential

Here's what a neuron does in response to a stimulus. If a stimulus has enough intensity to excite a neuron, the cell reaches its **threshold**—the minimum amount of stimulus a neuron needs to respond. This creates what we call an **action potential**, that is, a change in the membrane potential that produces a nerve impulse. The action potential is an **all-or-none response**—it doesn't fire "part way."

Depolarization

At the point where the axon connects to the cell body, tiny gated sodium ion channels open up and allow sodium ions to rush into the cell. So many sodium ions rush in that the cell now becomes more positive inside than outside. This change is known as **depolarization**: The interior of the cell has "switched" its polarity from a negative to a positive charge. For now, just remember that an action potential makes the cell depolarize.

The net change is substantial. The charge has shifted from the –70 mV we saw earlier to about +35 mV:

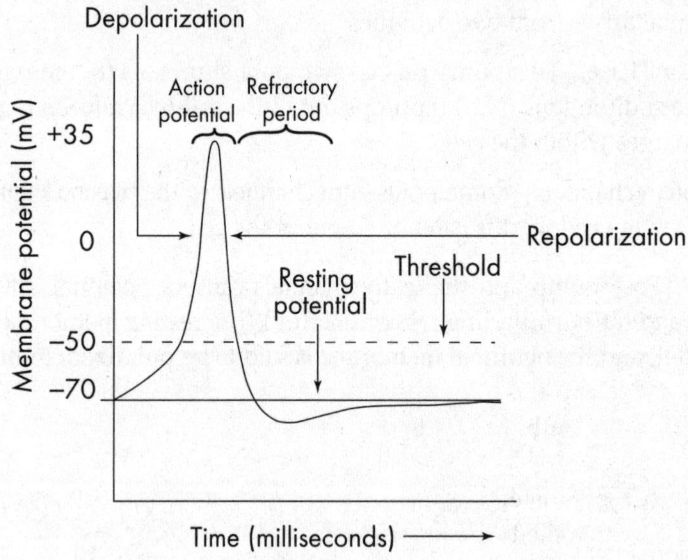

Let's recap what's happened so far. In an action potential:

- The cell's tiny "gates" open up.
- Sodium ions rush in.
- The polarity of the cell changes: The axon is now positive on the inside and negative on the outside.

Repolarization

Once sodium ions have flooded the neuron, the sodium channels close. At this point, the potassium channels open. The potassium ions, which are on the inside of the axon, now rush out. As the potassium ions move out of the cell, the electrical charges reverse again. The inside of the cell becomes more negative than the outside of the cell.

We can now say that this section of the neuron has been **repolarized**. In other words, the charge has returned to its original polarization.

The Refractory Period

Here's one thing you should remember: Although the charge has returned to its original state, at the end of the action potential the ions are now on the wrong side of the axonal membrane. Sodium ions are on the inside and potassium ions are on the outside of the axonal membrane. Originally, sodium ions were on the outside and potassium ions were on the inside. The neuron reestablishes the order of the ions, and this process is carried out by the **sodium-potassium pump**.

This pump reestablishes the original ion distribution by kicking three sodium ions out of the cell for every two potassium ions it brings into the cell. The period after an action potential is known as the **refractory period**.

During this period, the sodium channels are now reset and are able to open, but the cell membrane potential is further from the threshold. A greater stimulus is required to reach the threshold, so it is more difficult to initiate another action potential.

To summarize:

- A "resting" neuron is polarized; that is, it is more negative on the inside than on the outside.
- When an action potential comes along, the neuron transmits an impulse down its axon.
- First, voltage-gated sodium channels open, allowing sodium ions to rush in. This is known as *depolarization*: The neuron becomes more positive on the inside and more negative on the outside.
- Sodium channels close and potassium channels open, which restores its negative charge. This is known as *repolarization*.
- The neuron enters a refractory period.
- The neuron reestablishes the ion distribution thanks to the sodium-potassium pump.

When one small area is depolarized, it causes a "domino effect." The action potential spreads to the rest of the axon. The impulse is transmitted down the axonal membrane until it reaches the end of the axon called the **axon bulb**. Now, the neuron wants to pass the impulse to the next neuron. How does it manage this?

When an impulse reaches the end of an axon, the axon releases a chemical called a **neurotransmitter** into the space between the two neurons. This space is called a **synapse**. The neurotransmitter diffuses across the synaptic cleft and binds to receptors on the dendrites of the next neuron:

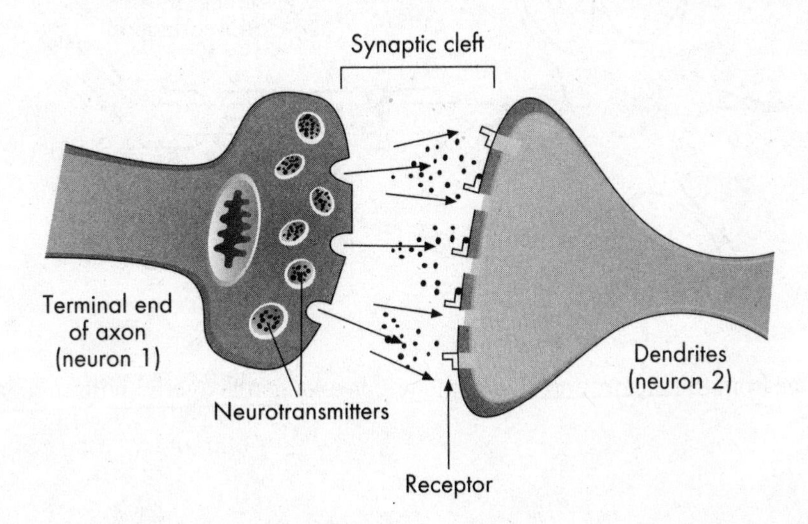

This usually triggers an action potential in the second neuron if the synaptic membrane is excited. Now, the impulse moves along the second neuron from dendrites to axon:

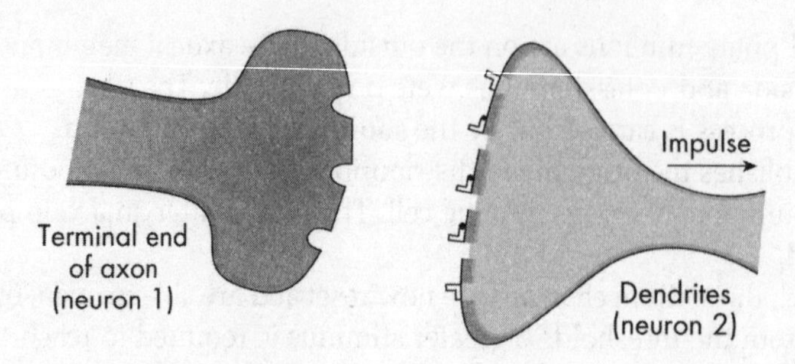

There are many neurotransmitters, but the most important one for the AP Biology Exam is called **acetylcholine**. Acetylcholine is a neurotransmitter that

- is released from the end of an axon, when Ca^{2+} moves into the terminal end of the axon
- is picked up almost instantly by the dendrites of the next neuron
- can stimulate muscles to contract or inhibit postsynaptic potential
- is released between neurons in the parasympathetic system, which we will discuss shortly

The extra acetylcholine in the synaptic cleft is broken down by the enzyme **acetylcholinesterase**.

Other important neurotransmitters include **norepinephrine** and **GABA**. Norepinephrine is a peptide neurotransmitter that is released between neurons within the central nervous system. GABA is secreted in the central nervous system and acts as an inhibitor.

Speed of an Impulse

Sometimes a neuron has supporting cells that wrap around its axon. These cells are called **Schwann cells**. Schwann cells produce a substance called the **myelin sheath**, which insulates the axon:

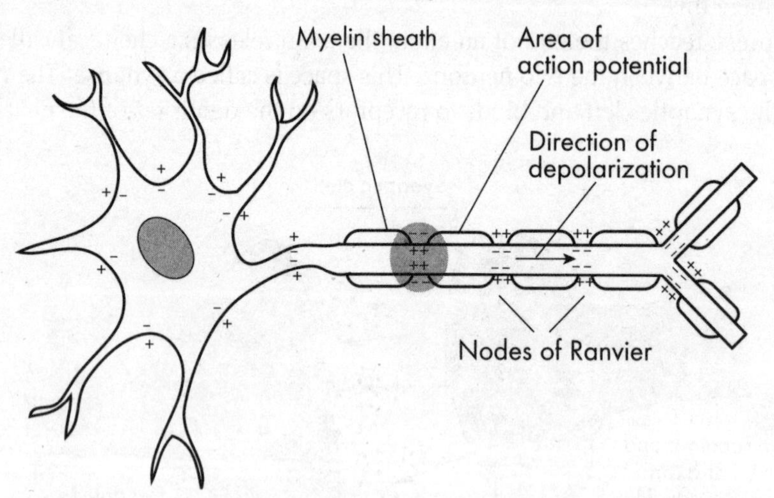

As you can see from the illustration above, the whole axon isn't covered with myelin sheaths. The spaces between myelin sheaths—the exposed regions of the axon—are called the **nodes of Ranvier**. Myelin sheaths speed up the propagation of an impulse. Instead of the standard "domino effect" that occurs during an action potential, the impulse can now jump from node to node. This form of conduction is called **saltatory conduction**.

Thanks to the myelin sheath, the neuron can transmit an impulse down the axon far more rapidly than it could without its help.

PARTS OF THE NERVOUS SYSTEM

The nervous system can be divided into two parts: the central nervous system and the peripheral nervous system.

Central Nervous System

All of the neurons within the brain and spinal cord make up the central nervous system. All of the other neurons lying outside the brain and the spinal cord—in our skin, our organs, and our blood vessels—are collectively part of the peripheral nervous system. Although both of these systems are really part of one system, we still use the terms *central* and *peripheral*.

So keep them in mind:

- The **central nervous system** includes the neurons in the brain and spinal cord.
- The **peripheral nervous system** includes all the rest.

Peripheral Nervous System

The peripheral nervous system is further broken down into the somatic nervous system and the autonomic nervous system.

- The **somatic nervous system** is the part that controls voluntary activities. For example, the movement of your eyes across the page as you read this line is under the control of your somatic nervous system.
- The **autonomic nervous system** is the part that controls involuntary activities. Your heartbeat and your digestive system, for example, are under the control of the autonomic nervous system.

The interesting thing about these two systems is that they sometimes overlap. For instance, you can control your breathing if you choose to. Yet most of the time you do not think about it: Your somatic system hands control of your respiration over to the autonomic system.

The autonomic system is broken down even further to the **sympathetic nervous system** and the **parasympathetic nervous system**. These two systems actually work antagonistically.

The sympathetic system controls the **"fight-or-flight" response**, which occurs when an organism confronted with a threatening situation prepares to fight or flee. To get ready for a quick, effective action, whether that be brawling or bolting, the sympathetic nervous system raises your heart and respiration rates, causes your blood vessels to constrict, increases the levels of glucose in your blood, and produces "goose bumps" on the back of your neck. It even reroutes your blood sugar to your skeletal muscles in case you need to make a break for it. After the threat has passed, the parasympathetic nervous system brings the body back to **homeostasis**—that is, back to normal. It lowers your heart and respiratory rates and decreases glucose levels in the blood.

The flow chart below will give you a nice overview of the different parts of the nervous system:

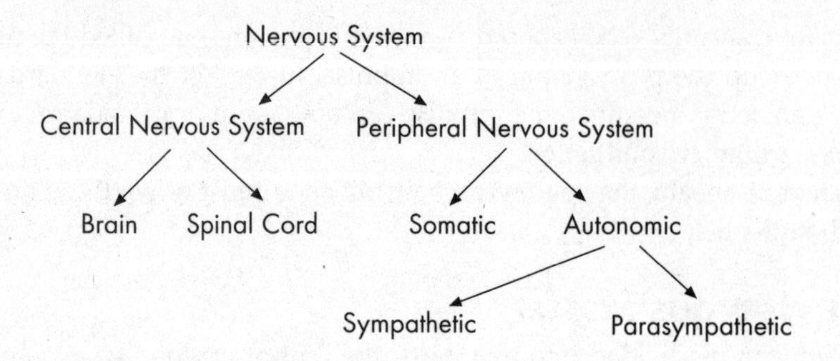

Parts of the Brain

The brain can also be divided into parts. Here's a summary of the major divisions within the brain.

DIVISIONS WITHIN THE BRAIN	
Parts of the Brain	Function
Cerebrum	Controls all voluntary activities; receives and interprets sensory information; largest part of human brain
Cerebellum	Coordinates muscle activity and refinement of movement
Hypothalamus	Regulates homeostasis and secretes hormones; regulates pituitary gland
Medulla	Controls involuntary actions such as breathing, swallowing, heartbeat, and respiration
Pons	Connects parts of the brain with one another and contains respiratory center
Midbrain	Center for visual and auditory reflexes (pupil reflex and blinking)
Thalamus	Main sensory relay center for conducting information between the spinal cord and cerebrum

The cerebrum consists of outer gray matter (the **cerebral cortex**) and inner white matter. One structure that is often mentioned on the AP biology test is the **corpus callosum**. The corpus callosum is a thick band of nerve fibers of the white matter that enable the right and left side of the cerebral hemispheres to communicate.

KEY WORDS

nerve net
ganglia
neurons
cell body
dendrites
axon
sensory neurons
motor (effector) neurons
interneurons
polarized
threshold
action potential
all-or-none response
depolarization
repolarized
sodium-potassium pump
refractory period
axon bulb
neurotransmitter
synapse
acetylcholine
acetylcholinesterase
norepinephrine
GABA
Schwann cells
myelin sheath
nodes of Ranvier
saltatory conduction
central nervous system
peripheral nervous system
somatic nervous system
autonomic nervous system
sympathetic nervous system
parasympathetic nervous system
"fight-or-flight" response
homeostasis
cerebrum
cerebellum
hypothalamus
medulla
pons
midbrain
thalamus
cerebral cortex
corpus callosum

NERVOUS SYSTEM REVIEW QUESTIONS

Answers can be found in Chapter 15.

1. The image below depicts an action potential occurring in an active neuron. Which of the following best explains why the membrane potential depolarizes at -50 mV?

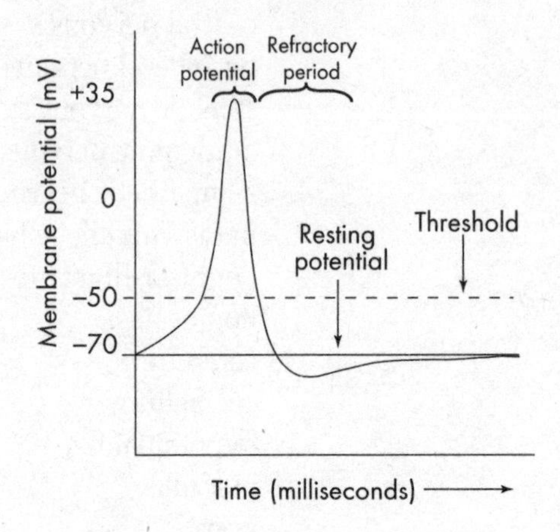

(A) At -50 mV, voltage-gated potassium channels open allowing potassium ions (K^+) to rush into the cell.

(B) At -50 mV, voltage-gated sodium channels open allowing sodium ions (Na^+) to rush into the cell.

(C) At -50 mV, voltage-gated potassium channels open allowing potassium ions (K^+) to rush out of the cell.

(D) At -50 mV, voltage-gated sodium channels open allowing sodium ions (Na^+) to rush out of the cell.

2. Fugu is a Japanese delicacy made from the preparation of pufferfish. Pufferfish produce a powerful toxin called tetrodotoxin, which binds to and blocks voltage-gated sodium channels. Specially trained Japanese chefs must carefully remove the toxin during preparation of fugu to prevent inadvertent poisoning by ingestion. Which of the following would likely occur to a neuron exposed to tetrodotoxin?

(A) The neuron would be able to send action potentials, but would not be able to release neurotransmitters to neighboring neurons.

(B) The neuron would be able to send action potentials, but would not be able to receive neurotransmitters from a presynaptic neuron.

(C) The neuron would not be able to send action potentials and would be unable to continue a neural response.

(D) The neuron would not be able to send action potentials, but will still be depolarized by the influx of sodium ions.

3. Jane has recently stumbled upon two bear cubs while hiking a trail in the forest. Upon their discovery, the mother of the cubs came running out of the woods and charged towards Jane. Which of the following best describes the neural response in Jane?

   (A) The sympathetic part of her autonomic nervous system was activated.
   (B) The sympathetic part of her somatic nervous system was activated.
   (C) The parasympathetic part of her autonomic nervous system was activated.
   (D) The parasympathetic part of her somatic nervous system was activated.

4. An ER patient has recently been in a car accident and is having major difficulty coordinating walking and motion. However, all of their muscles appear to be functionally active. Which of the following parts of the brain was likely injured by the accident?

   (A) Cerebrum
   (B) Cerebellum
   (C) Pons
   (D) Hypothalamus

THE MUSCULOSKELETAL SYSTEM

Most organisms need some form of support. Many animals wear their support on the *outside*. They have an **exoskeleton**—a hard covering or shell. Insects, for example, have an exoskeleton made of chitin. All **vertebrates** (animals with backbones) possess an **endoskeleton**—their entire skeleton is on the *inside*. In addition to ourselves, fish, amphibians, reptiles, birds, and all other mammals are considered vertebrates and therefore have endoskeletons.

THE HUMAN SKELETAL SYSTEM

In humans, the supporting skeleton is made of **cartilage** and **bone**. Cartilage is found in the embryonic stages of all vertebrates. It is later replaced by bone, except in your external ear or the tip of your nose. Here's one thing you should remember: Bone is a connective tissue that contains nerves and blood vessels. Cartilage, on the other hand, lacks nerves and blood vessels.

Bones

Bone is made up of two substances: **collagen** and **calcium salts**. Bone is a dynamic tissue that changes shape when **osteoblasts** (bone-building cells) and **osteoclasts** (bone-breaking cells) remodel it. Bones are held together by **joints**, like the ball-and-socket joint in your shoulder. What holds the joints together? They're held together by tough connective tissues called **ligaments**. Just remember that ligaments attach bone to bone. Bones not only serve as support but together with muscles also help us move about. The connective tissues that attach muscles to bones are called **tendons**.

Muscles

There are three kinds of muscle tissue: **skeletal**, **smooth**, and **cardiac**. For the AP Biology Exam, you'll need to know the differences among the types of muscles.

Skeletal muscles control voluntary movements. You'll notice that they have stripes called **striations**. They are also multinucleated. Let's look at a detailed skeletal muscle:

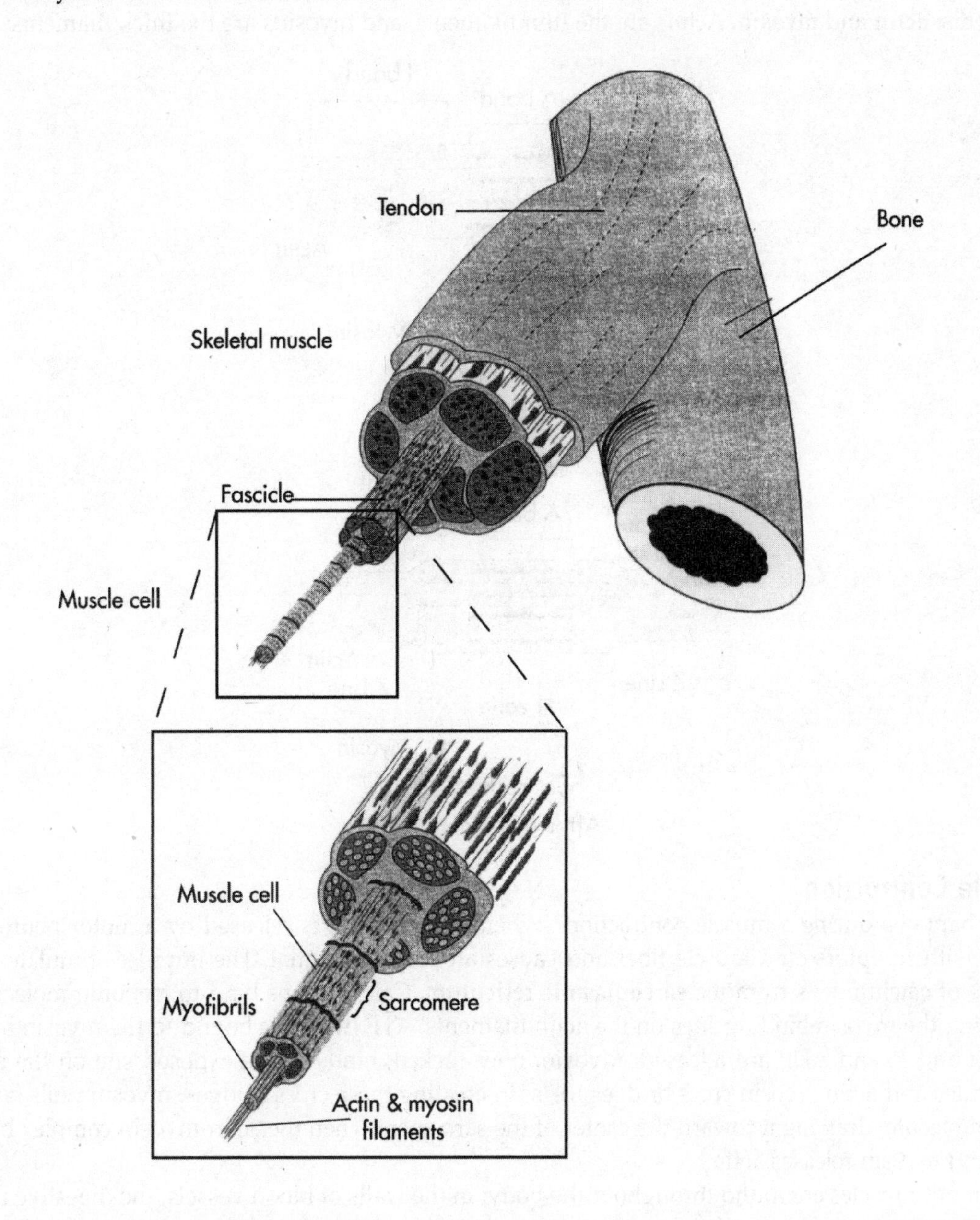

Organization of Skeletal Muscle

Muscles are made up of **muscle bundles**, which subdivide into **muscle fascicles**. Within each muscle fascicle are units called **muscle fiber cells**. Within each muscle fiber are contractile fibrils called **myofibrils**. A single myofibril is subdivided, by Z lines, into **sarcomeres** or contractile units.

The functional unit in a muscle cell is the sarcomere. Inside a sarcomere, there are two protein filaments: **actin** and **myosin**. Actins are the thin filaments, and myosins are the thick filaments:

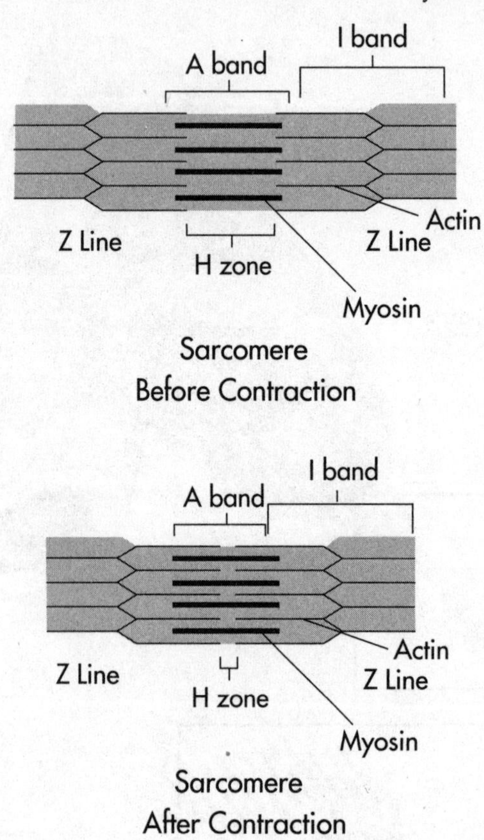

Muscle Contraction

What happens during a muscle contraction? When acetylcholine is released by a motor neuron, it binds with receptors on a muscle fiber and causes an action potential. The impulse stimulates the release of calcium ions from the **sarcoplasmic reticulum**. Calcium ions bind to troponin molecules, exposing the myosin-binding sites on the actin filaments. ATP (which is bound to the myosin head) is split and P_i and ADP are released. Myosin, now cocked, binds to the exposed site on the actin molecules and actin-myosin cross bridges form. In creating these cross bridges, myosin pulls on the actin molecule, drawing it toward the center of the sarcomere. Then the actin-myosin complex binds ATP and myosin releases actin.

Smooth muscles are found throughout the body: in the walls of blood vessels, the digestive tract, and internal organs. They are long and tapered, and each cell has a single nucleus. They contain actin and myosin but are not as well organized as skeletal muscles. This explains why they appear smooth. Smooth muscles are responsible for *involuntary* movements. Compared to those of skeletal muscles, the contractions in smooth muscles are slow.

Cardiac muscles are so-called because they're found in the heart. They have characteristics of both smooth and skeletal muscles. Cardiac muscles are striated, just like skeletal muscles, yet they are under *involuntary* control, like smooth muscles. One unique feature about cardiac muscle cells is that they are held together by special junctions called **intercalated discs**. Contractions in cardiac muscles are spontaneous and automatic. This simply means that the heart can beat on its own. Here's one more thing to remember: Both the smooth muscle and the cardiac muscle get their nerve impulses from the autonomic nervous system.

How does a muscle contract? Let's review the events that occur during muscle contraction. A muscle contraction begins with a neural impulse:

1. A nerve impulse is sent to a skeletal muscle.
2. The neuron sending the impulse releases a neurotransmitter onto the muscle cell.
3. The muscle depolarizes.
4. Depolarization causes the sarcoplasmic reticulum to release calcium ions.
5. These calcium ions cause the actin and myosin filaments to slide past each other.
6. The muscle contracts.

Let's compare the types of muscle tissues:

TYPES OF MUSCLE TISSUES			
	Skeletal	**Smooth**	**Cardiac**
Location	Attached to skeleton	Wall of digestive tract, inside the blood vessels	Wall of heart
Type of control	Voluntary	Involuntary	Involuntary
Striations	Yes	No	Yes
Multinucleated	Yes	No	No
Speed of contraction	Rapid	Slowest	Intermediate

KEY WORDS

exoskeleton
vertebrates
endoskeleton
cartilage
bone
collagen
calcium salts
osteoblasts
osteoclasts
joints
ligaments
tendons
skeletal muscles
smooth muscles
cardiac muscles
striations
muscle bundles
muscle fascicles
muscle fiber cells
myofibrils
sarcomere
actin
myosin
sarcoplasmic reticulum
intercalated discs

MUSCULOSKELETAL SYSTEM REVIEW QUESTIONS

Answers can be found in Chapter 15.

1. The sarcomere of a muscle cell is shown below prior to contraction. Following contraction of the muscle, which of the following will shorten?

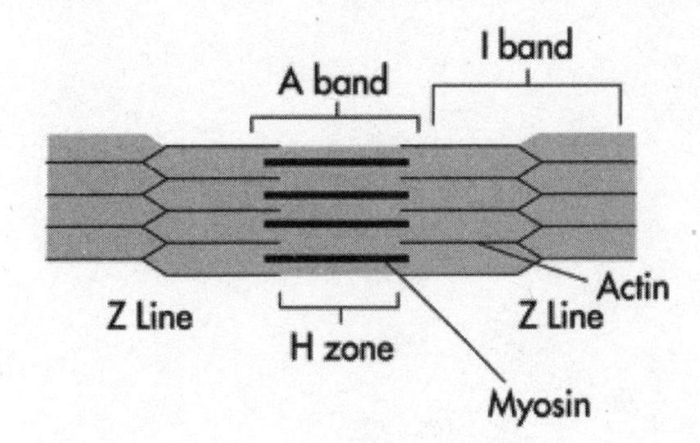

   (A) Z line
   (B) A band
   (C) H zone
   (D) Actin filaments

2. A key difference between the skeletal muscle and those of the heart and digestive tract is which of the following?

   (A) Skeletal muscle is under autonomic control, whereas cardiac and smooth muscle are under voluntary control.
   (B) Skeletal muscle is under voluntary control, whereas cardiac and smooth muscle are under autonomic control.
   (C) Skeletal muscle is striated, whereas cardiac and smooth muscle are not.
   (D) Cardiac and smooth muscle are striated, whereas skeletal muscle is not.

3. During a muscle contraction, calcium ions bind to troponin exposing the myosin-binding sites on the actin filaments. From where does the calcium ions originate?

   (A) They enter the cell through voltage-gated calcium channels.
   (B) They are released from the sarcoplasmic reticulum.
   (C) They are released from endocytic vesicles associated with phagocytosis.
   (D) They are released from the nucleus through voltage-gated channels.

4. Tendons and ligaments are both examples of connective tissues that are important for maintaining the shape and stability of the skeletal system. What is the role of these structures?

   (A) Tendons attach muscles to bones, whereas ligaments attach bones to bones.
   (B) Ligaments attach muscles to bones, whereas tendons attach bones to bones.
   (C) They both attach muscles to bones, however ligaments are specific to joints.
   (D) They both generate osteoblasts for generating bone, however ligaments are specific to joints.

THE ENDOCRINE SYSTEM

Chemical messengers can be produced in one region of the body to act on target cells in another part. These chemicals, known as **hormones**, are produced in specialized organs called **endocrine glands**. Hormones have a number of functions including regulating growth, behavior, development, and reproduction. For example, the hormone **ecdysone** promotes molting and the metamorphosis of a larva to a butterfly. Ecdysone is stimulated to release when targeted by another insect hormone, **brain hormone**, in the prothoracic glands. Another hormone, **juvenile hormone**, causes larvae to retain their characteristics. Other chemical messengers are used for communication. For example, **pheromones** help animals to communicate with members of their species and attract the opposite sex.

An endocrine gland releases hormones directly into the bloodstream, which carries them throughout the body. Take a look at the endocrine glands in the human body:

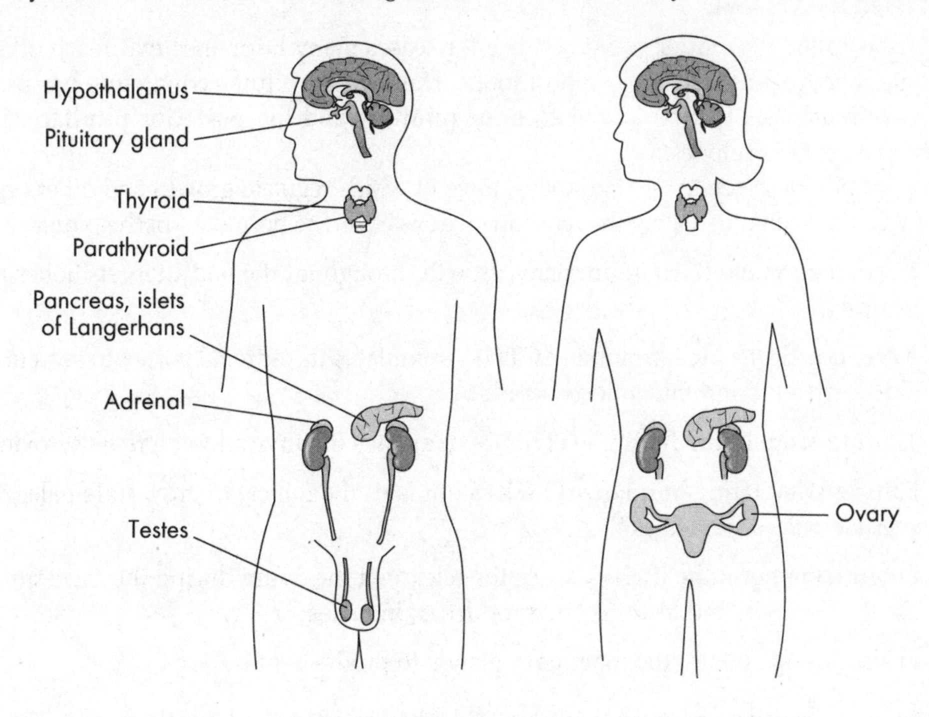

Before we launch into a review of the different hormones in the body, let's talk about how hormones work. Although hormones flow in your blood, they affect only specific cells. The cells that a hormone affects are known as the **target cells**. Suppose, for example, that gland X makes hormone Y. Hormone Y, in turn, has some effect on organ Z. We would then say that organ Z is the *target organ* of hormone Y.

Hormones also operate by a **negative feedback system**. That is, an excess of the hormone will signal the endocrine gland to temporarily shut down production. For example, when hormone Y reaches a peak level in the bloodstream, the organ secreting the hormone, gland X, will get a signal to stop producing hormone Y. Once the levels of hormone Y decline, the gland can resume production of the hormone.

THE PITUITARY GLAND

The **pituitary** is called the master gland because it releases many hormones that reach other glands and stimulate them to secrete their own hormones. The anterior pituitary therefore has many *target* organs. The pituitary has two parts: the **anterior pituitary** and the **posterior pituitary**. Each part secretes its own set of hormones.

The anterior pituitary secretes six hormones, three of which regulate growth and other organs. The other three are involved in regulating the reproductive systems. The hormones of the pituitary are:

- **Growth hormone (GH)**—stimulates growth throughout the body, targets bones and muscles
- **Adrenocorticotropic hormone (ACTH)**—stimulates the adrenal cortex to secrete glucocorticoids and mineralocorticoids
- **Thyroid-stimulating hormone (TSH)**—stimulates the thyroid to secrete thyroxine
- **Follicle-stimulating hormone (FSH)**—stimulates the follicle to grow in females, and spermatogenesis in males
- **Luteinizing hormone (LH)**—causes the release of the ovum during the menstrual cycle in females, and testosterone production in males
- **Prolactin**—stimulates the mammary glands to produce milk

The pituitary works in tandem with a part of the brain called the **hypothalamus**. The pituitary sits just below the hypothalamus:

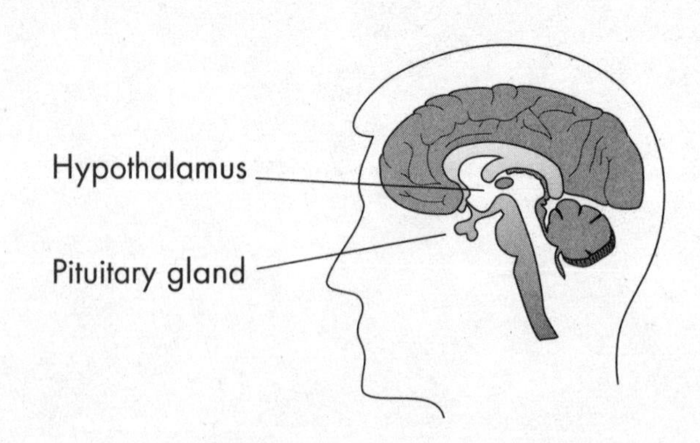

The hypothalamus regulates the anterior pituitary by secreting neurohormones that can stimulate or inhibit the actions of the anterior pituitary. The other part of the pituitary, the posterior pituitary, secretes two hormones:

- Antidiuretic hormone (or vasopressin)—regulates water intake by nephrons
- **Oxytocin**—stimulates contraction of uterus and ducts of mammary glands

These hormones are actually made in the hypothalamus but are *stored* in the posterior pituitary. How about a mnemonic? When you think of the **pit**uitary, think of the GATOR pit:

Growth hormone

ACTH

Thyroid-stimulating hormone

Oxytocin

R, for antidiu**R**etic hormone (or vasopressin)

(All of which come from the **pit**uitary gland.)

And, to make things easier, the first three come first, or are "anterior" to the last two, which are "posterior." For us, that means that G, A, and T come from the anterior pituitary, while O and R come from the posterior pituitary.

As for the other three hormones, think of "FLAP." FSH, LH, and prolactin are all hormones that have to do with the reproductive system. Together, these two mnemonics should help you keep the different hormones of the pituitary gland straight. Now let's move on to the target organs.

The Pancreas

We already know that the pancreas produces enzymes that it releases into the small intestine via the pancreatic duct. The pancreas also secretes two hormones, **glucagon (alpha cell)** and **insulin (beta cell)**, both of which are produced in clusters of cells called the **islets of Langerhans**. The target organs for these hormones are the liver and muscle cells. Glucagon, produced by α cells, stimulates the liver to convert **glycogen** into glucose and to release that glucose into the blood. Glucagon therefore *increases* the levels of glucose in the blood. Insulin has precisely the opposite effect that glucagon does.

When the blood has too much glucose floating around, insulin, produced by β cells, allows body cells to remove glucose from the blood. Consequently, insulin *decreases* the level of glucose in the blood. Insulin is particularly effective on muscle and liver cells. In short:

- *Insulin* lowers the blood sugar level.
- *Glucagon* raises the blood sugar level.

THE ADRENAL GLANDS

The adrenal glands contain two separate endocrine glands. One is called the **adrenal cortex**, and the other is called the **adrenal medulla**. Although they are part of the same organ, these two endocrine glands have very different effects on the body. Let's start by discussing the adrenal cortex.

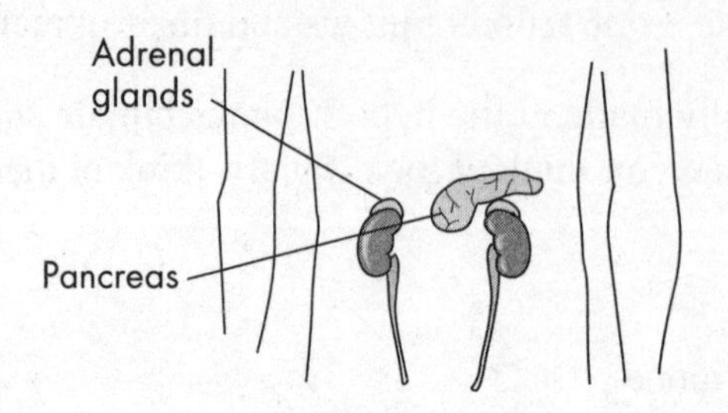

The Adrenal Cortex

Earlier we mentioned that adrenocorticotropic hormone, or simply ACTH, targets the adrenal cortex. When ACTH is released from the pituitary, it stimulates the adrenal cortex to produce and secrete its different hormones.

One group of hormones released by the adrenal cortex is the **glucocorticoids**. They increase the blood's concentration of glucose and help the body adapt to stress. In fact, glucocorticoids accomplish the same thing as glucagon, but in a slightly different way. Glucocorticoids promote the conversion of amino acids and fatty acids to glucose.

Another set of hormones is the **mineralocorticoids**. They help the body retain Na^+ and water in the kidneys. They accomplish this by promoting the reabsorption of sodium (Na^+) and chlorine (Cl^-), which get together to form common salt (NaCl). When salt is retained, water soon follows.

So then, just remember that the adrenal cortex releases two types of hormones.

- Glucocorticoids target the liver and promote the release of glucose.
- Mineralocorticoids target the kidney and promote the retention of water.

The Adrenal Medulla

The other adrenal gland, the adrenal medulla, is often referred to as the "emergency gland." It secretes two hormones: **epinephrine** and norepinephrine. These are the hormones involved in the fight-or-flight response. Both epinephrine and norepinephrine "kick in" under extreme stress. They increase your heart rate, metabolic rate, blood pressure, and give you a quick boost of energy.

The Thyroid

The **thyroid gland**, which is located in the neck, is the target organ of the thyroid-stimulating hormone (TSH):

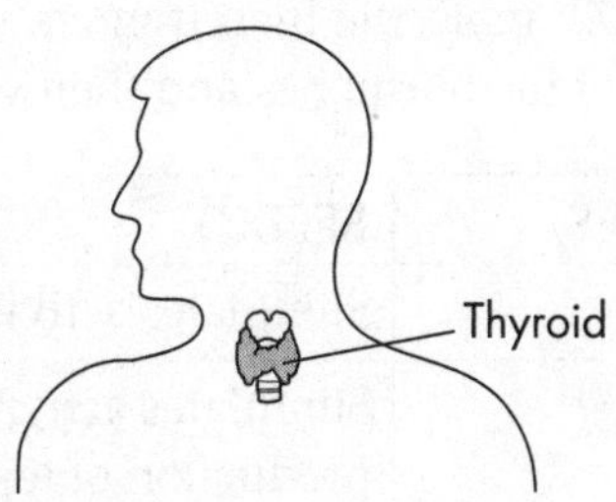

When the thyroid is stimulated by TSH, it releases the hormone **thyroxine**. Thyroxine, which contains iodine, is responsible for regulating the metabolic rate in your body tissues. Two conditions are associated with thyroid hormones. **Hyperthyroidism** occurs in individuals who regularly release too much thyroxine. They have a fast metabolic rate and tend to be irritable and nervous. On the other hand, individuals who suffer from **hypothyroidism** have too little thyroxine circulating in their bloodstream. They exhibit a slow metabolic rate and tend to be sluggish and overweight. To summarize:

> An individual with *hypo*thyroidism has a slow metabolic rate, while an individual with *hyper*thyroidism has a fast metabolic rate.

The thyroid also secretes another hormone called **calcitonin**. This hormone decreases your blood's concentration of calcium by concentrating free-floating calcium in the bones. You'll recall from our discussion of bone cells that bones contain collagen and calcium salts. Calcitonin is responsible for depositing these calcium molecules in the bones.

The Parathyroids

The **parathyroids** are four little pea-shaped organs that rest on the thyroid. They secrete **parathyroid hormone**. Parathyroid hormone increases your blood calcium levels. Consequently, parathyroid hormone has the opposite effect that calcitonin does. If your blood needs more calcium, parathyroid hormone releases calcium ions stored in the bones. This process of building or breaking down bones to store and release calcium is called **bone remodeling**.

The Sex Hormones

Three hormones that we'll discuss in detail in the chapter on reproduction are **estrogen, progesterone**, and **testosterone**. Estrogen and progesterone are hormones released by the ovaries and they regulate the menstrual cycle. Testosterone is the male hormone responsible for promoting spermatogenesis, the production of sperm. In addition, these hormones maintain secondary sex characteristics.

How Hormones Work

How do hormones trigger the activities of their target cells? That all depends on whether the hormone is a steroid (lipid soluble) or a protein, peptide, or **amine** (not lipid soluble). If the hormone is a steroid, then the hormone can diffuse across the membrane of the target cell. It then binds to a receptor protein in the nucleus and activates specific genes contained in the DNA, which in turn make proteins.

However, if the hormone is a protein, peptide, or amine, it can't get into the target cell by means of simple diffusion. Remember: "Like dissolves like." The hormone binds to a receptor protein on the *cell membrane* of the target cell. This protein in turn stimulates the production of a second messenger called **cyclic AMP (cAMP)**. The cAMP molecule then triggers various enzymes, leading to specific cellular changes. Here's a summary of the hormones and their effects on the body.

ORGAN	HORMONES	EFFECT
Anterior Pituitary	FSH	Stimulates activity in ovaries and testes
	LH	Stimulates activity in ovary (release of ovum) and production of testosterone
	ACTH	Stimulates the adrenal cortex
	Growth Hormone	Stimulates bone and muscle growth
	TSH	Stimulates the thyroid to secrete thyroxine
	Prolactin	Causes milk secretion
Posterior Pituitary	Oxytocin	Causes uterus to contract
	Vasopressin	Causes kidney to reabsorb water
Thyroid	Thyroid Hormone	Regulates metabolic rate
	Calcitonin	Lowers blood calcium levels
Parathyroid	Parathyroid Hormone	Increases blood calcium concentration
Adrenal Cortex	Aldosterone	Increases Na^+ and H_20 reabsorption in kidneys
Adrenal Medulla	Epinephrine Norepinephrine	Increase blood glucose level and heart rate
Pancreas	Insulin	Decreases blood sugar concentration
	Glucagon	Increases blood sugar concentration
Ovaries	Estrogen	Promotes female secondary sex characteristics and thickens endometrial lining
	Progesterone	Maintains endometrial lining
Testes	Testosterone	Promotes male secondary sex characteristic and spermatogenesis

While the nervous system and the endocrine system work in close coordination, there are significant differences between the two:

- The nervous system sends nerve impulses using neurons, whereas the endocrine system secretes hormones.
- Nerve impulses control rapidly changing activities, such as muscle contractions, whereas hormones deal with long-term adjustments.

KEY WORDS

hormones
endocrine glands
ecdysone
brain hormone
juvenile hormone
pheromones
target cells
negative feedback system
pituitary
anterior pituitary
posterior pituitary
growth hormone (GH)
adrenocorticotropic hormone (ACTH)
thyroid-stimulating hormone (TSH)
follicle-stimulating hormone (FSH)
luteinizing hormone (LH)
prolactin
hypothalamus
oxytocin
glucagon (alpha cell)
insulin (beta cell)
islets of Langerhans
glycogen
adrenal cortex
adrenal medulla
glucocorticoids
mineralocorticoids
epinephrine
thyroid gland
thyroxine
hyperthyroidism
hypothyroidism
calcitonin
parathyroids
parathyroid hormone
bone remodeling
estrogen
progesterone
testosterone
amine
cyclic AMP (cAMP)

ENDOCRINE SYSTEM REVIEW QUESTIONS

Answers can be found in Chapter 15.

1. Consider an experiment in which African green monkeys are injected with extract from the anterior pituitary gland. Which of the following would probably not be observed?

   (A) Increased production of ACH
   (B) Increased blood levels of glucocorticoids
   (C) Increased production of thyroxine
   (D) Increased production of vasopressin

2. Many hormones regulate functions by having antagonistic roles. Which of the following pairs of hormones result in opposite effects?

   (A) FSH and LH on testosterone production
   (B) Prolactin and oxytocin on milk production
   (C) Insulin and aldosterone on blood sugar concentrations
   (D) Calcitonin and parathyroid hormone (PTH) on blood calcium concentrations

3. All of the following hormones are produced by the thyroid or parathyroid hormones EXCEPT

   (A) thyroxine
   (B) thyroid stimulating hormone (TSH)
   (C) parathyroid hormone (PTH)
   (D) calcitonin

4. The hypothalamus is a critical structure for regulating hormonal levels. Which of the following is true of the hypothalamus?

   (A) It secretes thyroid-stimulating hormone (TSH).
   (B) It secretes luteinizing hormone (LH).
   (C) It is an extension of the pituitary gland.
   (D) It produces neurosecretory hormones.

THE REPRODUCTIVE SYSTEM AND EMBRYONIC DEVELOPMENT

Reproduction in animals involves the production of eggs and sperm.

Before we launch into reproduction, let's take a look at something that ties in very nicely with everything we've just discussed about hormones and the endocrine system: the menstrual cycle.

Since the ovaries release hormones, they are considered endocrine glands. The ovaries have two main responsibilities:

- They manufacture **ova**.
- They secrete estrogen and progesterone, sex hormones that are found in females.

The hormones secreted by the ovaries are involved in the menstrual cycle.

The Menstrual Cycle

Phase 1: The Follicular Phase

In phase 1, the anterior pituitary secretes two hormones: **follicle-stimulating hormone (FSH)** and **luteinizing hormone (LH)**. The FSH stimulates several follicles in the ovaries to grow. Eventually, one of these follicles gains the lead and dominates the others, which soon stop growing. The one growing follicle now takes command.

Because the follicle is growing in this phase, the phase itself is known as the **follicular phase.** Remember that during all this time the follicle is releasing estrogen. Estrogen helps the uterine lining to thicken and eventually causes the pituitary to release LH. This increase in estrogen causes a sudden surge in luteinizing hormone. This release of LH is known as a **luteal surge**. LH triggers **ovulation**—the release of the follicle from the ovary.

There are thus three hormones associated with the follicular phase:

- Follicle-stimulating hormone (FSH)—originates in the pituitary gland.
- Estrogen—originates in the follicle.
- Luteinizing hormone (LH)—originates in the pituitary gland.

The luteal surge makes the follicle burst and release the ovum. The ovum then begins its journey into the **fallopian tube**, which is also known as the **oviduct**. This is a crucial event in the female menstrual cycle and is known as ovulation. Once the ovum has been released, the follicular phase ends and the ovum is ready to move on to the next phase.

In addition to the growth of the follicle, the follicular phase involves the thickening of the **uterine walls**, or **endometrium**. This happens in preparation for the implantation of a fertilized cell. The entire follicular phase lasts about 10 days.

Phase 2: The Luteal Phase

By the end of the follicular phase, the ovum has moved into the fallopian tube and the follicle has been ruptured and left behind in the ovary. However, the ruptured follicle (now a fluid-filled sac) continues to function in the menstrual cycle. At this stage, it condenses into a little yellow blob called the **corpus luteum**, which is Latin for "yellow body."

The corpus luteum continues to secrete estrogen. In addition, it now starts producing the other major hormone involved in female reproduction, progesterone. Progesterone is responsible for readying the body for pregnancy. It does this by promoting the growth of glands and blood vessels in the endometrium. Without progesterone, a fertilized ovum cannot latch onto the uterus and develop into an embryo. We can therefore think of progesterone as the hormone of pregnancy.

After about 13 to 15 days, if fertilization and implantation have not occurred, the corpus luteum shuts down. Once it has stopped producing estrogen and progesterone, the final phase of the menstrual cycle begins.

Phase 3: The Flow Phase, or Menstruation

Once the corpus luteum turns off, the uterus can no longer maintain its thickened walls. It starts to reabsorb most of the tissue that the progesterone encouraged it to grow. However, since there is too much to reabsorb, a certain amount is shed. This "sloughing off," or bleeding, is known as **menstruation.**

With the end of menstruation, the cycle starts all over again, readying the body for fertilization. Let's recap some of the major steps:

1. In the follicular phase, the pituitary releases FSH, causing the follicle to grow.
2. The follicle releases estrogen, which helps the endometrium to grow.
3. Estrogen causes the pituitary to release LH, resulting in a luteal surge.
4. This excess LH causes the follicle to burst, releasing the ovum during ovulation.
5. The shed follicle becomes the corpus luteum, which produces progesterone.
6. Progesterone, the "pregnancy hormone," enhances the endometrium, causing it to thicken with glands and blood vessels.
7. If fertilization does not occur after about two weeks, the corpus luteum dies, leading to menstruation—the sloughing off of uterine tissue.

If pregnancy occurs, the extraembryonic tissue of the fetus releases **human chorionic gonadotropin (HCG)**, which helps maintain the uterine lining. The AP Biology Exam is likely to contain questions dealing with the events that occur during the menstrual cycle and fertilization and where those events take place.

Take a look at the following diagram. Familiarize yourself with the parts of the female reproductive system and pay special attention to the different sites of the stages we've just discussed.

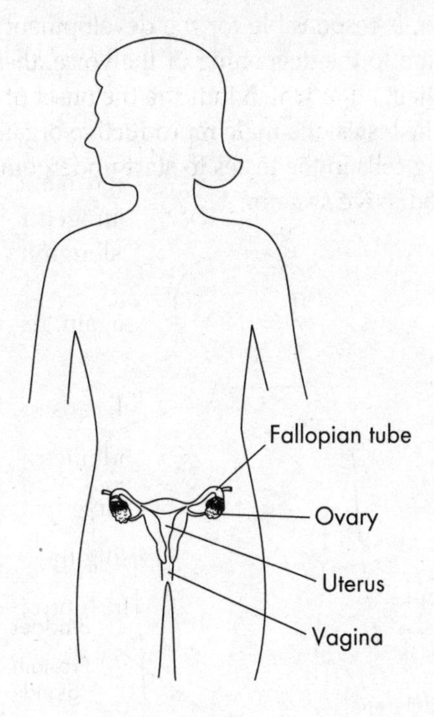

Remember:

- The follicles (and thus the ova) are contained in the ovaries.
- Hormones are released from the ovaries and pituitary gland.
- Fertilization occurs in the fallopian tube.
- The fertilized ovum implants itself in the uterus.

THE MALE REPRODUCTIVE SYSTEM

Now let's discuss the hormones in the male reproductive system. Testosterone, along with the cortical sex hormones we saw earlier, is responsible for the development of the sex organs and secondary sex characteristics. In addition to the deepening of the voice, these characteristics include body hair, muscle growth, and facial hair, all of which indicate the onset of **puberty**. Testosterone also has another function. It stimulates the testes, the male reproductive organs, to manufacture **sperm cells**. Testosterone does this by causing cells in the testes to start undergoing meiosis.

Take a look at the male reproductive system.

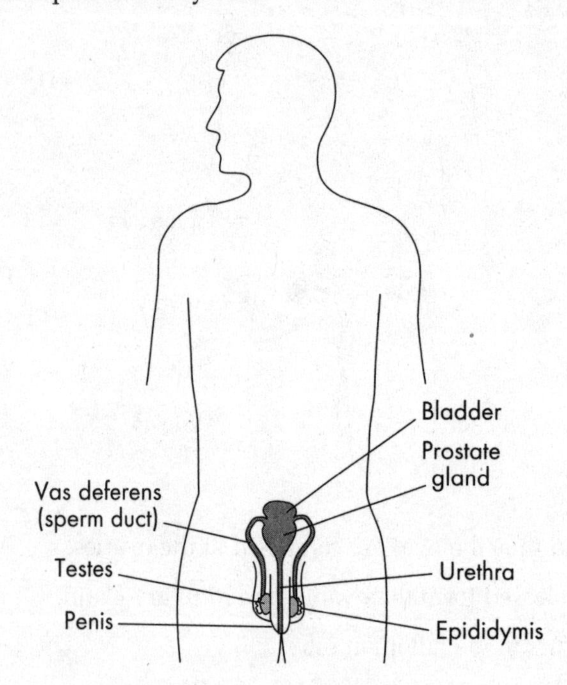

Sperm and male hormones are produced in the testes. The main tissues of the testes, called the **seminiferous tubules**, are where spermatogonia undergo meiosis. The spermatids then mature in the **epididymis**. The **interstitial cells**, which are supporting tissue, produce testosterone and other androgens. Sperms then travel through the **vas deferens** and pick up fluids from the **seminal vesicles** (which provides them with fructose for energy) and the **prostate gland** (which provides an alkaline fluid that neutralizes the vagina's acidic fluids). Semen is transported to the vagina by the penis.

Unlike the female reproductive system, the male reproductive system continues to secrete hormones throughout the life of the male. FSH targets the seminiferous tubules of the testes, where it stimulates sperm production. LH stimulates interstitial cells to produce testosterone.

Embryonic Development

How does a tiny, single-celled egg develop into a complex, multicellular organism? By dividing, of course. The cell will change shape and organization many times by going through a succession of stages. This process is called **morphogenesis**. In order for the human sperm to fertilize an egg it must dissolve the corona radiata, a dense covering of follicle cells that surrounds the egg. Then the sperm must penetrate the zona pellucida, the zone below the corona radiata.

When an egg is fertilized by a sperm, it forms a diploid cell called a **zygote**:

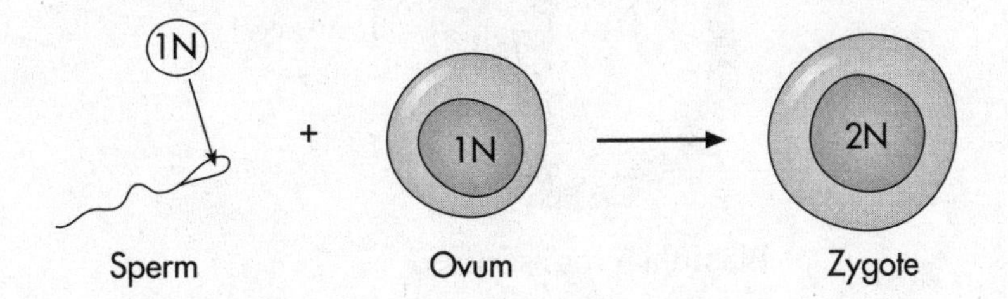

Fertilization triggers the zygote to go through a series of rapid cell divisions called **cleavage**. What's interesting at this stage is that the embryo doesn't grow. The cells just keep dividing to form a solid ball called a **morula**:

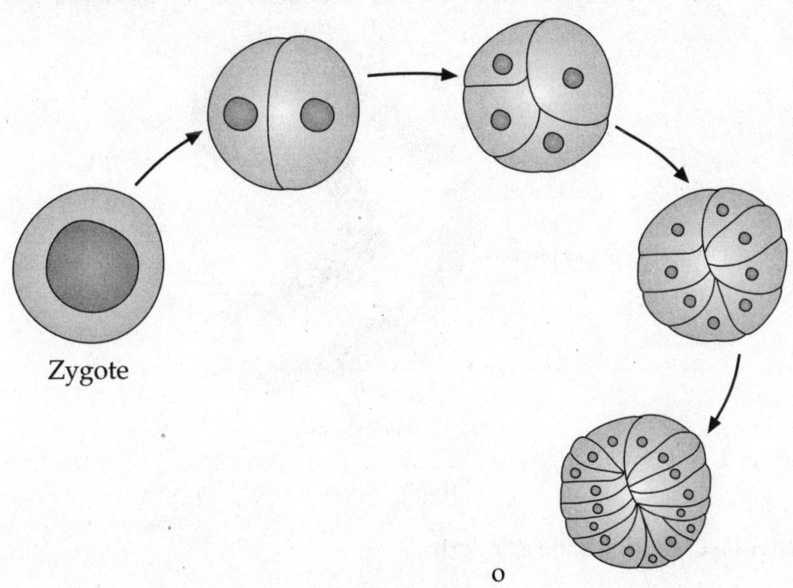

o

One cell becomes two cells, two cells become four cells, and so on.

Blastula

The next stage is called **blastula**. As the cells continue to divide, they "press" against each other and produce a fluid-filled cavity called a **blastocoel**:

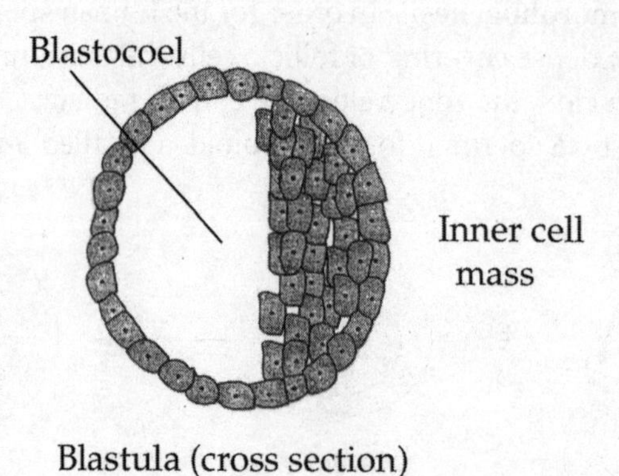

Blastula (cross section)

Gastrula

During **gastrulation**, the zygote begins to change its shape. Cells now migrate into the blastocoel and differentiate to form three germ layers: the ectoderm, mesoderm, and endoderm:

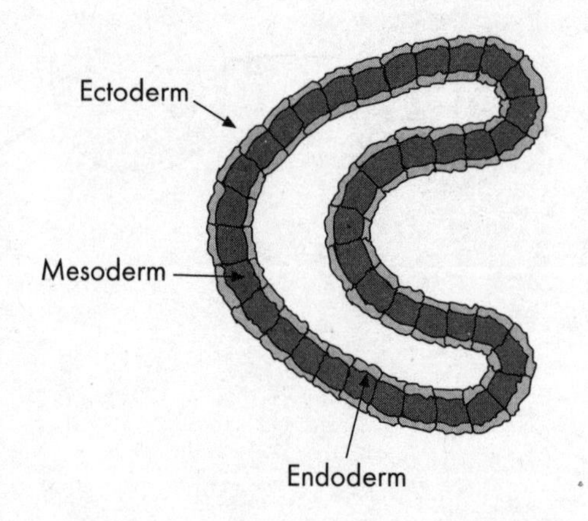

Gastrula (cross-section)

- The outer layer becomes the ectoderm.
- The middle layer becomes the mesoderm.
- The inner layer becomes the endoderm.

Each germ layer gives rise to various organs and systems in the body. Here's a list of the organs that develop from each germ layer.

- The ectoderm produces the epidermis (the skin), the eyes, and the nervous system.
- The endoderm produces the inner linings of the digestive tract and respiratory tract, as well as accessory organs such as the pancreas, gall bladder, and liver. These are called "accessory" organs because they are offshoots of the digestive tract, as opposed to the channels of the tract itself.
- The mesoderm gives rise to everything else. This includes bones and muscles as well as the excretory, circulatory, and reproductive systems.

Organogenesis

The **neurula** stage begins with the formation of two structures; the **notochord**, a rod-shaped structure running beneath the nerve cord, and the **neural tube** cells, which develop into the central nervous system. By the end of this stage, we're well on our way to developing a nervous system:

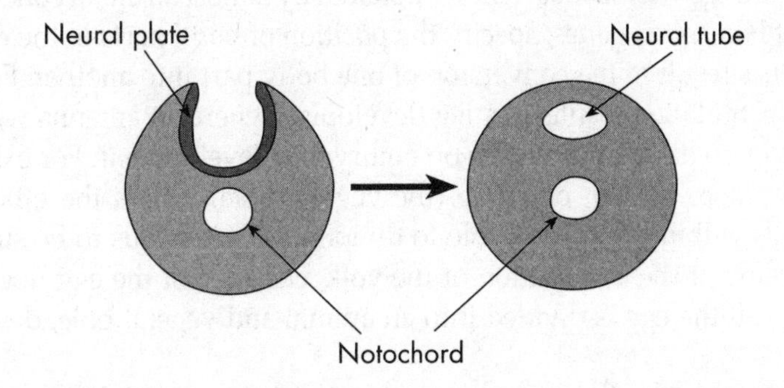

As far as the AP Biology Exam is concerned, the *order* of the stages and the various events is extremely important. For our purposes, think of the order of embryological development in this way:

Zygote → Cleavage → Blastula → Gastrula → Organogenesis

What About Chicken Embryos?

In addition to the primary germ layers, some animals have **extraembryonic membranes**. Your average developing chicken, for example, possesses these membranes.

There are basically four extraembryonic membranes: the **yolk sac**, **amnion**, **chorion**, and **allantois**. These extra membranes are common in birds and reptiles.

For the AP Biology Exam, you should be familiar with these membranes and their functions.

FUNCTIONS OF EXTRAEMBRYONIC MEMBRANES	
Extraembryonic membrane	**Function**
Yolk sac	Provides food for the embryo
Amnion	Forms a fluid-filled sac that protects the embryo
Allantois	Membrane involved in gas exchange; stores uric acid
Chorion	Outermost membrane that surrounds all the other extraembryonic membranes

Fetal Embryo

The fetal **embryo** also has extraembryonic membranes during development: the amnion, chorion, allantois, and yolk sac. The **placenta** and the **umbilical cord** are outgrowths of these membranes. The placenta is the organ that provides the **fetus** with nutrients and oxygen and gets rid of the fetus's wastes. The placenta develops from both the chorion and the uterine tissue of the mother. The umbilical cord is the organ that connects the embryo to the placenta.

Development of an Embryo

During embryonic development, some tissues determine the fate of other tissues in a process called **induction**. Certain cells, called **organizers**, release a chemical substance (a **morphogen**) that moves from one tissue to the target tissue. It is now known that development involves many episodes of embryonic induction.

Homeotic genes control the development of the embryo. Some homeotic genes, called **homeobox genes**, consist of homeoboxes (short, nearly identical DNA sequences) that encode proteins that bind to DNA; these proteins tell cells in various segments of the developing embryo what type of structures to make. The process by which a less specialized cell becomes a more specialized cell type is called **differentiation**. Interestingly, homeobox genes are shared by almost all eukaryotic species. **Hox genes,** which are a subset of homeobox genes, specify the position of body parts in the developing embryo. Mutations in Hox genes result in the conversion of one body part into another. For example, in Drosophila, a specific Hox mutation results in a leg developing where an antenna would normally be.

The cytoplasm can also have an influence on embryonic development. For example, chicken and frog embryos contain more yolk in one pole (the vegetal pole) versus the other pole (the animal pole). This causes cells within the animal pole to divide more, and thus to be smaller than those of the vegetal pole. Because of the distribution of the yolk, cleavage of the egg does not produce eggs that develop normally. If the egg is divided into an animal and vegetal pole, development does not proceed normally.

Apoptosis, or programmed cell death, plays a crucial role in normal differentiation and development. For example, in a human embryo apoptosis allows for the removal of tissue between newly developing fingers and toes to allow for separate digits. If this process is absent or incomplete, an individual is born with syndactyly, which is webbed hands or feet.

KEY WORDS

morphogenesis
zygote
fertilization
cleavage
morula
blastula
blastocoel
gastrulation
neurula
notochord
neural tube
extraembryonic membranes
yolk sac
amnion
chorion
allantois
embryo
placenta
umbilical cord
fetus
induction
organizers
morphogen
homeotic genes
homebox genes
differentiation
Hox genes
apoptosis
ova
follicle-stimulating hormone (FSH)
luteinizing hormone (LH)
follicular phase
luteal surge
ovulation
fallopian tube
oviduct
uterine walls
endometrium
corpus luteum
menstruation
human chorionic gonadotropin (HCG)
puberty
sperm cells
seminiferous tubules
epididymis
interstitial cells
vas deferens
seminal vesicles
prostate gland

REPRODUCTIVE SYSTEM AND EMBRYONIC DEVELOPMENT REVIEW QUESTIONS

Answers can be found in Chapter 15.

1. Shown below is the cross-section structure of a gastrula. Which of the following anatomical structures will ultimately be derived from the ectoderm?

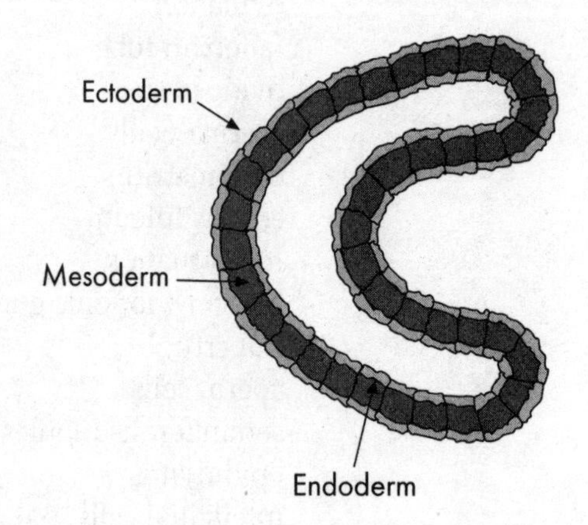

   (A) GI tract
   (B) Brain
   (C) Heart
   (D) Pancreas

2. The menstrual cycle is critical for maintaining a state of readiness to prepare for impregnation and implantation of a developing embryo. Which of the following structures is essential for producing progesterone, which protects and enhances the endometrium?

   (A) Ova
   (B) Corpus luteum
   (C) Uterus
   (D) Pituitary gland

3. The male prostate gland performs an important role in producing a fluid, which is added to sperm before being released. What primary function does the prostate secretions perform?

   (A) Providing nutrients such as fructose for energy
   (B) Providing growth factors which are required for the soon-to-be developing embryo
   (C) Providing alkaline salts which neutralize the acidic conditions in the vagina and uterus
   (D) Providing a chemical signal to the testes to promote spermatogenesis

11 Evolution

All of the organisms we see today arose from earlier organisms. This process, known as **evolution**, can be described as a change in a population over time. Interestingly, however, the driving force of evolution, **natural selection**, operates on the level of the individual. In other words, evolution is defined in terms of populations but occurs in terms of individuals.

NATURAL SELECTION

What is the basis of our knowledge of evolution? Much of what we now know about evolution is based on the work of **Charles Darwin**. Darwin was a nineteenth-century British naturalist who sailed the world in a ship named the HMS *Beagle*. Darwin developed his theory of evolution based on natural selection after studying animals in the Galápagos Islands and other places.

Darwin concluded that it was impossible for the finches and tortoises of the Galápagos simply to "grow" longer beaks or necks. Rather, the driving force of evolution must have been natural selection. Quite simply put, this means that nature would "choose" which organisms survive on the basis of their fitness. For example, on the first island Darwin studied, there must once have been short-necked tortoises. Unable to reach the higher vegetation, these tortoises eventually died off, leaving only those tortoises with longer necks. Consequently, evolution has come to be thought of as "the survival of the fittest": Only those organisms most fit to survive will survive and reproduce.

Darwin elaborated his theory in a book entitled *On the Origin of Species*. In a nutshell, here's what Darwin observed:

- Each species produces more offspring than can survive.
- These offspring compete with one another for the limited resources available to them.
- Organisms in every population vary.
- The fittest offspring, or those with the most favorable traits are the most likely to survive and therefore produce a second generation.

Lamarck and the Long Necks

Darwin was not the first to propose a theory explaining the variety of life on earth. One of the most widely accepted theories of evolution in Darwin's day was that proposed by **Jean-Baptiste de Lamarck**.

In the eighteenth century, Lamarck had proposed that *acquired* traits were inherited and passed on to offspring. For example, in the case of our tortoises, Lamarck's theory said that the tortoises had long necks because they were constantly reaching for higher leaves while feeding. This theory is referred to as the "law of use and disuse," or, as we might say now, "use it or lose it." According to Lamarck, tortoises have long necks because they constantly *use* them.

We know now that Lamarck's theory was wrong: Acquired changes—that is, changes at a "macro" level in somatic (body) cells—cannot be passed on to germ cells. For example, if you were to lose one of your fingers, your children would not inherit this trait.

Evidence for Evolution

In essence, nature "selects" which living things survive and reproduce. Today, we find support for the theory of evolution in several areas:

- **Paleontology**, or the study of fossils. Paleontology has revealed to us both the great variety of organisms (most of which, including trilobites, dinosaurs, and the woolly mammoth, have died off) and the major lines of evolution.
- **Biogeography**, or the study of the distribution of **flora** (plants) and **fauna** (animals) in the environment. Scientists have found related species in widely separated regions of the world. For example, Darwin observed that animals in the Galápagos have traits similar to those of animals on the mainland of South America. One possible explanation for these similarities is a common ancestor. As we'll see below, there are other explanations for similar traits. However, when organisms share multiple traits, it's pretty safe to say that they also shared a common ancestor.
- **Embryology**, or the study of the development of an organism. If you look at the early stages in vertebrate development, all the embryos look alike! All vertebrates—including fish, amphibians, birds, and even humans—show fishlike features called gill slits.

- **Comparative anatomy**, or the study of the anatomy of various animals. Scientists have discovered that some animals have similar structures that serve different functions. For example, a human's arm, a dog's leg, a bird's wing, and a whale's fin are all the same appendages, though they have evolved to serve different purposes. These structures, called **homologous structures**, also point to a common ancestor.

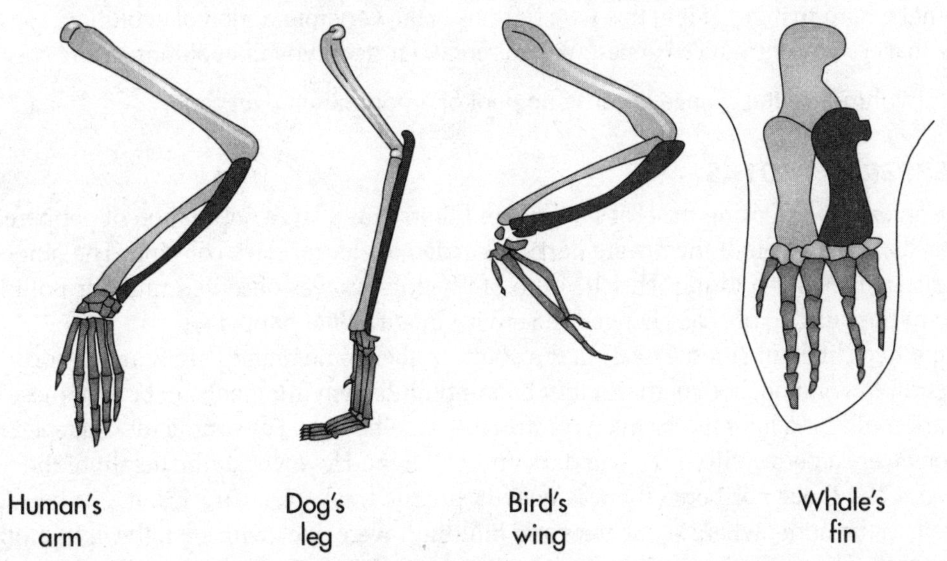

  In contrast, sometimes animals have features with the same function but that are structurally different. A bat's wing and an insect's wing, for example, are both used to fly. They therefore have the same function, but have evolved totally independently of one another. These are called **analogous structures**. Another classic example of an analogous structure is the eye. Though scallops, insects, and humans all have eyes, these three different types of eyes are thought to have evolved entirely independently of one another. They are therefore analogous structures.

- **Molecular biology**. Perhaps the most compelling proof of all is the similarity at the molecular level. Today, scientists can examine the nucleotide and amino acid sequences of different organisms. From these analyses, we've discovered that organisms that are closely related have a greater proportion of sequences in common than distantly related species. For example, most of us don't look much like chimpanzees. However, by some estimates, as much as 99% of our genetic code is identical to that of a chimp.

GENETIC VARIABILITY

In Chapter 8, we saw how traits are passed from parents to offspring. You'll recall from our discussion of heredity that different alleles are passed from parents to their progeny. For example, you might have an allele for brown eyes from your mother and an allele for blue eyes from your father. Since brown is dominant, you'll wind up with brown eyes. We also saw how these alleles are in fact just different forms of the same gene.

As you know, no two individuals are identical. The differences in each person are known as **genetic variability**. All this means is that no two individuals in a population have identical sets of alleles (except, of course, identical twins). In fact, the survival of a species is dependent on this genetic variation; it allows a species to survive in a changing environment. How did all this wonderful variation come about? Through **random mutation**.

It might be hard to think of it in this way, but this is the very foundation of evolution, as we'll soon see. Now that we've reintroduced genes, we can refine our definition of evolution. More specifically:

Evolution is the change in the gene pool of a population over time.

The Peppered Moths

Let's take an example. During the 1850s in England there was a large population of peppered moths. In most areas, exactly half of them were dark, or carried alleles for dark coloring. The other half carried the alleles for light coloring. This 1:1 ratio of phenotypes was observed until air pollution, due primarily to the burning of coal, changed the environment. What happened?

Imagine two different cities: City 1, in the south of the country, and City 2 in the north. Prior to the Industrial Revolution, both of these cities had unpolluted environments. In both of these environments, dark moths and light moths lived comfortably side-by-side. For simplicity's sake, let's say our proportions were a perfect fifty-fifty, half dark and half light. However, at the height of the Industrial Revolution, City 2, our northern city, was heavily polluted, whereas City 1, our southern city, was unchanged. In the north, where all the trees and buildings were thick with soot, the light moths didn't stand a chance. They were impossible for a predator to miss! As a result, the predators gobbled up light-colored moths just as fast as they could reproduce, sometimes even before they reached an age where they *could* reproduce. However, the dark moths were just fine. With all the soot around the predators couldn't even see them; they continued doing their thing—above all, *reproducing*. And when they reproduced, they had more and more offspring carrying the dark allele.

After a few generations, the peppered moth gene pool in City 2 changed. Although our original moth gene pool was 50 percent light and 50 percent dark, excessive predation changed the population's genetic makeup. By about 1950, the gene pool reached 90 percent dark and only 10 percent light. This occurred because the light moth didn't stand a chance in an environment where it was so easy to spot. The dark moths, on the other hand, multiplied just as fast as they could.

In the southern city, you'll remember, there was very little pollution. What happened there? Things remained pretty much the same. The gene pool was unchanged, and the population continued to have roughly equal proportions of light moths and dark moths.

CAUSES OF EVOLUTION

Natural selection, the evolutionary mechanism that "selects" which members of a population are best suited to survive and which are not, works both "internally" and "externally": *internally* through random mutations and *externally* through environmental pressures.

To see how this process unfolds in nature, let's return to the moth case. Why did the dark moths in the north survive? Because they were dark-colored. But how did they become dark-colored? The answer is, through random mutation. One day, a moth was born with dark-colored wings. As long as a mutation does not kill an organism before it reproduces (most mutations, in fact, do), it may be passed on to the next generation. Over time, this one moth had offspring. These, too, were dark. The dark- and the light-colored moths lived happily side by side until something from the outside—in our example, the environment—changed all that.

The initial variation came about by chance. This variation gave the dark moths an edge. However, the edge did not become apparent until something made it apparent. In our case, that something was the intensive pollution due to the burning of coal. The abundance of soot made it easier for predators to spot the light-colored moths, thus effectively removing them from the population. Therefore, dark color is an adaptation, a variation favored by natural selection.

Eventually, over long stretches of time, these two different populations might change so much that they could no longer reproduce together. At that point, we would have two different species, and we could say, definitively, that the moths had evolved. As a consequence of random mutation and the pressure put on the population by an environmental change, evolution occurred.

Types of Selection

The situation with our moths is an example of **directional selection**. One of the phenotypes was favored at one of the extremes of the normal distribution.

In other words, directional selection "weeds out" one of the phenotypes. In our case, dark moths were favored and light moths were practically eliminated. Here's one more thing to remember: Directional selection can happen only if the appropriate allele—the one that is favored under the new circumstances—is already present in the population. Two other types of selection are **stabilizing selection** and **disruptive selection**.

Stabilizing selection means that organisms in a population with extreme traits are eliminated. This type of selection favors organisms with common traits. It "weeds out" the phenotypes that are less adaptive to the environment. A good example is birth weight in human babies. An abnormally small baby has a higher chance of having birth defects; conversely an abnormally large baby will have a challenge in terms of a safe birth delivery.

Disruptive selection, on the other hand, does the reverse. It favors both the extremes and selects *against* common traits. For example, females are "selected" to be small and males are "selected" to be large in elephant seals. You'll rarely find a female or male of intermediate size.

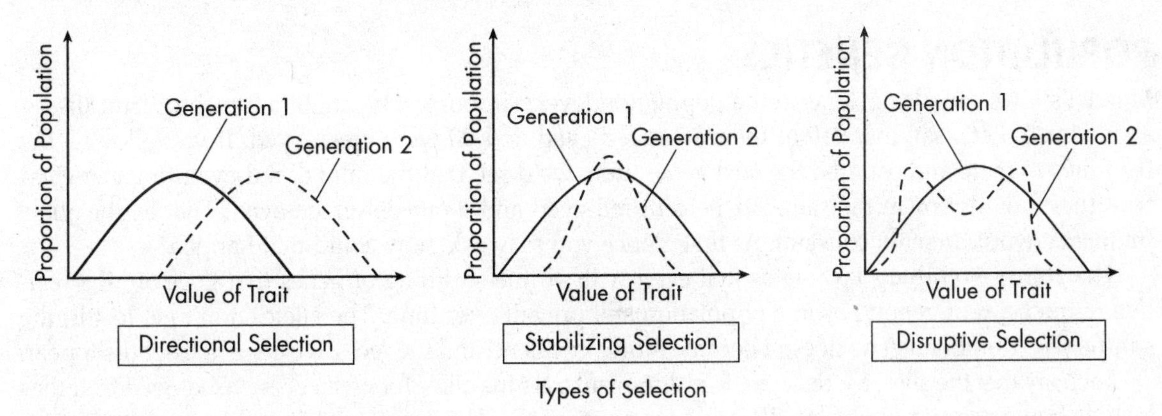

Types of Selection

SPECIES

A dog and a bumblebee obviously cannot come together to produce offspring. They are therefore different **species**. However, a poodle and a Great Dane could reproduce (at least in theory). We would not say that they are different species; they are merely different breeds.

Let's get back to our moths. We said above that evolution occurred when they could no longer reproduce. In fact, this is simply the endpoint of that particular cycle of evolution: **speciation**. Speciation refers to the emergence of new species. The type of evolution that our peppered moths underwent is known as **divergent evolution**.

Divergent evolution results in closely related species with different behaviors and traits. As with our example, these species often originate from a common ancestor. More often than not, the "engine" of evolution is cataclysmic environmental change, such as pollution in the case of the moths. Geographical barriers, new stresses, disease, and dwindling resources are all factors in the process of evolution. Pre and post-zygotic barriers also prevent organisms of two different species from mating to produce viable offspring. Pre-zygotic barriers prevent fertilization. Examples of this kind of barrier include temporal isolation, or when two species reproduce at different times of the year. A post-zygotic barrier is related to the inability of the hybrid to produce offspring. For example, a horse and a donkey can mate to produce a mule, but mules are sterile and therefore cannot produce a second generation.

Convergent evolution is the process in which two unrelated and dissimilar species come to have similar (analogous) traits, often because they have been exposed to similar selective pressures. Examples of convergent evolution include aardvarks, anteaters, and pangolins. They all have strong, sharp claws and long snouts with sticky tongues to catch insects, yet they evolved from three completely different mammals.

There are two types of speciation: **allopatric speciation** and **sympatric speciation**. Allopatric speciation simply means that a population becomes separated from the rest of the species by a geographical barrier so that they can't interbreed. An example would be a mountain that separates two populations of ants. In time, the two populations might evolve into different species. If, however, new species form without any geographic barrier, it is called sympatric speciation. This type of speciation is common in plants. Two species of plants may evolve in the same area without any geographic barrier.

POPULATION GENETICS

Mendel's laws can also extend to the population level. Suppose you caught a bunch of fruit flies—about 1,000. Let's say that 910 of them were red-eyed and 90 were green-eyed. If you allowed the fruit flies to mate and counted the next generation, we'd see that the ratio of red-eyed to green-eyed fruit flies would remain the same: 91 percent red-eyed and 9 percent green-eyed. That is, the allele frequency would remain constant. At first glance you may ask, how could that happen?

The **Hardy-Weinberg law** states that even with all the shuffling of genes that goes on, the relative frequencies of genotypes in a population still prevail over time. The alleles don't get lost in the shuffle. The dominant gene doesn't become more prevalent, and the recessive gene doesn't disappear.

Let's say that the allele for red eyes, R, is dominant over the allele for green eyes, r. Red-eyed fruit flies include homozygous dominants, RR, and heterozygous, Rr. The green-eyed fruit flies are recessive, rr.

HARDY-WEINBERG EQUATIONS

The frequency of each allele is described in the equation below. The allele must be either R or r. Let "p" represent the frequency of the R allele and "q" represent the frequency of the other allele in the population.

$$p + q = 1$$

This sum of the frequencies must add up to one. If you know the value of one of the alleles, then you'll also know the value of the other allele.

We can also determine the frequency of the *genotypes* in a population using another equation:

$$p^2 + 2pq + q^2 = 1$$

In this equation, p^2 represents the homozygous dominants, $2pq$ represents the heterozygotes and q^2 represents the homozygous recessives.

So how do we use these equations? Use the proportions in the population to figure out both the allele and genotype frequencies. Let's calculate the frequency of the genotype for green-eyed fruit flies. If 9 percent of the fruit flies are green-eyed, then the *genotype* frequency, q^2, is 0.09. You can now use this value to figure out the frequency of the recessive allele in the population. The allele frequency for green eyes is equal to the square root of 0.09—that's 0.3. If the recessive allele is 0.3, the dominant allele must be 0.7. That's because 0.3 + 0.7 equals 1.

Using the second equation, you can calculate the genotypes of the homozygous dominants and the heterozygotes. The frequency for the homozygous dominants, p^2, is 0.7 × 0.7, which equals 0.49. The frequency for the heterozygotes, $2pq$, is 2 × 0.3 × 0.7, which equals 0.42. If you include the frequency of the recessive genotype—0.09—the numbers once again add up to 1.

Hardy-Weinberg Equilibrium

The **Hardy-Weinberg law** says that a population will be in genetic equilibrium only if it meets these five conditions: 1) a large population, 2) no mutations, 3) no immigration or emigration, 4) random mating, and 5) no natural selection.

Violations of the Hardy-Weinberg Law

When these five conditions are met, the gene pool in a population is pretty stable. Any departure from them results in changes in allele frequencies in a population. For example, if a small group of your fruit flies moved to a new location, the allele frequency may be altered and result in evolutionary changes. That's an example of **genetic drift** called the founder effect. In other words, the gene frequency may differ from the original gene pool. Genetic drift often occurs in new colonies with small populations.

KEY WORDS

evolution
natural selection
Charles Darwin
Jean-Baptiste de Lamarck
paleontology
biogeography
flora
fauna
embryology
comparative anatomy
homologous structures
analogous structures
molecular biology
genetic variability
random mutation
directional selection
stabilizing selection
disruptive selection
species
speciation
divergent evolution
convergent evolution
adaptation
pre-zygotic barriers
post-zygotic barriers
allopatric speciation
sympatric speciation
Hardy-Weinberg law
genetic drift

CHAPTER 11 REVIEW QUESTIONS

Answers can be found in Chapter 15.

1. The eye structures of mammals and cephalopods such as squid evolved independently to perform very similar functions and have similar structures. This evolution is an example of which of the following?

   (A) Allopatric speciation
   (B) Sympatric association
   (C) Divergent evolution
   (D) Convergent evolution

Questions 2 – 5 refer to the following graph and paragraph.

During the industrial revolution, a major change was observed in many insect species due to the mass production and deposition of ash and soot around cities and factories. One of the most famous instances was within the spotted moth populations. An ecological survey was performed where the number of spotted moths and longtail moths were counted in 8 different urban settings over a square kilometer in 1802. A repeat experiment was performed 100 years later in 1902. The results of the experiment are shown below.

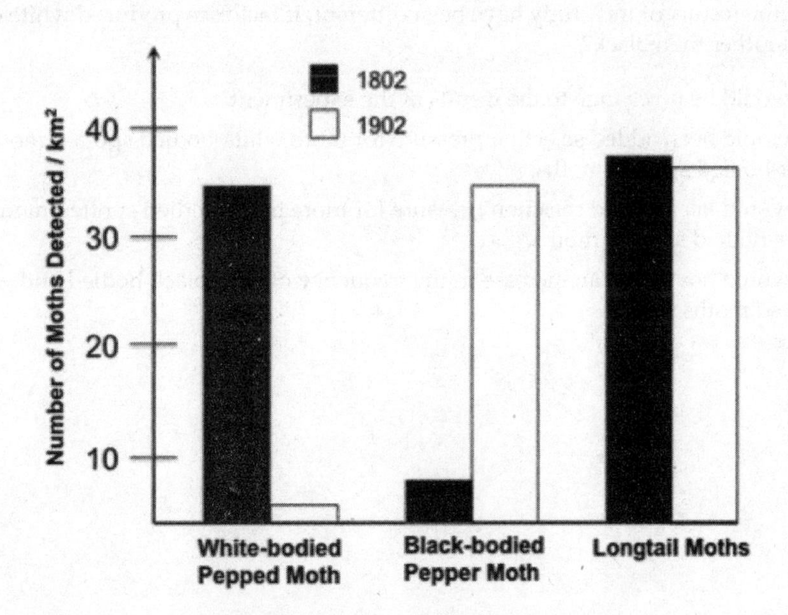

2. What type of selection is represented by the results of this study?

   (A) Stabilizing selection
   (B) Directional selection
   (C) Disruptive selection
   (D) Divergent selection

3. Which of the following statements best explains the data?

(A) As time passed from 1802 to 1902, the frequency of white-bodied pepper moths increased and black-bodied pepper moths decreased.

(B) As time passed from 1802 to 1902, the frequency of white-bodied pepper moths decreased and black-bodied pepper moths increased.

(C) As time passed from 1802 to 1902, the frequency of white-bodied pepper moths and black-bodied pepper moths both increased.

(D) As time passed from 1802 to 1902, the frequency of white-bodied pepper moths and black-bodied pepper moths both decreased.

4. Why was the population of longtail moths also surveyed in this study?

(A) Variations in the environment were expected to alter the population of Longtail moths.

(B) Longtail moths were included as a control since they were not expected to change appreciably due to changes associated with the Industrial Revolution.

(C) As time passed from 1802 to 1902, the frequency of white-bodied pepper moths and black-bodied pepper moths both increased.

(D) As time passed from 1802 to 1902, the frequency of white-bodied pepper moths and black-bodied pepper moths both decreased.

5. How would the results of this study have been different, if factories produced white or light gray ash and soot rather than black?

(A) There would be no change to the results of the experiment.

(B) There would been added selection pressure for more white-bodied spotted moths and against black-bodied spotted moths.

(C) There would been added selection pressure for more black-bodied spotted moths and against white-bodied spotted moths.

(D) There would have been an increase in the frequency of both black-bodied and white-bodied spotted moths.

Question 6 represents a question requiring a numeric answer.

6. A recessive allele of a gene has a calculated frequency of 0.3 in a population. Assuming the population is in Hardy-Weinberg equilibrium, what percentage of the population is expected to be heterozygous for the gene? Give your answer to the nearest hundredths place.

	/	/	/	
.	.	.	.	.
0	0	0	0	0
1	1	1	1	1
2	2	2	2	2
3	3	3	3	3
4	4	4	4	4
5	5	5	5	5
6	6	6	6	6
7	7	7	7	7
8	8	8	8	8
9	9	9	9	9

12

Behavior and Ecology

In the preceding chapters, we looked at individual organisms and the ways in which they solve life's many problems: acquiring nutrition, reproducing, etc. Now let's turn to how organisms deal with their environments. We can divide the discussion of organisms and their environments into two general categories: **behavior** and **ecology**.

BEHAVIOR

Some animals behave in a programmed way to specific stimuli while others behave according to some type of learning. We humans can do both. For the AP Biology Exam, you'll have to know a little about the different types of behavior.

All behavior means, basically, is how organisms cope with their environments. This chapter looks at the general types of behavior: instinct, imprinting, classical conditioning, operant conditioning, and insight. Let's start with instinct.

INSTINCTS

Instinct is an inborn, unlearned behavior. Sometimes the instinctive behavior is triggered by environmental signals called releasers. The releaser is usually a small part of the environment that is perceived. For example, when a male European robin sees another male robin, the sight of a tuft of red feathers on the male is a releaser that triggers fighting behavior. In fact, because instinct underlies all other behavior, it can be thought of as the circuitry that guides behavior.

For example, hive insects such as bees and termites never learn their roles; they are born knowing them. On the basis of this inborn knowledge, or instinct, they carry out all their other behaviors: A worker carries out "worker tasks," a drone "drone tasks," and a queen "queen tasks." Another example is the "dance" of the honey bee, which is used to communicate the location of food to other members of the beehive.

There are other types of instinct that last for only a part of an animal's life and are gradually replaced by "learned" behavior. For example, human infants are born with an ability to suck from a nipple. If it were not for this instinctual behavior, the infant would starve. Ultimately, however, the infant will move beyond this instinct and learn to feed itself. What exactly, then, is instinct?

For our purposes:

Instinct is the inherited "circuitry" that directs and guides behavior.

A particular type of innate behavior is a **fixed action pattern**. These behaviors are not simple reflexes and yet they are not conscious decisions. An example is the egg-rolling behavior exhibited by a graylag goose. If the egg is removed from the goose, it will continue to make the same movements. That is, the innate movements are independent of the environment.

LEARNING

Another form of behavior is **learning**. Learning refers to a change in a behavior brought about by an experience (which is what you're doing this very moment). Animals learn in a number of ways. For the AP Biology Exam, we'll take a look at the key types of learning.

Imprinting

Have you ever seen a group of goslings waddling along after their mother? How is it that they recognize her? Well, the mother arrives and gives out a call that the goslings "recognize." The goslings, hearing the call, know that this is their mother, and follow her around until they are big enough to head out on their own.

Now imagine the same goslings, newly hatched. If the mother is absent, they will accept the first moving object they see as their mother. This process is known as **imprinting**.

Animals undergo imprinting within a few days after birth in order to recognize members of their own species. While there are different types of imprinting—including parent, sexual, and song imprinting—they all occur during a **critical period**—a window of time when the animal is sensitive to certain aspects of the environment.

Remember that:

> Imprinting is a form of learning that occurs during a brief period of time, usually early in an organism's life.

Classical Conditioning

If you have a dog or a cat, you know that every time you hit the electric can opener, your cat or dog comes running. This is a form of **classical conditioning**. To feed your pet, you need to open its can of food. For your pet, the sound of the opener has come to be associated with eating: Every time it hears the opener, it thinks that it is about to be fed. We can say that your pet has been "conditioned" to link the buzz of the can opener and its evening meal.

The classic experiment demonstrating conditioning was done by a Russian scientist named Ivan Pavlov, who made his dogs salivate by ringing a bell. He did this the same way you make your dog come running with the can opener. Each time he fed his dogs, Pavlov rang a bell. Eventually, the dogs came to "associate" the bell with the food. By the end of the experiment, Pavlov had merely to ring the bell to start the dogs salivating. This type of learning is now known as **associative learning**.

Pavlov's experiments demonstrated what many of us know already: We can learn through conditioning, or repeated instances of an event.

Operant Conditioning

Another type of associative learning is **operant conditioning** (or **trial-and-error learning**).

In operant conditioning, an animal learns to perform an act in order to receive a reward. This type of behavior was extensively studied by psychologist B. F. Skinner. He put a rat in an experimental cage and watched to see if it would randomly press different levers. Through trial and error, the animal figured out that one lever in particular would always produce food from a dispenser. Over time, the animal made an "association" between pressing the lever and getting food (the reward). Skinner even detected that some rats were so "conditioned" that they started to "hang out" near the lever. These same rats were also subjected to a negative form of operant conditioning, where touching the bar led to an electric shock. As you could imagine, rats quite quickly learned not to touch the bar that gave them the shock.

If the behavior is not reinforced, the conditioned response will be lost. This is called extinction. Here's one thing to remember:

> In operant conditioning, the animal's behavior determines whether it gets the reward or the punishment.

Habituation is another form of learning. It occurs when an animal learns not to respond to a stimulus. For example, if an animal encounters a stimulus over and over again without any consequences, the response to it will gradually lessen and may altogether disappear. One example is how a marine worm, *Nerels*, withdraws into its protective tube if a shadow passes over. With repeated exposure to the stimulus, however, the response decreases.

Insight

This is the highest form of learning and is exercised only by higher animals. **Insight** means the ability to figure out a behavior that generates a desired outcome. It is sometimes referred to as the "aha experience." As far as the animal world is concerned, human beings tend to be pretty good at using insight, or **reasoning**, to solve problems. However, we are not the only ones who reason. Chimpanzees, for example, have been known to use rudimentary tools, such as twigs and stones, to get their food.

To recap, there are four basic types of learning:

- Imprinting occurs early in life and helps organisms recognize members of their own species.
- Classical conditioning involves learning through association.
- Operant conditioning occurs when a response is associated with new stimuli (also a form of associative learning).
- Insight involves "reasoning" or problem solving.

Internal Clocks: The Circadian Rhythm

There are other instinctual behaviors that occur in both animals and plants. One such behavior deals with time. Have you ever wondered how roosters always know when to start crowing? The first thought that comes to mind is that they've caught a glimpse of the sun. Yet many crow even before the sun has risen.

Roosters do have internal alarm clocks. Plants have them as well. These internal clocks, or cycles, are known as **circadian rhythms**.

If you've ever flown overseas, you know all about these. They're the basis of jet lag. Our bodies tell us it's one time while our watches tell us it's another. The sun may be up, but our body's internal clock is crying "Sleep!" This sense of time is purely instinctual: You don't need to know how to tell time in order to feel jet lag.

Circadian rhythms are yet another example of instinct. When an organism does something on a daily basis, we say it acts according to its circadian rhythm. But how do we know for certain that it's instinctual?

In a famous experiment, an American scientist took a bunch of plants and animals to the South Pole and put them on a turntable set to rotate at exactly the same speed as the earth but in the opposite direction of the earth's rotation. As a result, the organisms had absolutely no indication of day or night. Yet all of them continued to carry out their regular 24-hour cycles. This proved that the cycles have nothing to do with sunlight and everything to do with the internal clock.

Watch out though: Seasonal changes, like the loss of leaves by deciduous trees or the hibernation of mammals, are not examples of circadian rhythms. *Circadian* refers only to daily rhythms. Need a mnemonic? Just think how bad your jet lag would be after a trip around the world. In other words: Circling the globe screws up your circadian rhythm.

How Animals Communicate

Some animals use signals as a way of communicating with members of their species. These signals, which can be chemical, visual, electrical or tactile, are often used to influence mating and social behavior.

Chemical signals are one of the most common forms of communication among animals. **Pheromones**, for instance, are chemical signals between members of the same species that stimulate olfactory receptors and ultimately affect behavior. For example, when female insects give off their pheromones they attract males from great distances.

Visual signals also play an important role in the behavior observed among members of a species. For example, fireflies produce pulsed flashes that can be seen by other fireflies far away. The flashes are sexual displays that help male and female fireflies identify and locate each other in the dark.

Other animals use electrical channels to communicate. For example, some fish generate and receive weak electrical fields. Finally, tactile signals are found in animals that have mechanoreceptors in their skin to detect prey. For instance, cave-dwelling fishes use mechanoreceptors in their skin for communicating with other members and detecting prey.

Social Behavior

Many animals are highly social species, and they interact with each other in complex ways. Social behaviors can help members of the species survive and reproduce more successfully. Several behavioral patterns for animal societies are summarized below:

- **Agonistic behavior** is aggressive behavior that occurs as a result of competition for food or other resources. Animals will show aggression toward other members that tend to use the same resources. A typical form of aggression is fighting between competitors.
- **Dominance hierarchies** (or pecking orders) occur when members in a group have established which members are the most dominant. The more dominant male will often become the leader of the group and will usually have the best pickings of the food and females in the group. Once the dominance hierarchy is established, competition and tension within the group is reduced.
- **Territoriality** is a common behavior when food and nesting sites are in short supply. Usually the male of the species will establish and defend his territory (called a home range) within a group in order to protect important resources. This behavior is typically found among birds.
- **Altruistic behavior** is defined as unselfish behavior that benefits another organism in the group at the individual's expense because it advances the genes of the group. For example, when ground squirrels give warning calls to alert other squirrels of the presence of a predator, the calling squirrel puts itself at risk of being found by the predator.

PLANT BEHAVIOR

Plants have also evolved specific ways to respond to their environment. The plant behaviors covered on the AP Biology test are photoperiodism and tropisms. Plants flower in response to changes in the amount of daylight and darkness. This is called **photoperiodism**. Although you'd think that plants bloom based on the amount of sunlight they receive, they actually flower according to the amount of uninterrupted darkness.

Tropisms

Plants need light. Notice that all the plants in your house tip towards the windows. This movement of plants toward the light is known as phototropism. As you know, plants generally grow up and down: The branches grow upward, while the roots grow downward into the soil, seeking water. This tendency to grow toward or away from the earth is called gravitropism. All of these tropisms are examples of behavior in plants.

A **tropism** is a turning in response to a stimulus.

There are three basic tropisms in plants. They're easy to remember because their prefixes the stimuli to which plants react:

- **Phototropism** refers to how plants respond to sunlight. For example, bending towards light.
- **Gravitropism** refers to how plants respond to gravity. Stems exhibit negative gravitropism (i.e., they grow up, away from the pull of gravity), whereas roots exhibit positive gravitropism (i.e., they grow downward into the earth).
- **Thigmotropism** refers to how plants respond to touch. For example, ivy grows around a post or trellis.

ECOLOGY

The study of the interactions between living things and their environments is known as ecology. We've spent most of our time discussing individual organisms. However, in the "real world," organisms are in constant interaction with other organisms and the environment. The best way for us to understand the various levels of ecology is to progress from the big picture, the biosphere, down to the smallest ecological unit, the population.

Just as in taxonomy, there is a hierarchy within the "ecology world." Each of the following terms represents a different level of ecological interaction.

- **Biosphere**—The entire part of the earth where living things exist. This includes soil, water, light, and air. In comparison to the overall mass of the earth, the biosphere is relatively small. If you think of the earth as a basketball, the biosphere is equivalent to a coat of paint over its surface.
- **Ecosystem**—The interaction of living and nonliving things.
- **Community**—A group of populations interacting in the same area.
- **Population**—A group of individuals that belong to the same species and that are interbreeding.

Biosphere

The biosphere can be divided into large regions called **biomes**. Biomes are massive areas that are classified mostly on the basis of their climates and plant life. Because of the different climates and terrains on the earth, the distribution of living organisms varies. For the AP Biology Exam, you're expected to know both the names of the different biomes and their characteristic flora (plant life) and fauna (animal life).

Here's a summary of the major biomes:

Major Biomes

Tundra
Regions—northernmost regions
Plant life—few, if any, trees; primarily grasses and wildflowers
Characteristics—contains permafrost (a layer of permanently frozen soil); has a short growing season
Animal life—includes lemmings, arctic foxes, snowy owls, caribou, and reindeer

Taiga
Region—northern forests
Plant life—wind-blown conifers (evergreens), stunted in growth, possess modified spikes for leaves
Characteristics—very cold, long winters
Animal life—includes caribou, wolves, moose, bear, rabbits, and lynx

Temperate Deciduous Forest
Regions—northeast and middle eastern United States, western Europe
Plant life—deciduous trees that drop their leaves in winter
Characteristics—moderate precipitation; warm summers, cold winters
Animal life—includes deer, wolves, bear, small mammals, birds

Grasslands
Regions—American Midwest, Eurasia, Africa, South America
Plant life—grasses
Characteristics—hot summers, cold winters; unpredictable rainfall
Animal life—includes prairie dogs, bison, foxes, ferrets, grouse, snakes, and lizards

Deserts
Regions—western United States
Plant life—sparse, includes cacti, drought-resistant plants
Characteristics—arid, low rainfall; extreme diurnal temperature shifts
Animal life—includes jackrabbits (in North America), owls, kangaroo rats, lizards, snakes, tortoises

Tropical Rain Forests
Regions—South America
Plant life—high biomass; diverse types
Characteristics—high rainfall and temperatures; impoverished soil
Animal life—includes sloths, snakes, monkeys, birds, leopards, and insects

Remember that the biomes tend to be arranged along particular latitudes. For instance, if you hiked from Alaska to Kansas, you would pass through the following biomes: tundra, taiga, temperate deciduous forests, and grasslands.

Ecosystem

Ecosystems are self-contained regions that include both living and nonliving factors. For example, a lake, its surrounding forest, the atmosphere above it, and all the organisms that live in or feed off the lake would be considered an ecosystem. As you probably know, there is an exchange of materials between the components of an ecosystem. Take a look at the flow of carbon through a typical ecosystem:

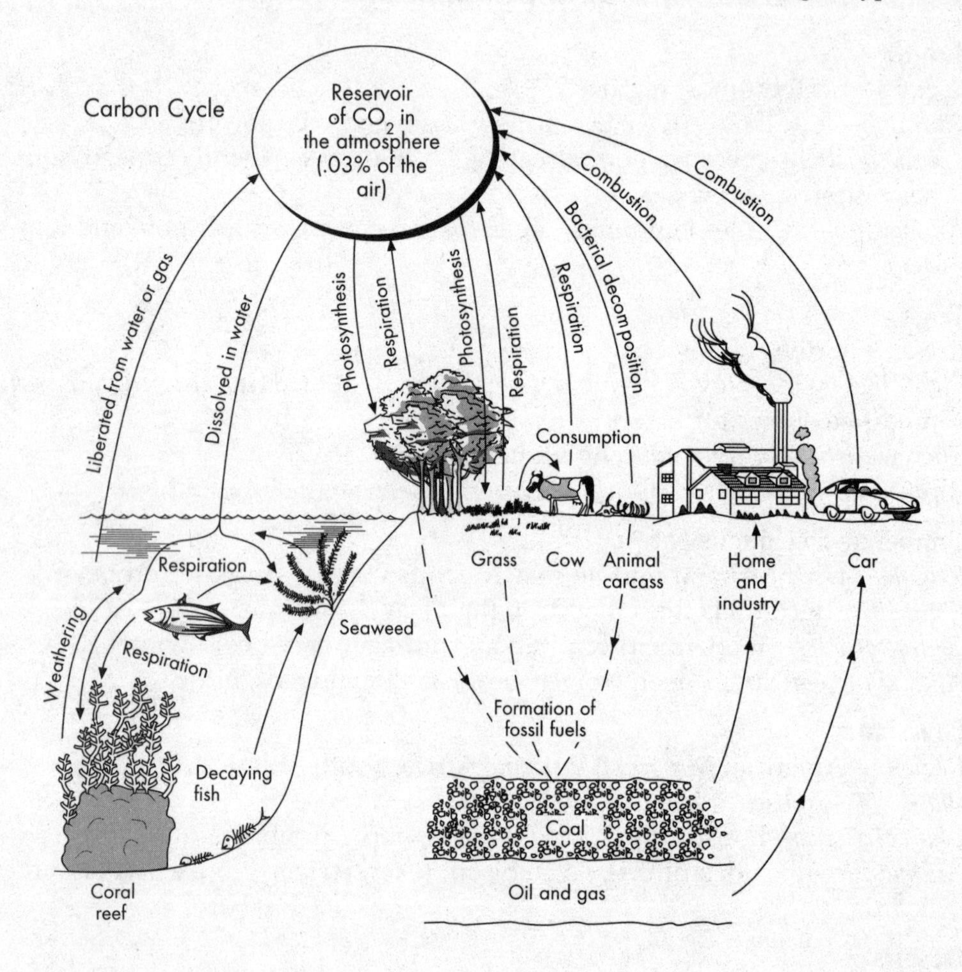

You'll notice how carbon is recycled throughout the ecosystem—this is called the **carbon cycle**. In other words, carbon flows through ecosystems. Other substances have similar cycles.

Community

The next smaller level is the community. A community refers to a group of interacting plants and animals that show some degree of interdependence. For instance, you, your dog, and the fleas on your dog are all members of the same community. All organisms within a community fill one of the following roles: producers (or autotrophs), consumers (or heterotrophs), or decomposers.

Producers

Producers, or autotrophs, have all of the raw building blocks to make their own food. From water and the gases that abound in the atmosphere, and with the aid of the sun's energy, autotrophs convert light energy to chemical energy. They accomplish this through photosynthesis.

Consumers

Consumers, or heterotrophs, are forced to find their energy sources in the outside world. Basically, heterotrophs digest the carbohydrates of their prey into carbon, hydrogen, and oxygen, and use these molecules to make organic substances.

The bottom line is: Heterotrophs, or consumers, get their energy from the things they consume.

Decomposers

All organisms at some point must finally yield to decomposers. Decomposers are the organisms that break down organic matter into simple products. Generally, fungi and bacteria are the decomposers. They serve as the "garbage collectors" in our environment.

NICHE

Each organism has its own **niche**—its position or function in a community. Because every species occupies a niche, it's going to have an effect on all the other organisms. These connections are shown in the **food chain**. A food chain describes the way different organisms depend on one another for food. There are basically four levels to the food chain: producers, primary consumers, secondary consumers, and tertiary consumers.

Autotrophs produce all of the available food. They make up the first trophic (feeding) level. They possess the highest biomass (the total weight of all the organisms in an area) and the greatest numbers. Did you know that plants make up about 99 percent of the earth's total biomass?

Primary consumers are organisms that directly feed on producers. A good example is a cow. These organisms are also known as **herbivores**. They make up the second trophic level.

The next level consists of organisms that feed on primary consumers. They are the **secondary consumers**, and they make up the third trophic level. Above these are **tertiary consumers**.

So now you have it. We've got our four complete levels of the food chain.

- Producers make their own food.
- Primary consumers (herbivores) eat producers.
- Secondary consumers (carnivores and omnivores) eat producers and primary consumers.
- Tertiary consumers eat all of the above.

The 10% Rule

In a food chain, only about 10 percent of the energy is transferred from one level to the next—this is called the **10% rule**. The other 90 percent is used for things like respiration, digestion, running away from predators—in other words, it's used to power the organism doing the eating! The producers have the most energy in an ecosystem; the primary consumers have less energy than producers; the secondary consumers have less energy than the primary consumers; and the tertiary consumers will have the least energy of all.

The energy flow, biomass, and numbers of members within an ecosystem can be represented in an **ecological pyramid**. Organisms that are "higher up" on the pyramid have less biomass and energy, and fewer numbers.

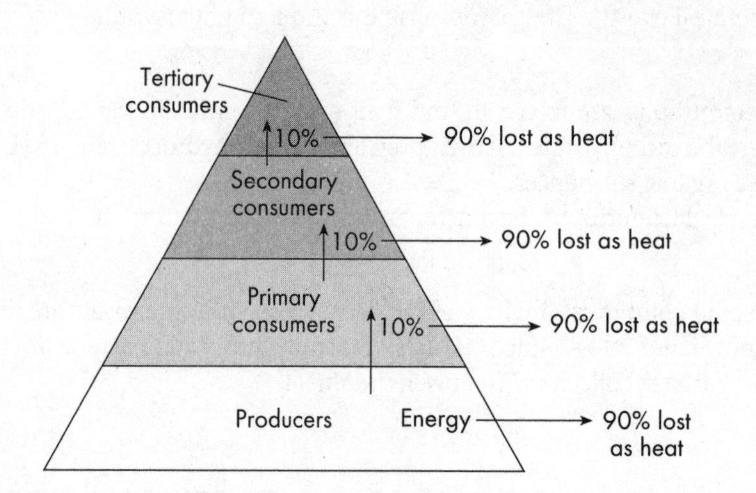

So, for example, primary consumers have less biomass and are fewer in number than producers. But what about decomposers? Where do they fit in on the food chain? They don't. They are not really considered part of the food chain. Decomposers are usually placed just below the food chain to show that they can decompose any organism.

In addition to the distinctions drawn above, remember:

> Toxins in an ecosystem are more concentrated and thus more dangerous for animals further up the pyramid.

This simply means that if a toxin is introduced into an ecosystem, the animals most likely to be affected are those at the top of the pyramid. This occurs because of the increasing concentration of such toxins. The classic example of this phenomenon is DDT, an insecticide initially used to kill mosquitoes. Although the large-scale spraying of DDT resulted in a decrease in the mosquito population, it also wound up killing off ospreys.

Ospreys are aquatic birds whose diet consists primarily of fish. The fish that ospreys consumed had, in turn, been feeding on contaminated insects (bioaccumulation). Because fish eat thousands of insects, and ospreys hundreds of fish, the toxins grew increasingly concentrated (biomagnification). Though the insecticide seemed harmless enough, it resulted in the near-extinction of certain osprey populations. What no one knew was that in sufficient concentrations, DDT weakened the eggshells of ospreys. Consequently, eggs broke before they could hatch, killing the unborn ospreys.

Such environmental tragedies still occur. All ecosystems, small or great, are intricately woven, and any change in one level invariably results in major changes at all the other levels.

Symbiotic Relationships

Many organisms that coexist exhibit some type of symbiotic relationship. These include remoras, or "sucker fish," which attach themselves to the backs of sharks, and lichen, the fuzzy, moldlike stuff that grows on rocks. Lichen appears to be one organism, when in fact it is two organisms—a fungus and an alga or photosynthetic bacterium—living in a complex symbiotic relationship.

Overall, there are three basic types of symbiotic relationships:

- **Mutualism**—in which both organisms win (for example, the lichen).
- **Commensalism**—in which one organism lives off another with no harm to the "host" organism (for example, the remora).
- **Parasitism**—in which the organism actually harms its host.

POPULATION ECOLOGY

Population ecology is the study of how populations change. Whether these changes are long-term or short-term, predictable or unpredictable, we're talking about the growth and distribution patterns of a population.

When studying a population, you need to examine four things: the size (the total number of individuals), the density (the number of individuals per area), the distribution patterns (how individuals in a population are spread out), and the age structure.

One way to understand the growth pattern and make predictions about the population growth of a country is by examining age structure histograms. For example, in underdeveloped countries, where the population increase is high, the base of the histogram is very wide compared to countries that show moderate growth.

Another way to study the changes in a population is by looking at survivorship curves. These curves are graphs of the numbers of individuals surviving to different ages, indicating the probability of any individual living to a given age.

For example, the graph below shows there is a high death rate among the young of oysters, but those that survive do well. On the other hand, there is a low death rate among the young of humans, but, after age 60, the death rate is high.

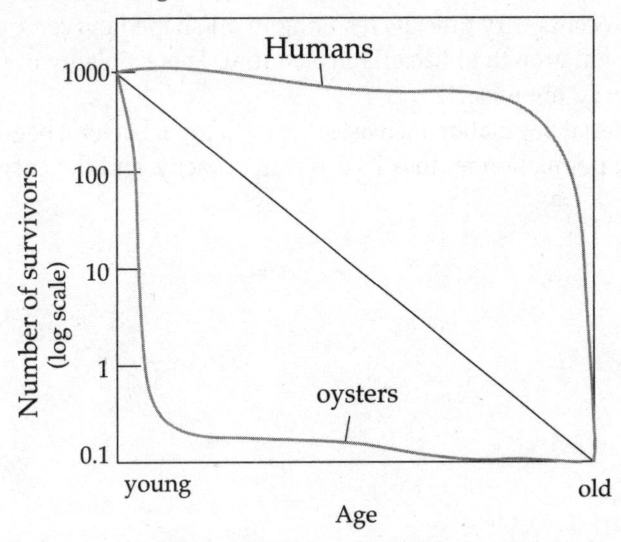

Survivorship Curve

The growth of a population can be represented as the number of crude births minus the number of crude deaths divided by the size of the population:

r = (births – deaths)/N

(r is the reproductive rate, and N is the population size)

Each population has a **carrying capacity**—the maximum number of individuals of a species that a habitat can support. Most populations, however, don't reach their carrying capacity because they're exposed to limiting factors.

One important factor is **population density**. The factors that limit a population are either density-independent or density-dependent. **Density-independent factors** are factors that affect the population regardless of the density of the population. Some examples are severe storms and extreme climates. On the other hand, **density-dependent factors** are those with effects that depend on population density. Resource depletion, competition, and predation are all examples of density-dependent factors. In fact, these effects become even more intense as the population density increases.

Exponential Growth

The growth rates of populations also vary greatly. There are two types of growth: exponential growth and logistic growth. **Exponential growth** occurs when a population is in an ideal environment. Growth is unrestricted because there are lots of resources, space, and no disease or predation. Here's an example of exponential growth. Notice that the curve arches sharply upward—the exponential increase.

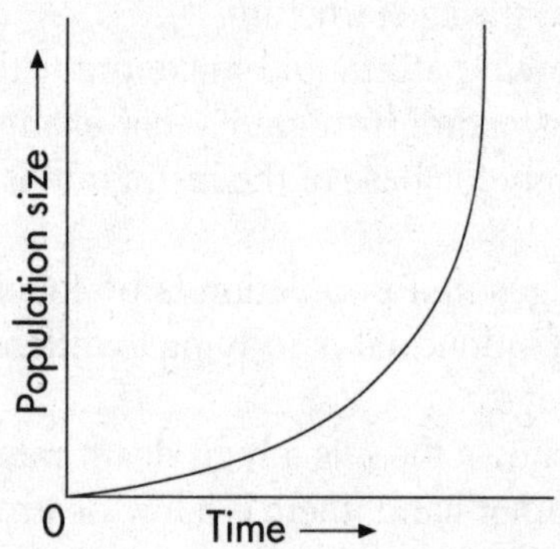

Exponential growth occurs very quickly, resulting in a J-shaped curve. A good example of exponential growth is the initial growth of bacteria in a culture. There's plenty of room and food, so they multiply rapidly—every 20 minutes.

However, as the bacterial population increases, the individual bacteria begin to compete with each other for resources. The population reaches its carrying capacity, and the curve levels off.

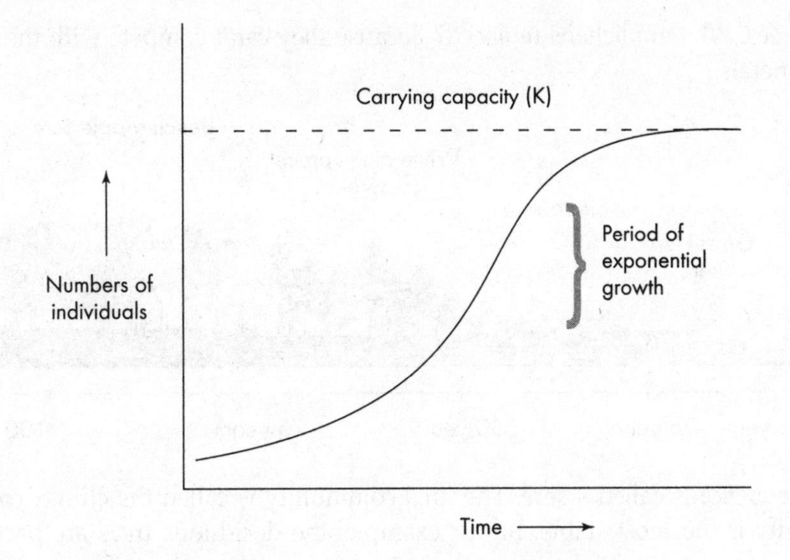

The population becomes restricted in size because of limited resources. This is referred to as **logistic growth**. Notice that the growth forms an S-shaped curve. These growth patterns are associated with two kinds of life-history strategies: r-selected species and k-selected species.

We've already mentioned that organisms that grow exponentially approach the carrying capacity. These organisms tend to thrive in areas that are barren or uninhabited. Once they colonize an area, they reproduce as quickly as possible. Why? They know they've got to multiply before competitors arrive on the scene! The best way to ensure their survival is to produce lots of offspring. These organisms are known as **r-strategists**. Typical examples are common weeds, dandelions, and bacteria.

At the other end of the spectrum are the **k-strategists**. These organisms are best suited for survival in stable environments. They tend to be large animals such as elephants, with long lifespans. Unlike r-strategists, they produce few offspring. Given their size, k-strategists usually don't have to contend with competition from other organisms.

Ecological Succession

Communities of organisms don't just spring up on their own; they develop gradually over time. **Ecological succession** refers to the predictable procession of plant communities over a relatively short period of time (decades or centuries).

Centuries may not seem like a short time to us, but if you consider the enormous stretches of time over which evolution occurs, hundreds of thousands or even millions of years, you'll see that it is pretty short.

The process of ecological succession where no previous organisms have existed is called **primary succession**.

The Pioneers

How does a new habitat full of bare rocks eventually turn into a forest? The first stage of the job usually falls to a community of lichens. Lichens are hardy organisms. They can invade an area, land on bare rocks and erode the rock surface, and over time turn it into soil. Lichens are considered **pioneer organisms**.

Once lichens have made an area more habitable, they've set the stage for other organisms to settle in. Communities establish themselves in an orderly fashion. Lichens are replaced by mosses and ferns, which in turn are replaced by tough grasses, then low shrubs, then evergreen trees, and finally, the deciduous trees. Why are lichens replaced? Because they can't compete with the new plants for sunlight and minerals.

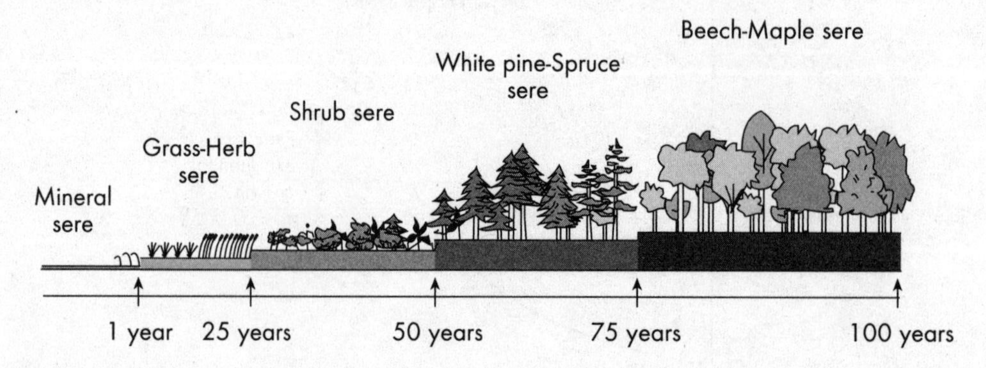

The entire sequence is called a **sere**. The final community is called the **climax community**. The climax community is the most stable. In our example, the deciduous trees are part of the climax community.

Now what happens when a forest is devastated by fire? The same principles apply, but the events occur much more rapidly. The only exception is that the first invaders are usually not lichens but grasses, shrubs, saplings, and weeds. When a new community develops where another community has been destroyed or disrupted, this event is called **secondary succession**.

HUMAN IMPACT ON THE ENVIRONMENT

Unfortunately, humans have disturbed the existing ecological balance, and the results are far-reaching. Soils have been eroded and various forms of pollution have increased. The potential consequences on the environment are summarized below:

- **Greenhouse effect**—The increasing atmospheric concentrations of carbon dioxide through the burning of fossil fuels and forests have contributed to the warming of the earth. Higher temperatures may cause the polar ice caps to melt and flooding to occur. Other potential effects of global warming include changes in precipitation patterns, changes in plant and animal populations, and detrimental changes in agriculture.

- **Ozone depletion**—Pollution has also led to the depletion of the atmospheric ozone layer by such chemicals as chlorofluorocarbons (CFCs), which are used in aerosol cans. Ozone (O_3), forms when UV radiation reacts with O_2. Ozone protects the earth's surface from excessive ultraviolet radiation. Its loss could have major genetic effects and could increase the incidence of cancer.

- **Acid rain**—The burning of fossil fuels produces pollutants such as sulfur dioxide and nitrogen dioxide. When these compounds react with droplets of atmospheric water in clouds they form sulfuric and nitric acids, respectively. The rain that falls from these clouds is weakly acid and is called acid rain. Acid rain lowers the pH of aquatic ecosystems and soil which damages water systems, plants and soil. For example, the change in soil pH causes calcium and other nutrients to leach out, which damages plant roots and stunts their growth. Furthermore, useful microorganisms that release nutrients from decaying organic matter into the soil are also killed, resulting in less nutrients being available for the plants. Low pH also kills fish, especially those that have just hatched.

- **Desertification**—When land is overgrazed by animals, it turns grasslands into deserts and reduces the available habitats for organisms.

- **Deforestation**—When forests are cleared (especially by the slash and burn method), erosion, floods, and changes in weather patterns can occur.

- **Pollution**—Another environmental concern is the toxic chemicals in our environment. One example is DDT, a pesticide used to control insects. DDT was overused at one time and later found to damage plants and animals worldwide. DDT is particularly harmful because it resists chemical breakdown and today it can still be found in the tissues of nearly every living organism. The danger with toxins such as DDT is that as each trophic level consumes DDT, the substance becomes more concentrated by a process called **biomagnification**.
- **Reduction in biodiversity**—As different habitats have been destroyed, many plants and animals have become extinct. Some of these plants could have provided us with medicines and products that may have been beneficial.

KEY WORDS

behavior
ecology
instinct
fixed action pattern
learning
imprinting
critical period
classical conditioning
associative learning
operant conditioning (or trial-and-error learning)
habituation
insight
reasoning
circadian rhythms
pheromones
agonistic behavior
dominance hierarchy
territoriality
altruistic behavior
photoperiodism
tropism
phototropism
gravitropism
thigmotropism
biosphere
ecosystem
community
population
biomes
carbon cycle
niche
food chain
primary consumers
herbivores
secondary consumers
tertiary consumers
10% rule
ecological pyramid
mutualism
commensalism
parasitism
carrying capacity
population density
density-independent factors
density-dependent factors
exponential growth
logistic growth
r-strategists
k-strategists
ecological succession
primary succession
pioneer organisms
sere
climax community
secondary succession
greenhouse effect
ozone depletion
acid rain
desertification
deforestation
pollution
biomagnification

CHAPTER 12 REVIEW QUESTIONS

Answers can be found in Chapter 15.

1. A chimpanzee stacks a series of boxes on top of one another to reach a bunch of bananas suspended from the ceiling. This is an example of which of the following behaviors?

    (A) Operant learning
    (B) Imprinting
    (C) Instinct
    (D) Insight

2. Viruses are obligate intracellular parasites, requiring their host cells for replication. Consequently, viruses generally attempt to reproduce as efficiently and quickly as possible in a host. Below is a graph depicting the initial growth pattern of a bacteriophage within a population of E. coli. This reproductive strategy is most similar to which of the following?

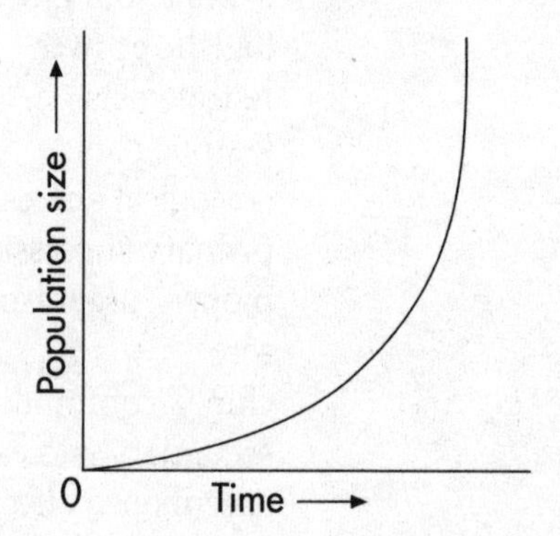

    (A) An r-strategist, because it aims to produce a large abundance of offspring to ensure survival
    (B) A k-strategist, because it aims to produce a large abundance of offspring to ensure survival
    (C) An r-strategist, because it is best suited to thrive in stable environments and over a long life-span
    (D) A k-strategist, because it is best suited to thrive in stable environments and over a long life-span

3. In a pond ecosystem, spring rains trigger an expansion of species at levels of the food chain. Runoff from nearby hills brings nutrients which when combine with warming temperatures trigger an algae bloom. The populations of small protozoans such as plankton expand by ingesting the algae. These plankton subsequently are consumed by small crustaceans such as crayfish, which ultimately become prey for fish such as catfish or carp. In this ecosystem, which of the following accurately describes the crayfish?

    (A) They are producers.
    (B) They are primary consumers.
    (C) They are secondary consumers.
    (D) They are tertiary consumers.

Questions 4 – 6 refer to the following table, chart and paragraph.

Table 1: Minimum pH Tolerance of Common Aquatic Organisms

Animal	pH Minimum
Brook Trout	4.9
American Bullfrog	3.8
Yellow Perch	4.6
Tiger Salamander	4.9
Crayfish	5.4
Snails	6.1
Clams	6.0

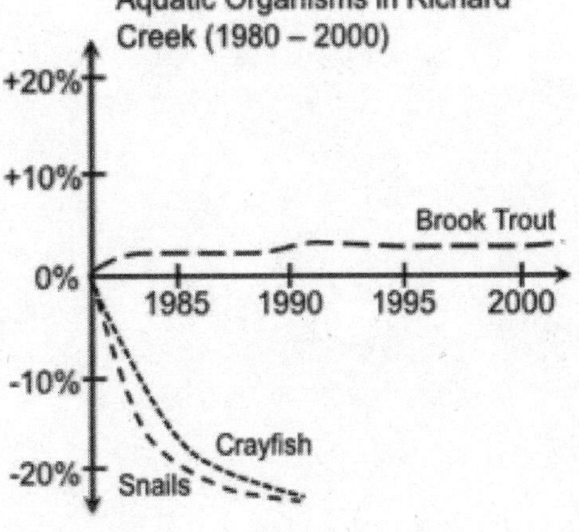

Over the period of 1980 to 2000, the average pH in Richard Creek changed drastically. An ecological survey was performed to evaluate the effect of detectable decreases in pH on the aquatic life of the creek. Four times a year, ecological surveys were performed to identify the amount of snails (a primary consumer), crayfish (a secondary consumer), and brook trout (a tertiary consumer) present at five different locations. The percent change relative to 1980 is shown in **Chart 1** above. Many aquatic organisms cannot live in low pH conditions. The minimum pH necessary for common aquatic organisms to sustain life is show in **Table 1**.

4. According to the data in Chart 1 and Table 1, the average pH in the creek is most nearly which of the following?

   (A) 5.9
   (B) 5.5
   (C) 5.1
   (D) 4.7

5. Which of the following is the most likely explanation for the abrupt decrease in pH in Richard Creek?

   (A) Greenhouse effect
   (B) Deforestation
   (C) Acid rain
   (D) Pollution

6. In order for snails to return to Richard Creek, the pH of the creek must exceed

   (A) 5.4
   (B) 5.7
   (C) 6.0
   (D) 6.1

13

Sample Free-Response Questions

THE ESSAYS

We discussed the essay portion of the test in Chapter 2. Let's see if you can write a good essay using your free-response techniques. Take 21 minutes to write a response to the sample essay.

1. All organisms need nutrients to survive. Angiosperms and vertebrates have each developed various methods to obtain nutrients from their environment.

   a. Discuss the ways angiosperms and vertebrates procure their nutrients.
   b. Discuss two structures used for obtaining nutrients among angiosperms. Relate structure to function.
   c. Discuss two examples of symbiotic relationships that have evolved between organisms to obtain nutrients.

GRADING CHECKLIST

To help you grade this sample essay, we've put together a checklist that you can use to calculate the number of points that should be assigned to each part of this question. We'll first explain the important points on the checklist, and then give you sample essays to show you how test reviewers would evaluate them.

ESSAY CHECKLIST

Part a: Types of Nutritional Requirements—4 Points Maximum

Angiosperms:
1 point—Autotrophs—if defined
1 point—They require H_2O, CO_2, and minerals.

Vertebrates:
1 point—Heterotrophs—if defined
1 point—They require organic compounds (sugars, proteins, fats), H_2O, vitamins, and minerals.

Part b: Structures in Angiosperms—4 Points Maximum

First two structures only

(1 point each) **Structure**	**(1 point each)** **Function described**
Cuticles	Prevent desiccation
Stomata	Pores in leaf that regulate gas exchange
Lenticels	Pores in stems that regulate gas exchange
Palisade mesophyll	Loosely packed cells in a leaf; contains chloroplasts
Roots	Structures that absorb water and minerals
Root hairs	Extensions of the root that increase the surface area for water absorption
	(1 point)

Carnivorous plants (such as Venus flytrap) have modified leaves to trap insects.

Part c: Symbiotic Relationships—4 Points Maximum

First 2 relationships only

(1 point each)	(1 point each)
Relationship	**Benefits to Each**
Nitrogen-fixing bacteria	Bacteria get food/Plants get usable nitrogen
Mycorrhizae	Fungi get food/Plants get water and minerals
Intestinal bacteria	Humans get Vitamin K/Bacteria have a place to live
Ruminants flora	Bacteria digest the cellulose for cows/Bacteria have a place to live
Wood-eating termites	Protists digest wood/Termites obtain the wood

EXPLANATION OF CHECKLIST

For Part a, you'll receive points if you demonstrate a clear understanding of the terms autotroph *and* heterotroph. You must define the terms and describe the necessary nutrients for both angiosperms and vertebrates. For angiosperms, you have to mention the raw materials for photosynthesis (CO_2 + H_2O) and minerals. For vertebrates, you have to mention organic compounds and additional vitamins or minerals.

For Part b, you have to relate structure to function. Which structures are involved in obtaining nutrients? You're asked to discuss *two* structures in angiosperms. One point is given for the structure and another point is given if you show a clear understanding of the function of the structure.

For Part c, you are asked to give *two* examples of symbiotic relationships. Since they ask for only two examples, you earn points only for the first two examples. An additional point is also given if you describe the benefits to each organism.

Notice that you don't have to know everything about biology to get a high score on the essay section. However, you do have to present the proper information with a certain level of detail. Additionally, you can earn a maximum of 4 points for each part. However, you can't earn more than 10 points total.

Before you use this checklist to assign points to your essay, let's evaluate two sample essays.

SAMPLE ESSAYS

Let's read an essay written by Joe Bloggs, an average student, which was given 4 points. (All of Joe's grammar and spelling mistakes were retained.)

JOE BLOGGS'S ESSAY

Angiosperms are autotrophs which make their own food. They have chloroplasts within the leaves which captures sunlight. Angiosperms also get nutrients through their roots which absorb water and bring it up the leaf. Food, water and minerals are then transport in the plant by xylem and phloem. Vertebrates can not make their own food so they must get their nutrients from other organisms. Some vertebrates only eat plants while other vertebrates eat both plants and animals.

Two structures that angiosperms use to obtain nutrients from the environment are leaves and roots. Plants have leaves which are shaped to absorb the maximum amount of sunlight. Plants also have root hairs which increase the surface area so that more water and minerals can be taken up the leaf.

An example of a symbiotic relationship is the bacteria found in the root nodules of plants. Plants need nitrogen to make plant proteins but they can't use the nitrogen from the atmosphere. Plants therefore work with nitrogen-fixing bacteria which convert nitrogen to nitrates. Both organisms benefit from this relationship.

Explanation

Joe Bloggs would probably get a 4 for this essay. For Part a, he clearly defined the term autotroph and received 1 point. However, he did not list the starting materials ($CO_2 + H_2O$) to make food in plants. He also failed to define heterotrophs. Consequently, he didn't receive any points for his explanation of heterotrophs. Notice that for this paragraph, he wasn't penalized for any grammar mistakes.

For Part b, Joe received 2 points for discussing root hairs and their function. Unfortunately, his explanation of leaves was not detailed enough, so he did not receive any additional points.

For Part c, Joe received 1 point for his example of a symbiotic relationship—nitrogen-fixing bacteria and plants. However, he did not specifically describe the benefits to *each* organism. Although he mentioned the benefits to plants, he neglected to discuss the benefits to bacteria. Had he completed his thought, he would have earned another point. Overall, although his essay was a little choppy, Joe was not penalized for it.

Now let's read an essay written by Josephine Bloggs that was given 9 points.

JOSEPHINE BLOGGS'S ESSAY

Organisms have developed a variety of ways to obtain nutrients. Angiosperms, which are flowering plants, are autotrophs. This means that they are capable of making their own food via photosynthesis. The reaction for photosynthesis is:

$$6CO_2 + 12H_2O \xrightarrow{\text{sunlight}} C_6H_{12}O_6 + 6O_2 + 6H_2O + ATP$$

Plants use carbon dioxide and water to produce glucose and ATP. Angiosperms also use their roots to absorb water and essential minerals from the soil.

Vertebrates, however, are heterotrophs. They are unable to make their own food. They must therefore rely on other organisms to obtain energy. They require organic compounds, such as proteins and carbohydrates, and break them down to simpler molecules that can be absorbed into the body. Vertebrates also require other essential substances such as vitamins, which serve as coenzymes in crucial biological reactions.

Two structural adaptations used by angiosperms to obtain nutrients are root hairs and stomates. Plants contain numerous root hairs which are tiny projections from a plant's root that increase the surface area for the uptake of water and minerals. Plants also have stomates which regulate the intake of CO_2. The carbon dioxide that enters the stomates can then be used to make glucose.

There are many examples of symbiotic relationships. A common example is nitrogen fixation. Certain plants, called legumes, require nitrogen to synthesize plant amino acids. These plants have nitrogen-fixing bacteria which live in their root nodules and convert nitrogen to nitrates. The plant uses the nitrates to make plant proteins. The bacteria, in turn, gets energy from the carbohydrates of the plant.

Another example of a symbiotic relationship is bacteria in the stomach of cows. Cows are herbivores that can not digest cellulose. They rely on bacteria in their stomach to digest some of the cellulose found in their stomach. Cows, therefore, get some of their energy from fermenting bacteria. A third example of symbiosis is the bacteria found in vertebrates, such as humans. Humans have bacteria that reside in their large intestine. These microorganisms break down undigested food particles and produce Vitamin K. The bacteria in turn have a place to live.

Explanation

This essay received 9 points. Josephine clearly demonstrated a mastery of the material. In Part a, she defined and discussed the terms autotroph and heterotroph. She also mentioned the important raw materials that are involved in each case. Notice that she presented and explained the equation for photosynthesis. She received 4 points—the maximum—for this section.

In Part b, Josephine presented two structures and their function. Her coverage of the material was fairly thorough. She talked about two important structures: root hairs and stomates. She received 2 points, one for mentioning each structure, and 1 point for her discussion of root hairs. She didn't mention the structure of stomates—that they're pores in the epidermis—so she did not receive a point for the function of stomates.

For Part c, Josephine presented three types of symbiotic relationships. The reader reviewed only the first two examples to assign points. Josephine received 2 points for her discussion of nitrogen fixation but only 1 point for her explanation of bacteria in the cow's stomach. In this case, she forgot to mention the benefits to the bacteria.

HOW TO USE OUR PRACTICE ESSAYS

The most important advice for this section of the test is *PRACTICE, PRACTICE, PRACTICE!* No matter how well you think you know the material, it's important to practice formulating your thoughts on paper. Before you take our practice tests in the back of the book, make sure you review our techniques for the essay portion. Then use our checklist—which is put together just like ETS's—to assign points.

To give you as much practice as possible, we've compiled a few sample AP biology questions. In fact, most of these questions refer to themes in biology that crop up time and time again. More often than not, you'll see one of these questions (or something similar) on your exam.

SAMPLE ESSAY QUESTIONS

1. Describe how the modern theory of evolution is supported by evidence from the following areas:

   a. Comparative anatomy
   b. Biogeography
   c. Embryology

2. Describe the chemical nature of genes. Discuss the replicative process of DNA in eukaryotic organisms. What are the various types of gene mutations that can occur during replication?

3. Describe the process of photosynthesis. What are the major plant pigments involved in photosynthesis? Design an experiment to measure the rate of photosynthesis.

SAMPLE SHORT-FORM FREE-RESPONSE QUESTIONS

1. Describe the structure of a generalized eukaryotic cell.

2. Discuss the following responses in plants and give one example of each.

   a. Photoperiodism
   b. Phototropism

3. Select one of the following pairs of hormones and discuss the concept of negative feedback.

   a. Thyroid-stimulating hormone (TSH) and thyroxine
   b. Parathyroid hormone and calcitonin
   c. Cortisol and ACTH

The biology you need to answer all of these essay questions is in this book. After you have completed your essays, go back to the chapters that cover these topics and check to see if you have presented all the relevant points. Use our Essay Checklist as a guideline to give yourself (or have someone give you) an approximate score. Good luck!

14

Laboratory

All AP biology courses have a laboratory component that gives students hands-on experience regarding some of the biology topics covered in class. Through these laboratory exercises you can learn the scientific method, lab techniques, and problem-solving skills. College Board suggest that you complete at least 8 of the following 13 lab experiences, in order to reinforce some of the most important topics in biology.

How do these labs relate for the AP biology exam? Approximately 10 percent of the questions on the multiple-choice section and one essay question will refer to certain aspects of these 13 lab exercises. ETS incorporates questions concerning the labs to determine two things: (1) how well you understand the key concepts in biology and (2) how well you think analytically. ETS wants to find out if you can design experiments, interpret data, and draw conclusions from these experiments.

There are basically two ways to approach a review of the laboratory exercises for this exam. We could review every single experiment, including the procedures and results. Not only would this take up more than half the book, it wouldn't necessarily improve your score.

We at The Princeton Review have a better approach. Since schools are not required to finish every single lab, we'd rather not overwhelm you with all the boring details about each and every experiment. Instead, we're going to tell you the objective of each lab. That way, even if you haven't completed all 13 labs, you will still be aware of the important concepts and results related to each experiment. How should you use this chapter to review for the AP Biology Exam? Read each summary and focus on the concepts stressed in each experiment. Now, let's get cracking!

LAB 1: ARTIFICIAL SELECTION

This lab explores artificial selection, the process by which humans decide traits to enhance or diminish in other species by crossing individuals with the desired phenotype. The principles that you need to understand are the basic principles of natural selection and how natural selection drives evolution, which are:

- Variation is present in any population.
- Natural selection, or differential reproduction in a population, is a major mechanism in evolution.
- Natural selection acts on phenotypic variations in populations. Some organisms will have traits that are more favorable to the environment and will survive to reproduce more than other individuals, causing the genetic makeup of the population to change over time.

This lab specifically deals with Wisconsin Fast Plants, and in order to artificially select for certain traits, plants with the desired traits can be crossed, changing the genetic makeup of the population. For example, if you wanted to select for height, you could cross only the tallest plants with one another. The new population will have a higher mean height than the previous generation.

LAB 2: MATHEMATICAL MODELING: HARDY-WEINBERG

In order to understand everything you need for the AP Biology exam, you need to be able to use the Hardy-Weinberg principles and equations to determine allele frequencies in a population. One way to study evolution is to study how the frequencies of alleles change from generation to generation.

- Know how to calculate the allele and genotype frequencies using the two Hardy-Weinberg equations: $p + q = 1$ and $p^2 + 2pq + q^2 = 1$. Don't forget: If the population obeys Hardy-Weinberg's ruled, these frequencies remain constant over time.
- Review the discussion of the Hardy-Weinberg principle in this book. Know the five conditions of the Hardy-Weinberg equilibrium: (1) large population, (2) no mutations, (3) no immigration or emigration, (4) random mating, and (5) no natural selection.
- Review natural selection and how it can lead to changes in the genetic makeup of a population.

LAB 3: COMPARING DNA SEQUENCES TO UNDERSTAND EVOLUTIONARY RELATIONSHIPS WITH BLAST

This laboratory uses BLAST (Basic Local Alignment Search Tool), a database that allows you to input a DNA sequence for a gene to look for similar or identical sequences present in other species. In order to understand everything you need for the AP Exam you need to be able to do the following:

- Be able to look at a phylogenic tree, which is a way to visually represent evolutionary relatedness. Endpoints of each branch correspond to a specific species and each junction on the tree represents a common ancestor. Species that are closer on a phylogenic tree are more closely related.
- Understand BLAST scoring. Most phylogenic trees are constructed by examining nucleotide sequences; the more identical two species sequences are for a specific gene, the more closely related they are. BLAST is able to analyze different sequences to tell you how similar they are to one another based on a score. The higher the score the closer the two sequences align.

LAB 4: DIFFUSION AND OSMOSIS

This lab investigates the process of diffusion and osmosis in a semipermeable membrane as well as the effects of solute concentration and water potential on these processes.

What are the general concepts you really need to know?

Fortunately, this lab covers the same concepts about diffusion and osmosis that are discussed in this book. Just remember that osmosis is the movement of water across a semipermeable membrane from a region of high water concentration to one of low water concentration from a hypotonic region (low solute concentration) to a hypertonic region (high solute concentration).

- Be familiar with the concept of osmotic potential, which is simply the free energy of water. It is a measure of the tendency of water to diffuse across a membrane. Water moves across a selectively permeable membrane from an area of higher osmotic potential to an area of lower water potential.
- Be familiar with the effects of water gain in animal and plant cells. In animals, the direction of osmosis depends on the concentration of solutes both inside and outside of the cell. In plants, osmosis is also influenced by turgor pressure—the pressure that develops as water presses against a cell wall. If a plant cell loses water, the cell will shrink away from the cell wall and plasmolyze.

Another important concept to understand is the importance of surface area and volume in cells. Cells maintain homeostasis by regulating the movement of solutes across the cell membrane. Small cells have a large surface area to volume ratio, however as cells become larger, this ratio becomes smaller, giving the cell relatively less surface area to exchange solutes. A cell is limited in size by the surface area to volume ratio. There are many organisms that have evolved strategies for increasing surface area, like root hairs on plants and villi in the small intestines of animals.

LAB 5: PHOTOSYNTHESIS

The chemical equation for photosynthesis is:

$$6CO_2 + 6H_2O \rightarrow 6CO_2 + C_6H_{12}O_6 + 6O_2$$

Because plants consume some of this energy during photosynthesis, measuring the oxygen produced by a plant can tell us about the net photosynthesis that is occurring. In this laboratory photosynthesis rates are measured using leaf discs that begin to float as photosynthesis is carried out, allowing you to see that photosynthesis is occurring.

There are several properties that affect the rates of photosynthesis including:

- Light intensity, color and direction
- Temperature
- Leaf color, size, type

Be able to hypothesize about the effects of these variables: for example, as light intensity increases, so does the rate of photosynthesis. Remember, plants and animals both contain mitochondria and carry out cell respiration!

LAB 6: CELLULAR RESPIRATION

In this lab the respiratory rate of germinating and nongerminating seeds, as well as small insects, is investigated. The equation for cellular respiration is:

$$C_6H_{12}O_6 + O_2 \rightarrow 6CO_2 + 6H_2O$$

Germinating seeds respire and need to consume oxygen in order to continue to grow. Non-germinating seeds do not respire actively. In this lab, the amount of oxygen consumed by these types of seeds was measured using a respirometer. The experiment was also conducted at two temperatures, 25°C and 10°C, because seeds consume more oxygen at higher temperatures. You should know the following concepts:

- Oxygen is consumed in cellular respiration.
- Germinating seeds have a higher respiratory rate than non-germinating seeds, which have a very low respiratory rate.
- Know how to design a study to determine the effect of temperature on cell respiration.
- Be able to explain the significance of a control using glass beads. A control is a condition held constant during an experiment. In this case glass beads are used as a control because they will not consume any oxygen.

LAB 7: MITOSIS AND MEIOSIS

This lab highlights the differences between mitosis and meiosis. In this lab, slides of onion root tips were prepared to study plant mitosis. The important information and skills to review in this lab are

- Mitosis produces two genetically identical cells, while meiosis produces haploid gametes.
- Cell division is highly regulated by checkpoints that depend, in part, on complexes of proteins called cyclins with other proteins called cyclin-dependent kinases. One such example is mitosis promoting factor (or MPF) that has its highest concentration during cell division and thought to usher a cell into mitosis.
- Nondisjunction, or the failure of chromosomes to separate correctly, can lead to an incorrect umber of chromosomes (too many or too few) in daughter cells.
- Know what each phase of the cell cycle looks like under a microscope (i.e. during metaphase the chromosomes are aligned on the metaphase plate.) Chapter 8 contains diagrams of each phase.

In one section of this lab, the sexual life cycle of the fungus Sordaria fimicola was examined. Sexual reproduction in this fungus involves the fusion of two nuclei: a (+) strain and a (-) strain- to form a diploid zygote. This zygote immediately undergoes meiosis to produe asci, which each contain eight haploid spores.

- During meiosis, crossing-over can occur to increase genetic variation. If crossing over, or recombination has occurred, different genetic combinations will be observed in the offspring when compared to the parent strain.
- These offspring with new genetic combinations are called recombinants. By examining the numbers of recombinants with the total number of offspring an estimate of the linkage map distance between two genes can be calculated using the following equation:

  Map distance (in map units) = [(# recombinants)/ (# total offspring)] × 100.

LAB 8: BIOTECHNOLOGY: BACTERIAL TRANSFORMATION

In this lab, the principles of genetic engineering are studied. Biotechnologists are able to insert genes into an organism's DNA in order to introduce new traits or phenotypes, like inserting genes into a corn genome that help the crops ward off pests. This process is very complicated in higher plants and animals, but relatively simple in bacteria. You are responsible for knowing the ways in which bacteria can accept fragments of foreign DNA.

- conjugation: the transfer of genetic material between bacteria via a pilus, a bridge between the two cells
- transformation: when bacteria take up foreign genetic material from the environment
- transduction: when a bacteriophage (a virus that infects bacteria) transfers genetic material from one bacteria to another

In addition, DNA can also be inserted into bacteria using plasmids, which are small circular DNA fragments that can serve as a vector to incorporate genes into the host's chromosome. Plasmids are key elements in genetic engineering. The concepts you need to know about plasmids are:

- One way to incorporate specific genes into a plasmid is to use restriction enzymes, which cut foreign DNA at specific sites, producing DNA fragments. A specific fragment can be mixed together with a plasmid, and this recombinant plasmid can then be taken up by *E. coli*.
- Plasmids can give a transformed cell selective advantage. For example, if a plasmid carries genes that confer resistance to an antibiotic like ampicillin, it can transfer these genes to the bacteria. These bacteria are then said to be transformed. That means if ampicillin is in the culture, only transformed cells will grow. This is a clever way scientists can find out which bacteria have taken up a plasmid.
- In order to make a bacteria cell take up a plasmid, you must (1) add $CaCl_2$, (2) heat shock the cells, and (3) incubate them in order to allow the plasmid to cross the plasma membrane.

LAB 9: BIOTECHNOLOGY: RESTRICTION ENZYME ANALYSIS OF DNA

This laboratory introduces you to the technique of electrophoresis. This technique is used in genetic engineering to separate and identify DNA fragments. You need to know the steps of this lab technique for the AP exam.

1. DNA is cut with various restriction enzymes.
2. The DNA fragments are loaded into wells on an agarose gel.
3. As electricity runs through the gel, the fragments move according to their molecular weights. DNA is negatively charges molecule, and therefore will migrate towards the positive electrode. The longer the DNA fragment, the slower it moves through the gel.
4. The distance of each fragment is recorded.

Restriction mapping allows scientists to distinguish between the DNA of different individuals. Since a restriction enzyme will only cut a specific DNA sequences, it will cause unique set of fragments for each individual called *restriction fragments length polymorphisms*, or RFLPs for short. This technology is used at crime scenes to help match DNA samples to suspects.

LAB 10: ENERGY DYNAMICS

This lab examines energy storage and transfer in ecosystems. Almost all organisms receive energy from the sun either directly or indirectly.

- Producers, or autotrophs are organisms that can make their own food using energy captured from the sun. They convert this into chemical energy that is stored in high-energy molecules like glucose.
- Consumers, or heterotrophs, must obtain their energy from organic molecules in their environments. They can then take this energy to make the organic molecules they need to survive.
- Biomass is measure of the mass of living matter in an environment. It can be used to estimate the energy present in an environment.

LAB 11: TRANSPIRATION

This lab investigates the mechanisms of transpiration, the movement of water from a plant to the atmosphere through evaporation. What do you need to take away from this lab?

- The special properties of water that allow it to move through a plant from the roots to the leaves include polarity, hydrogen bonding, cohesion, and adhesion
- The vascular tissues that are involved in transport in plants. Xylem transport water from roots to leaves, while phloem transport sugars made by photosynthesis in the leaves down to the stem and roots.
- Stomata are small pores present in leaves that allow CO_2 to enter for photosynthesis, and also are a major place that water can exits a plant through during transpiration.

LAB 12: FRUIT FLY BEHAVIOR

In this lab, fruit flies are given choices between two environments using a choice chamber, which allows fruit flies to move freely between the two. Typically fruit flies prefer an environment that supplies either food or a place to reproduce. They also respond to light and gravity.

- Taxis is the innate movement of an organism based on some sort of stimulus. Movement toward a stimulus is called positive taxis, while movement away from a stimulus is negative taxis. In this lab fruit flies exhibited a negative gravitaxis (or a movement opposite to the force of gravity, and positive phototaxis (a movement towards a light source).

LAB 13: ENZYME ACTIVITY

This lab demonstrated how an enzyme catalyzes a reaction, and what can influence rates of catalysis. In this lab the enzyme peroxidase was used to catalyze the conversion of hydrogen peroxide to water and oxygen.

$$H_2O_2 + \text{peroxidase} \rightarrow 2H_2O + O_2 + \text{peroxidase}$$

The following are the major concepts you need to understand for the AP exam:

- Enzymes are proteins that increase the rate of biological reactions. They accomplish this by lowering the activation energy of the reaction.
- Enzymes have active sites, which are pockets the substrates (reactants) can enter that are specific to one substrate or set of substrates.
- Enzymes have optimal temperature and pH ranges at which they catalyze reactions. Enzyme concentration and substrate concentration can also influence the rates of reaction.
- If you're asked to design an experiment to measure the effect of these four variables on enzyme activity, keep all the conditions constant except for the variable of interest. For example, to measure the effects of pH in an experiment, maintain the temperature, enzyme concentration, and substrate concentration as you change pH.

15

Answers and Explanations to the Chapter Review Questions

CHAPTER 3, THE CHEMISTRY OF LIFE

1. **C** Water has many unique properties, which favor life including a high specific heat (A), high surface tension, and cohesive properties (B), and high intermolecular forces due to hydrogen bonding. However, water is a very polar molecule and is a excellent solvent (making (C) inaccurate and the correct answer).

2. **B** The pH scale is logarithmically based, meaning that each difference of 1 on the log scale is indicative of a ten-fold difference in the hydrogen (H^+) ion concentration. According to the question, the pH values of the cytoplasm of cells and gastric juices are approximately 7.4 and 2.5, respectively. Therefore, the pH values vary by 5 and the hydrogen (H^+) ion concentrations much vary by 5 orders of ten (10^5) or 100,000-fold. Since lower pH values have more hydrogen ions, they are also more acidic.

3. **C** All amino acids share a carboxylic acid group (B; COOH), an amino-group (D; NH_2). a bound hydrogen atom to the central carbon (A). Differences in amino acids are defined by variation in the fourth position called the R group labeled C.

4. **B** The hypothesis that life may have arisen from formation of complex molecules from the primordial "soup" of Earth is not supported by the absence of nucleic acids. All life is DNA-based, yet no nucleic acid molecules were detected. The presence of complex carbon molecules (A), amino acids (C), and sugars (D), which are common compounds, which comprise life, supports the hypothesis.

5. **C** The amino acids cysteine and methionine contain the element sulfur (as indicated by the S in the amino acid structure shown). However, no sulfur-based compounds were included in the Miller-Urey experiment and therefore it was impossible to form these two amino acids under the conditions of the experiment.

6. **B** Silica is a mineral form of glass, is not a common component of life forms, and is largely chemically inert. Since oxygen is already present in several compounds included in the experiment, addition of this compound does not provide any additional elements or chemical substrates, which would permit generation of additional amino acids or synthesis of nucleic acids. Addition of sulfur compounds (A and C) and phosphorus (D) are necessary to generate some amino acids and all nucleic acids.

CHAPTER 4, CELLS

1. **C** Active transport is the movement of substances across a membrane against their concentration gradient through the use of energy (ATP). The sodium-potassium pump is a critical structure which uses ATP hydrolysis to move sodium and potassium ions against their respective concentration gradients. Diffusion of oxygen (A) and water (D) are examples of simple diffusion and can occur without the need of channels or pores. Movement of sodium ions by a voltage-gated ion channel (B) is an example of facilitated diffusion. Since the sodium ions are still undergoing diffusion from high to low concentrations without the need of ATP, this does not represent a form of active transport.

2. **C** Bacterial cells can be visualized using light microscopy. In fact, some of the earliest studies using primitive microscopes back in the 17th century recorded the shape and organization of bacteria. Viruses (A) and cell organelle structure (B and D) are too small to be observed using light microscopy and require electron microscopy.

3. **D** The cell wall is a structure that is present in bacteria, but absent in animal cells. Consequently, this structure is targeted by several leading classes of antibiotics and would be an effective target of therapeutics against *V. cholerae*. Flagella (A), plasma membrane (B), and ribosomes (C) are all structures that are present in animal cells.

4. **C** Based on the microscopy data, the organism has a cell wall and lacks mitochondria. The presence of a cell wall suggests that the organism is not a protozoan (B). The absence of mitochondria is indicative of prokaryotic structure since they lack organelles (eliminating A and D). The most likely conclusion is that this organism is a bacterial species.

5. **700** The graphed line of chlorophyll a has its highest peak at 700 nm.

CHAPTER 5, CELLULAR ENERGETICS

1. **B** The Krebs cycle mostly occurs in the matrix of the mitochondria. The inner membrane (A) and intermembrane space (C) are used in oxidative phosphorylation.

2. **D** Inhibitor Y is binding at a site outside of the active site and inducing a conformational change in the enzyme structure. By binding outside of the active site, it must be an allosteric inhibitor, (eliminating (A) and (C). Since the inhibitor is binding outside of the active site, it is not competing with the substrate for binding and therefore is considered a noncompetitive inhibitor.

3. **B** Enzymes are biological catalysts, which lower the activation energy (the energy threshold that must be met to proceed from reactant to product). The reaction coordinate diagram must reflect a decrease in the activation energy (eliminating C). Furthermore, the enzyme does not alter the energy of the reactants or products (eliminating A and D).

4. **B** Based on the pathway provided, consumption of 1 glucose and 2 ATP results in production of 4 ATP. In other words, each glucose results in a net gain of 2 ATP. Therefore, 2 glucose molecules would result in a net gain of 4 ATP.

5. **D** NAD^+ is a required cofactor for glycolysis and must be regenerated to permit glycolysis to continue to occur. In fermentation, pyruvic acid is converted into either ethanol or lactic acid. During this process, NADH is recycled into NAD^+.

6. **B** Glycolysis results in the production of ATP (energy) and therefore is considered an exergonic process.

CHAPTER 6, PHOTOSYNTHESIS

1. **C** Antenna pigments capture light and transfer its energy to the reaction centers where photosystems I and II exist.

2. **D** The light-independent reactions take CO_2 and NADPH and generate glucose. As a byproduct of this reaction, $NADP^+$ is generated which must be recycled into NADPH by the light dependent reactions. NAD^+ (B) is associated with mitochondrial activity. NADPH (C) and ATP (A) are generated by the light-dependent reactions.

3. **C** Carbon dioxide is used during the light-dependent reactions referred to as the Calvin cycle or C_3 pathway to generate glucose.

4. **B** Noncyclic photophosphorylation results in production of NADPH, which is subsequently used in the light-independent reactions to generate glucose. Cyclic photophosphorylation only produces ATP, not NADPH, and therefore is less efficient as a photophosphorylation process.

5. **D** Under conditions where ferredoxin is inhibited, photosystem II will still be active in the noncyclic photophosphorylation cascade. A byproduct of photosystem II is the production of free O_2 from the split of water using sunlight.

6. **A** A key fundamental difference between cyclic and noncyclic photophosphorylation is the photolysis of water. In cyclic photophosphorylation, water is not split and no NADPH is produced. However, in noncyclic photophosphorylation, photolysis occurs and water is split to release free O_2 gas.

CHAPTER 7, MOLECULAR GENETICS

1. **B** Okazaki fragments are generated during DNA replication when the DNA polymerase must create short DNA segments due to its requirement for 5′ to 3′ polymerization. Since the newly discovered yeast cell has 3′ to 5′ activity, there would be no lagging strand and likely no Okazaki fragments.

2. **C** Since the gene is much shorter than expected; a stop codon must have been introduced by mutagenesis. This is an example of a nonsense mutation.

3. **D** The order for DNA replication is helicase, RNA primase, DNA polymerase, and ligase.

4. **C** If an mRNA codon is UAC, the complementary segment on a tRNA anticodon is AUG.

5. **D** During post-translational modification, the polypeptide undergoes a conformational change. Intron excision (A) and poly(A) addition (B) are examples of post-transcriptional modifications, and formation of peptide bonds occurs during translation, not afterwards.

6. **B** If 21 nucleotides comprise a sequence and 3 nucleotides comprise each codon, there would be 7 codons and thus a maximum of 7 amino acids.

CHAPTER 8, CELL REPRODUCTION

1. **C** During anaphase, the chromatids are separated by shortening of the spindle fibers. Chemically blocking the shortening of these fibers would arrest the cell in metaphase. The cells are arrested in metaphase as indicated by the alignment of the chromosomes in the center of the cell and their attachment to spindle fibers, eliminating (A) and (D). The next phase (anaphase) involves separation of the chromatids during shortening of the spindle fibers. Therefore, the cells are not arrested due to dissociation of the kinetochore from the spindle fibers (eliminating B).

2. **A** Crossing-over occurs during prophase I and constitutes the key step which creates exchange of alleles during meiosis.

3. **C** Lactose is facilitating transcription by removing a molecule which blocks RNA polymerase activity; therefore it is an inducer. The molecular target of lactose is a repressor, which prevents the RNA polymerase from reading through a gene and synthesizing RNA.

4. **C** The synthesis or S phase of the cell cycle represents the step where the genetic material is duplicated. The only phase labeled in the experiment that represents an increase is phase B. Based on the time scale on the *x*-axis, this phase lasts approximately 30 min.

5. **D** Anaphase represents the cell division stage of the cell cycle and would be the phase that occurs right before the amount of genetic material should decrease. Phase D is the phase right before the genetic material would drop, and thus (D) is the correct answer.

6. **52** The organism has 13 chromosomes in gametic cells, which are haploid. Therefore, a diploid cell would normally have two copies of each chromosome or 26 chromosomes. During S phase, the cell must make a duplicate copy of all of its chromosomes, therefore the cell would contain 52 chromosomes.

CHAPTER 9, HEREDITY

1. **D** The father and mother both have an AB blood type. Since neither parent has a recessive allele, it is impossible for their child to be an O blood type.

2. **B** When the phenotype associated with two traits is mixed, this is considered an example of incomplete dominance. In this case, neither red nor blue color was dominant, and the resulting progeny exhibited a mixture of the traits (purple).

3. **B** Crossing the pea plant that is heterozygous for both traits (TtGg) with a plant that is recessive for both traits (ttgg) results in the following possible combinations, each of which should occur 25% of the time: TtGg (Tall and Green), Ttgg (Tall and Yellow), ttGg (Short and Green), and ttgg (Short and Yellow).

4. **C** Since the woman is a carrier, she must have one normal copy of the X chromosome and one diseased copy. Since the boy will receive an X chromosome from his mother, there is a 50% chance that he will receive a diseased copy. Since he doesn't have a second X chromosome, he must have the disease if he receives the diseased X chromosome.

5. **D** They both must have a normal copy of the X chromosome. It is possible that the woman may be affected with hemophilia, however that scenario is extremely unlikely since such a case would require two diseased copies of the X chromosome.

6. **D** Essentially, transmission of hemophilia to a girl born with Turner Syndrome would be very similar to the conditions by which a boy would receive the disease. Both boys and girls with Turner Syndrome only have one copy of the X chromosome. Therefore, they both would have a 50% chance of receiving the diseased copy of the X chromosome.

CHAPTER 10, ANIMAL STRUCTURE AND FUNCTION

DIGESTIVE SYSTEM

1. **C** Trypsin is an enzyme that used for the breakdown of proteins. Lacteals (A), Bile (B) and pancreatic lipase (D) are all involved in the digestion and absorption of lipids.

2. **C** Food that is swallowed and reaches the stomach is referred to as a bolus. Once digested in the stomach and mixed with gastric juices, food is then referred to as chyme. Feces (B) are the waste products generated after food leaves the large intestine and starches (A) are predigested foods.

3. **D** Colonic bacteria play no major role in the absorption of water and salts. They do however provide critical vitamins (such as Vitamin K), help break down food products (B) and limit the growth of potentially pathogenic bacteria (C).

4. **A** Carbohydrates begin digestion in the mouth following treatment with salivary amylase in the saliva. Some lipids also begin digestion in the mouth, though they require more complex digestive processes.

RESPIRATORY SYSTEM

1. **A** Air passes from the nose through the pharynx (throat), larynx (voicebox), trachea, bronchi, bronchioles, and lastly alveoli.

2. **A** Oxygen is predominately carried within erythrocytes (red blood cells) attached to hemoglobin molecules. A small fraction is also dissolved in the plasma of the blood.

3. **C** Carbon dioxide (CO_2) gas is combined with water to form carbonic acid (H_2CO_3) which dissolves to form bicarbonate ions (HCO_3^-). This is the primary mechanism by which carbon dioxide is carried in the blood. A small fraction is also carried on hemoglobin in erythrocytes and dissolved directly in the plasma of the blood.

4. **D** Immediately prior to reaching the lungs, blood is oxygen poor and carbon dioxide rich. The purpose of trafficking blood to the lungs is to exchange carbon dioxide, which is a waste product, for oxygen, which is necessary for cellular respiration.

CIRCULATORY SYSTEM

1. **D** The patient may receive blood from any blood group. Individuals with AB blood have both A and B antigens on their erythrocytes. Consequently, they have no antibodies against either antigen and would therefore not mount an immune response against blood type antigen.

2. **D** The right ventricle pumps blood via the pulmonary artery to the lungs to be oxygenated. A myocardial infarction leading to damage in the right ventricle would result in poor circulation to the lungs.

3. **A** All blood cells originate in the bone marrow. Leukocytes have nuclei and erythrocytes do not (B). Erythrocytes carry oxygen bound to hemoglobin, leukocytes do not (C). Leukocytes protect the body from foreign antigens, erythrocytes do not (D).

4. **B** Electrical impulses in the heart originate at the SA node, (also referred to as the pacemaker node) and travel to the AV node, followed by the Purkinje fibers, and lastly bundle of His.

Lymphatic and Immune Systems

1. **B** Lymph nodes harbor large quantities of lymphocytes. During active infections or cancer, mass expansion of lymphocytes causes enlargement and swelling of lymph nodes. This is why doctors often feel the areas around lymph nodes nearest to where a suspected infection is.

2. **C** The placebo strain was included as a control for the experiment. Since no active antibodies should be produced to the placebo strain, the scientists would be able to compare antibodies elicited specifically to the vaccine strain.

3. **B** B cells (or plasma cells) are responsible for creating antibodies. These are the primary cells associated with lasting memory to vaccinations.

4. **A** The most practical way to increase statistical significance is to increase the sample size. More control variables (B) will only ensure that they haven't introduced bias. Repeating a third challenge (C), would evaluate the length of potential protection, not the extent. Using a different statistical test (D) would simply manipulate the data and not provide true significance.

Excretory System

1. **C** Normal kidney function should not result in proteins being leaked into the urine. A common test for kidney failure is to look for proteins or cells in urine. Urea (A), water (B), and salts (D) are all common components of urine.

2. **D** The patient needs to take a drug which will result in increased urination. Only (D) describes a functional response that results in release of water. By inhibiting antidiuretic hormone (ADH), less water is reabsorbed meaning that more water is being excreted.

3. **C** Urea is a nitrogenous byproduct of protein metabolism. It must be excreted because it nitrogen compounds such as ammonia are toxic in high quantities in the blood.

4. **D** The glomeruli are the capillary beds within Bowman's capsules. Destruction of these structures will result in direct injury of the glomerulus.

Nervous System

1. **B** The membrane potential depolarizes at -50 mV so that the voltage-gated sodium channels can open. Note that the sodium channels open and close more quickly than the potassium channels.

2. **C** The blocking of voltage-gated sodium (Na^+) channels by tetrodotoxin would prevent an action potential in cells and therefore, the transmission of a neural response to the subsequent neurons.

3. **A** Upon seeing the bear charging towards her, Jane's sympathetic "fight or flight" response would be activated. Since this pathway is not voluntary, it is considered part of her autonomic nervous system.

4. **B** The cerebellum is responsible for coordinating the complex balance and motions of the body. Since all of the patient's muscles appear to be functionally active, the impairment would likely be limited to coordination rather than physical inhibition.

Musculoskeletal System

1. **C** During contraction, the Z lines move closer to one another as myosin and actin filaments overlap and coordinate a contraction. The H zone will shorten as the Z lines move closer. There would be no apparent difference in the length of the Z line (A), A band (B), or actin filaments (D).

2. **B** Cardiac and smooth muscles are not under voluntary control. However, you are unable to control motion of your skeletal muscles. Skeletal and cardiac muscle are striated (smooth is smooth).

3. **B** Muscle cells have a unique organelle component called the sarcoplasmic reticulum which harbors calcium ions (Ca^{2+}) until activated by a neuron for a muscular response.

4. **A** Tendons attach muscles to bones (recall that Achilles tendon which connects the calf muscles to the base of the foot) and ligaments attach bones to bones (an example is the ACL which holds the bones of your knee together).

Endocrine System

1. **D** Increased levels of vasopressin would not likely be observed since it is secreted by the posterior pituitary. ACH (A), glucocorticoids (B), and thyroxine (C) production would all be triggered by hormones produced by the anterior pituitary.

2. **D** Calcitonin and parathyroid hormone (PTH) have antagonistic effects on blood calcium concentrations. Calcitonin lowers blood calcium levels, while PTH increases blood calcium levels.

3. **B** Thyroid stimulating hormone (TSH) is produced by the anterior pituitary gland to increase production of thyroid hormones.

4. **D** The hypothalamus is the part of the brain that regulates the pituitary gland and produces neurosecretory hormones. TSH and LH are hormones secreted by the anterior pituitary and the hypothalamus is not an extension of the pituitary gland.

Reproductive System and Embryonic Development

1. **B** The ectoderm will ultimately become the epithelial tissues and neural components of the organism. The GI tract will be associated with the endoderm and the heart and pancreas will be associated with the mesoderm.

2. **B** Following ovulation, the corpus luteum is responsible for secreting progesterone to protect and enhance the endometrium for receiving a fertilized cell.

3. **C** The secretions of the prostate gland are alkaline and neutralize the acidic conditions of the vagina and uterus to prevent rapid death of sperm in the low pH conditions.

CHAPTER 11, EVOLUTION

1. **D** Mammals and cephalopods developed similar eye structures independently due to similar selective pressures. This is an example of convergent evolution.

2. **B** The data provided show a transition towards one extreme (black) and against another (white). This is an example of directional selection.

3. **B** Based on the data, the amount of white-bodied peppered moths decreased between 1802 and 1902 and the amount of black-bodied peppered moths increased during the same period.

4. **B** Longtail moths were included in the experiment as a control to compare the effects that are not associated with color.

5. **B** If the color of ash or soot produced by the Industrial Revolution were white or light gray, this would likely reverse the trend observe applying additional selection against the black moths.

6. **0.43** The frequency can be calculated as shown below:

$$p + q = 1$$

$$p + (0.3) = 1$$

$$p = 0.7$$

$$2pq = \text{Heterozygotes} = 2(0.7)(0.3) = 0.43$$

CHAPTER 12, BEHAVIOR AND ECOLOGY

1. **D** The behavior displayed by the chimpanzee represents insight because the chimpanzee has figured how to solve the problem without external influence or learning.

2. **A** Viruses would display reproductive strategies most similar to r-strategists since they aim to reproduce as fast as possible and create as many progeny as possible to increase their odds to transmission to other hosts.

3. **C** They would be considered secondary consumers since they consume the primary consumers (plankton) which consume algae (the producers).

4. **C** Since brook trout can tolerate pH values as low as 4.9 and do not appear to diminish, the pH of the creek must exceed 4.9, so we can eliminate (D). Since crayfish cannot tolerate pH levels lower than 5.4, the pH of the stream must have dropped lower than this value. Only (C) falls in this range.

5. **C** Only acid rain would directly explain why the pH would drop in the creek over the time period.

6. **D** Based on Table 1, the pH must exceed 6.1 for snails to be able to return to the creek ecosystem.

PART V

PRACTICE TESTS

16

Practice Test 1

AP® Biology Exam

SECTION I: Multiple-Choice Questions

DO NOT OPEN THIS BOOKLET UNTIL YOU ARE TOLD TO DO SO.

At a Glance

Total Time
1 hour and 30 minutes

Number of Questions
69

Writing Instrument
Pencil required

Instructions

Section I of this examination contains 69 multiple-choice questions. These are broken down into Part A (63 multiple-choice questions) and Part B (6 grid-in questions).

Indicate all of your answers to the multiple-choice questions on the answer sheet. No credit will be given for anything written in this exam booklet, but you may use the booklet for notes or scratch work. After you have decided which of the suggested answers is best, completely fill in the corresponding oval on the answer sheet. Give only one answer to each question. If you change an answer, be sure that the previous mark is erased completely. Here is a sample question and answer.

Sample Question	Sample Answer
Chicago is a	Ⓐ ● Ⓒ Ⓓ

(A) state
(B) city
(C) country
(D) continent

Use your time effectively, working as quickly as you can without losing accuracy. Do not spend too much time on any one question. Go on to other questions and come back to the ones you have not answered if you have time. It is not expected that everyone will know the answers to all the multiple-choice questions.

About Guessing

Many candidates wonder whether or not to guess the answers to questions about which they are not certain. Multiple choice scores are based on the number of questions answered correctly. Points are not deducted for incorrect answers, and no points are awarded for unanswered questions. Because points are not deducted for incorrect answers, you are encouraged to answer all multiple-choice questions. On any questions you do not know the answer to, you should eliminate as many choices as you can, and then select the best answer among the remaining choices.

This page intentionally left blank.

GO ON TO THE NEXT PAGE.

BIOLOGY
SECTION I
Time—1 hour and 30 minutes

Directions: Each of the questions or incomplete statements below is followed by four suggested answers or completions. Select the one that is best in each case and then fill in the corresponding oval on the answer sheet.

1. The resting membrane potential depends on which of the following?
   I. Active transport
   II. Selective permeability
   III. Differential distribution of ions across the axonal membrane

   (A) III only
   (B) I and II only
   (C) II and III only
   (D) I, II, and III

2. The Krebs cycle in humans occurs in the
   (A) mitochondrial matrix
   (B) inner mitochondrial membrane
   (C) outer mitochondrial membrane
   (D) intermembrane space

3. A heterotroph
   (A) obtains its energy from sunlight, harnessed by pigments
   (B) obtains its energy by oxidizing organic molecules
   (C) makes organic molecules from CO_2
   (D) obtains its energy by consuming exclusively autotrophs

4. Regarding meiosis and mitosis, one difference between the two forms of cellular reproduction is that in meiosis
   (A) there is one round of cell division, whereas in mitosis there are two rounds of cell division
   (B) separation of sister chromatids occurs during the second division, whereas in mitosis separation of sister chromatids occurs during the first division
   (C) chromosomes are replicated during interphase, whereas in mitosis chromosomes are replicated during prophase
   (D) spindle fibers form during prophase, whereas in mitosis the spindle fibers form during metaphase

5. A feature of amino acids NOT found in carbohydrates is the presence of
   (A) carbon atoms
   (B) oxygen atoms
   (C) nitrogen atoms
   (D) hydrogen atoms

6. Which of the following is NOT a characteristic of bacteria?
   (A) Circular double-stranded DNA
   (B) Membrane-bound cellular organelles
   (C) Plasma membrane consisting of lipids and proteins
   (D) Ribosomes that synthesize polypeptides

GO ON TO THE NEXT PAGE.

7. Which of the following best explains why a population is described as the evolutionary unit?
   (A) Genetic changes can only occur at the population level.
   (B) The gene pool in a population remains fixed over time.
   (C) Natural selection affects individuals, not populations.
   (D) Individuals cannot evolve, but populations can.

8. The endocrine system maintains homeostasis using many feedback mechanisms. Which of the following is an example of positive feedback?
   (A) Infant suckling causes a mother's brain to release oxytocin, which in turn stimulates milk production
   (B) An enzyme is allosterically inhibited by the product of the reaction it catalyzes
   (C) When ATP is abundant the rate of glycolysis decreases
   (D) When blood sugar levels decrease to normal after a meal, insulin is no longer secreted

9. A scientist carries out a cross between two guinea pigs, both of which have black coats. Black hair coat is dominant over white hair coat. Three quarters of the offspring have black coats, and one quarter have white coats. The genotypes of the parents were most likely
   (A) bb × bb
   (B) Bb × Bb
   (C) Bb × bb
   (D) BB × Bb

10. A large island is devastated by a volcanic eruption. Most of the horses die except for the heaviest males and heaviest females of the group. They survive, reproduce, and perpetuate the population. Since weight is highly heritable and the distribution of weights approximates a binomial distribution, the offspring of the next generation would be expected to have
    (A) a higher mean weight compared with their parents
    (B) a lower mean weight compared with their parents
    (C) the same mean weight as members of the original population
    (D) a higher mean weight compared with members of the original population

11. All of the following play a role in morphogenesis EXCEPT
    (A) apoptosis
    (B) homeotic genes
    (C) operons
    (D) inductive effects

12. During the period when life is believed to have begun, the atmosphere on primitive Earth contained abundant amounts of all the following gases EXCEPT
    (A) oxygen
    (B) hydrogen
    (C) ammonia
    (D) methane

13. Villi and microvilli are present in the small intestine and aid in reabsorption by
    (A) increasing the surface area of the small intestine
    (B) decreasing the surface area of the small intestine
    (C) making the small intestine more hydrophilic
    (D) making the small intestine more hydrophobic

14. Which of the following does NOT take place in the small intestine?
    (A) Pancreatic lipase breaks down fats to fatty acids and glycerol.
    (B) Pepsin breaks down proteins to amino acids.
    (C) Pancreatic amylase breaks down carbohydrates into simple sugars.
    (D) Bile emulsifies fats into smaller fat particles.

15. In animal cells, which of the following represents the most likely pathway that a secretory protein takes as it is synthesized in a cell?
    (A) Plasma membrane–Golgi apparatus–ribosome–secretory vesicle–rough ER
    (B) Ribosome–Golgi apparatus–rough ER–secretory vesicle–plasma membrane
    (C) Plasma membrane–Golgi apparatus–ribosome–secretory vesicle–rough ER
    (D) Ribosome–rough ER–Golgi apparatus-secretory vesicle–plasma membrane

GO ON TO THE NEXT PAGE.

16. All of the following statements are correct regarding alleles EXCEPT:
 (A) Alleles are alternative forms of the same gene.
 (B) Alleles are found on corresponding loci of homologous chromosomes.
 (C) A gene can have more than two alleles.
 (D) An individual with two identical alleles is said to be heterozygous with respect to that gene.

17. Once a plasmid has incorporated specific genes, such as the gene coding for the antibiotic ampicillin, into its genome, the plasmid may be cloned by
 (A) inserting it into a virus to generate multiple copies
 (B) treating it with a restriction enzyme in order to cut the molecule into small pieces
 (C) inserting it into a suitable bacterium in order to produce multiple copies
 (D) running it on a gel electrophoresis in order to determine the size of the gene of interest

18. Although mutations occur at a regular and predictable rate, which of the following statements is the LEAST likely reason the frequency of mutation appears to be low?
 (A) Some mutations produce alleles that are recessive and may not be expressed.
 (B) Some undesirable phenotypic traits may be prevented from reproducing.
 (C) Some mutations cause such drastic phenotypic changes that they are removed from the gene pool.
 (D) The predictable rate of mutation results in ongoing variability in a gene pool.

19. Which of the following adaptive features would most likely be found in an animal living in a hot arid environment?
 (A) Long loops of henle to maximize water reabsorption
 (B) Storage of water in fatty tissues
 (C) Large ears to aid in heat dispersion
 (D) Short loops of henle to maximize water secretion

20. Which of the following best accounts for the ability of legumes to grow well in nitrogen-poor soils?
 (A) These plants make their own proteins.
 (B) These plants have a mutualistic relationship with nitrogen-fixing bacteria.
 (C) These plants are capable of directly converting nitrogen gas into nitrates.
 (D) These plants do not require nitrogen to make plant proteins.

21. Which of the following is most correct concerning cell differentiation in vertebrates?
 (A) Cells in different tissues contain different sets of genes, leading to structural and functional differences
 (B) Differences in the timing and expression levels of different genes leads to structural and functional differences
 (C) Differences in the reading frame of mRNA leads to structural and functional differences
 (D) Differences between tissues result from spontaneous morphogenesis

Question 22 refers to the diagram below.

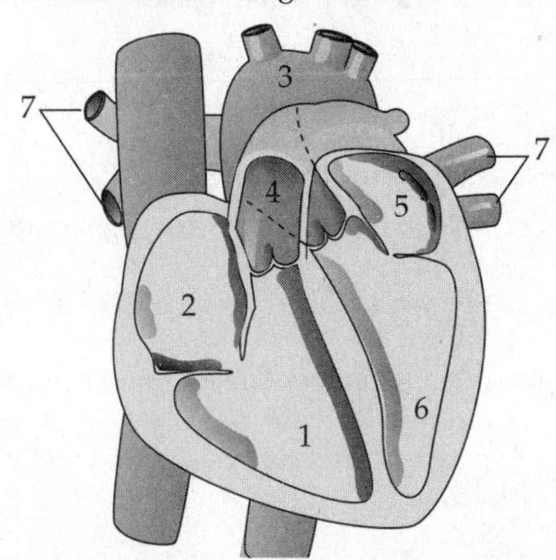

22. Which of the following chambers or vessels carry deoxygenated blood in the human heart?
 (A) 4 only
 (B) 1 and 2 only
 (C) 5 only
 (D) 1, 2 and 4

GO ON TO THE NEXT PAGE.

23. In chick embryos, the extraembryonic membrane that provides nourishment to the fetus is the

(A) amnion
(B) chorion
(C) placenta
(D) egg yolk

24. Some strains of viruses can change normal mammalian cells into cancer cells in vitro. This transformation of the mammalian cell is usually associated with the

(A) formation of a pilus between the mammalian cell and the virus
(B) incorporation of the viral genome into the mammalian cell's nuclear DNA
(C) conversion of the host's genome into the viral DNA
(D) release of spores into the mammalian cell

25. The major difference between cartilage and bone is that cartilage

(A) is a part of the skeletal system
(B) is composed of collagen and salts
(C) lacks blood vessels and nerves
(D) secretes a matrix

26. All of the following are examples of events that can prevent interspecific breeding EXCEPT:

(A) The potential mates experience geographic isolation.
(B) The potential mates experience behavioral isolation.
(C) The potential mates have different courtship rituals.
(D) The potential mates have similar breeding seasons.

27. Which of the following is NOT a characteristic of asexual reproduction in animals?

(A) Daughter cells have the same number of chromosomes as the parent cell.
(B) Daughter cells are identical to the parent cell.
(C) The parent cell produces diploid cells.
(D) The daughter cells fuse to form a zygote.

28. Which of the following is the correct characteristic of arteries?

(A) They are thin-walled blood vessels.
(B) They contain valves that prevent backflow.
(C) They always carry oxygenated blood.
(D) They carry blood away from the heart.

29. Transpiration is a result of special properties of water. The special properties of water include all of the following EXCEPT

(A) cohesion
(B) adhesion
(C) capillary action
(D) hydrophobicity

30. Crossing-over occurs during which of the following phases in meiosis?

(A) Prophase I
(B) Metaphase I
(C) Anaphase I
(D) Prophase II

31. Which of the following about meiosis is NOT true?

(A) Meiosis produces two haploid gametes.
(B) Homologous chromosomes join during synapsis.
(C) Sister chromatids separate during meiosis I.
(D) Crossing over increases genetic variation in gametes.

32. A plant grows in the opposite direction of the gravitational force. This is an example of

(A) positive thignotropism
(B) negative phototropism
(C) positive phototropism
(D) negative gravitropism

GO ON TO THE NEXT PAGE.

33. In most ecosystems, net primary productivity is important because it represents the
    (A) energy available to producers
    (B) total solar energy converted to chemical energy by producers
    (C) biomass of all producers
    (D) energy available to heterotrophs

34. Hawkmoths are insects that are similar in appearance and behavior to hummingbirds. Which of the following is LEAST valid?
    (A) These organisms are examples of convergent evolution.
    (B) These organisms were subjected to similar environmental conditions.
    (C) These organisms are genetically related to each other.
    (D) These organisms have analogous structures.

35. Which of the following describes a symbiotic relationship?
    (A) A tapeworm feeds off of its host's nutrients causing the host to lose large amounts of weight.
    (B) Certain plants grow on trees in order to gain access to sunlight, not affecting the tree.
    (C) Remora fish eat parasites off of sharks; the sharks stay free of parasites and the remora fish are protected from predators.
    (D) Meerkats sound alarm calls to warn other meerkats of predators.

36. Destruction of all beta cells in the pancreas will cause which of the following to occur?
    (A) Glucagon secretion will stop and blood glucose levels will increase.
    (B) Glucagon secretion will stop and blood glucose levels will decrease.
    (C) Glucagon secretion will stop and digestive enzymes will be secreted.
    (D) Insulin secretion will stop and blood glucose levels will increase.

37. All of the following are stimulated by the sympathetic nervous system EXCEPT
    (A) dilation of the pupil of the eye
    (B) constriction of blood vessels
    (C) increased secretion of the sweat glands
    (D) increased peristalsis in the gastrointestinal tract

38. The calypso orchid, *Calypso bulbosa*, grows in close association with mycorrhizae fungi. The fungi penetrate the roots of the flower and take advantage of the plant's food resources. The fungi concentrate rare minerals, such as phosphates, in the roots and make them readily accessible to the orchid. This situation is an example of
    (A) parasitism
    (B) commensalism
    (C) mutualism
    (D) endosymbiosis

39. Which of the following are characteristics of both bacteria and fungi?
    (A) Cell wall, DNA, and plasma membrane
    (B) Nucleus, organelles, and unicellularity
    (C) Plasma membrane, multicellularity, and Golgi apparatus
    (D) Cell wall, unicellularity, and mitochondria

40. A sustained decrease in circulating Ca^{2+} levels might be caused by decreased levels of which of the following substances?
    (A) Growth hormone
    (B) Parathyroid hormone
    (C) Thyroid hormone
    (D) Calcitonin

41. The synthesis of new proteins necessary for lactose utilization by the bacterium *E. coli* using the *lac* operon is regulated
    (A) by the synthesis of additional ribosomes
    (B) at the transcription stage
    (C) at the translation stage
    (D) by differential replication of the DNA that codes for lactose-utilizing mechanisms

42. Which of the following statements about trypsin is NOT true?
    (A) It is an organic compound made of proteins.
    (B) It is a catalyst that alters the rate of a reaction.
    (C) It is operative over a wide pH range.
    (D) The rate of catalysis is affected by the concentration of substrate.

GO ON TO THE NEXT PAGE.

43. In DNA replication, which of the following does NOT occur?

(A) Helicase unwinds the double helix.
(B) DNA ligase links the Okazaki fragments.
(C) RNA polymerase is used to elongate both chains of the helix.
(D) DNA strands grow in the 5' to 3' direction.

44. A change in a neuron membrane potential from +50 millivolts to –70 millivolts is considered

(A) depolarization
(B) repolarization
(C) hyperpolarization
(D) an action potential

45. The energy given up by electrons as they move through the electron transport chain is used to

(A) break down glucose
(B) make glucose
(C) produce ATP
(D) make NADH

46. If a photosynthesizing plant began to release $^{18}O_2$ instead of normal oxygen, one could most reasonably conclude that the plant had been supplied with

(A) H_2O containing radioactive oxygen
(B) CO_2 containing radioactive oxygen
(C) $C_6H_{12}O_6$ containing radioactive oxygen
(D) NO_2 containing radioactive oxygen

47. All of the following statements describe the unique characteristics of water EXCEPT:

(A) It is a polar solvent.
(B) It forms hydrogen bonds with disaccharides.
(C) It can dissociate into hydrogen ions and hydroxide ions.
(D) It is a hydrophobic solvent.

48. Chemical substances released by organisms that elicit a physiological or behavioral response in other members of the same species are known as

(A) auxins
(B) hormones
(C) pheromones
(D) enzymes

49. Homologous structures are often cited as evidence for the process of natural selection. All of the following are examples of homologous structures EXCEPT

(A) the wings of a bird and the wings of a bat
(B) the flippers of a whale and the arms of a man
(C) the pectoral fins of a porpoise and the flippers of a seal
(D) the forelegs of an insect and the forelimbs of a dog

50. The sliding action in the myofibril of skeletal muscle contraction requires which of the following?

I. Ca^{2+}

II. ATP

III. actin

(A) I only
(B) II only
(C) II and III
(D) I, II, and III

51. Certain populations of finches have long been isolated on the Galapagos Islands off the western coast of South America. Compared with the larger stock population of mainland finches, these separate populations exhibit far greater variation over a wider range of species. The variation among these numerous finch species is the result of

(A) convergent evolution
(B) divergent evolution
(C) disruptive selection
(D) stabilizing selection

52. Which of the following contributes the MOST to genetic variability in a population?

(A) Sporulation
(B) Binary fission
(C) Vegetative propagation
(D) Mutation

GO ON TO THE NEXT PAGE.

Directions: Each group of questions below concerns an experimental or laboratory situation or data. In each case, first study the description of the situation or data. Then choose the one best answer to each question following it and fill in the corresponding oval on the answer sheet.

Questions 53–55 refer to the following information and table.

A marine ecosystem was sampled in order to determine its food chain. The results of the study are shown below.

Type of Organism	**Number of Organisms**
Shark	2
Small crustaceans	400
Mackerel	20
Phytoplankton	1000
Herring	100

53. Which of the following organisms in this population are secondary consumers?

(A) Sharks
(B) Mackerels
(C) Herrings
(D) Small crusteceans

54. Which of the following organisms has the largest biomass in this food chain?

(A) Phytoplanktons
(B) Mackerels
(C) Herrings
(D) Sharks

55. If the herring population is reduced by predation, which of the following is most likely to occur in this aquatic ecosystem?

(A) The mackerels will be the largest predator in the ecosystem.
(B) The small crustacean population will be greatly reduced.
(C) The plankton population will be reduced over the next year.
(D) The small crustaceans will become extinct.

GO ON TO THE NEXT PAGE.

Questions 56–58 refer to the following information and diagram.

To understand the workings of neurons, an experiment was conducted to study the neural pathway of a reflex arc in frogs. A diagram of a reflex arc is given below.

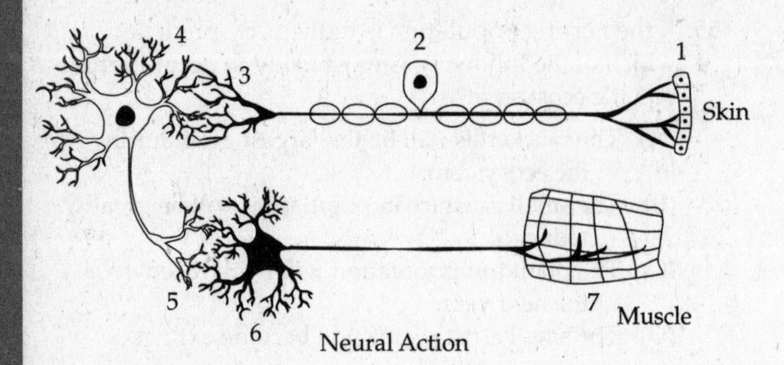

56. Which of the following represents the correct pathway taken by a nerve impulse as it travels from the spinal cord to effector cells?
(A) 1-2-3-4
(B) 6-5-4-3
(C) 2-3-4-5
(D) 4-5-6-7

57. The brain of the frog is destroyed. A piece of acid-soaked paper is applied to the frog's skin. Every time the piece of paper is placed on its skin, one leg moves upward. Which of the following conclusions is best supported by the experiment?
(A) Reflex actions are not automatic.
(B) Some reflex actions can be inhibited or facilitated.
(C) All behaviors in frogs are primarily reflex responses.
(D) This reflex action bypasses the brain.

58. A nerve impulse requires the release of neurotransmitters at the axonal bulb of a presynaptic neuron. Which of the following best explains the purpose of neurotransmitters, such as acetylcholine?
(A) They speed up the nerve conduction in a neuron.
(B) They open the sodium channels in the axonal membrane.
(C) They excite or inhibit the postsynaptic neuron.
(D) They open the potassium channels in the axonal membrane.

GO ON TO THE NEXT PAGE.

<u>Questions 59–61</u> refer to the figure and chart below.

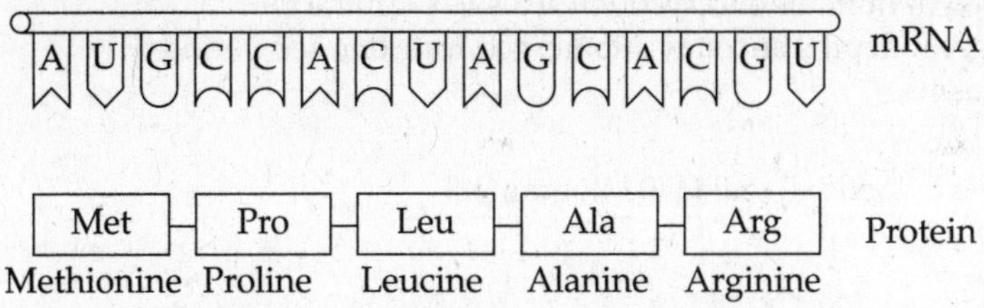

A U G C C A C U A G C A C G U mRNA

Met – Pro – Leu – Ala – Arg Protein

Methionine Proline Leucine Alanine Arginine

Formation of a Protein

		The Genetic Code: Condons of mRNA that Specify a Given Amino Acid			
		Third Position (3' end)			
First Position (5' end)	Second Position	U	C	A	G
U	U	UUU	UUC	UUA	UUG
		Phenylalanine		Lucine	
	C	UCU	UCC	UCA	UCG
		Serine			
	A	UAU	UAC	UAA	UAG
		Tytrosine			
	G	UGU	UGC	UGA	UGG
		Cysteine			Tryptophan
C	U	CUU	CUC	CUA	CUG
		Leucine			
	C	CCU	CCC	CCA	CCG
		Proline			
	A	CAU	CAC	CAA	CAG
		Histidine		Glutamine	
	G	CGU	CGC	CGA	CGG
		Arginine			
A	U	AUU	AUC	AUA	AUG
		Isoleucine			
	C	ACU	ACC	ACA	ACG
		Threonine			
	A	AAU	AAC	AAA	AAG
		Asparagine		Lysine	
	G	AGU	AGC	AGA	AGG
		Serine		Arginine	
G	U	GUU	GUC	GUA	GUG
		Valine			
	C	GCU	GCC	GCA	GCG
	A	GAU	GAC	GAA	GAG
		Aspartic Acid		Glutamic acid	
	G	GGU	GGC	GGA	GGG
		Glycine			

59. Which of the following DNA strands would serve as a template for the amino acid sequence shown above?

(A) 3′-ATGCGACCAGCACGT-5′
(B) 3′-AUGCCACUAGCACGU-5′
(C) 3′-TACGGTGATCGTGCA-5′
(D) 3′-UACGGUGAUCGUGCA-5′

60. If a mutation occurs in which uracil is deleted from the messenger RNA after methionine is translated, which of the following represents the resulting amino acid sequence?

(A) serine–histidine–serine–threonine
(B) methionine–proline–glutamine–histidine
(C) methionine–proline–leucine–alanine–arginine
(D) methionine–proline–alanine–arginine–arginine

61. The mRNA above was found to be much smaller than the original mRNA synthesized in the nucleus. This is due to the

(A) addition of a poly(A) tail to the mRNA molecule
(B) addition of a cap to the mRNA molecule
(C) excision of exons from the mRNA molecule
(D) excision of introns from the mRNA molecule

GO ON TO THE NEXT PAGE.

Questions 62–63 refer to the following information.

A scientist studies the storage and distribution of oxygen in humans and Weddell seals to examine the physiological adaptations that permit seals to descend to great depths and stay submerged for extended periods. The figure below depicts the oxygen storage in both organisms.

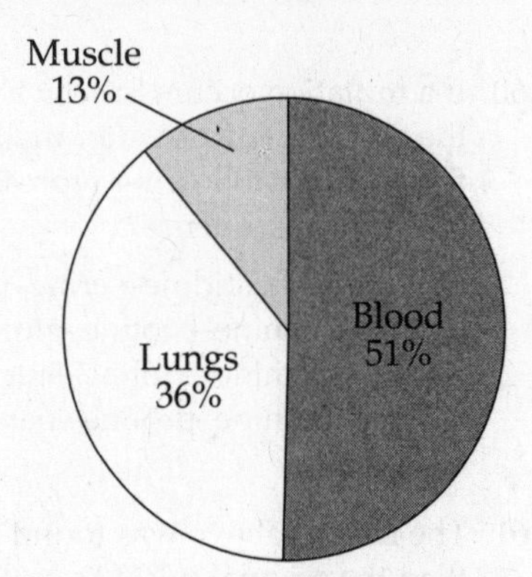

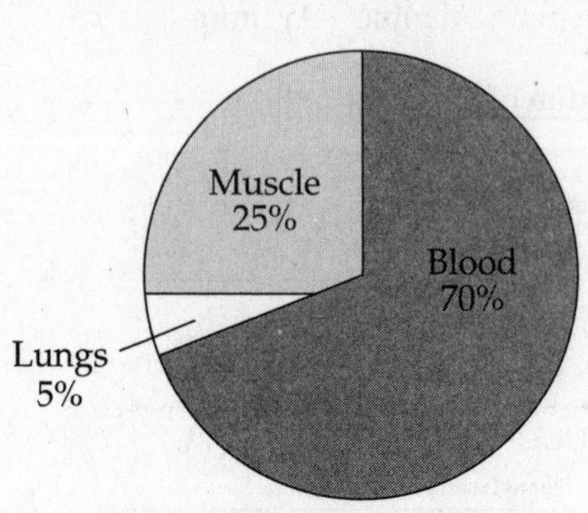

62. Compared with humans, approximately how many liters of oxygen does the Weddell seal store per kilogram of body weight?
    (A) The same amount of oxygen
    (B) Twice the amount of oxygen
    (C) Three times the amount of oxygen
    (D) Five times the amount of oxygen

63. During a dive, a Weddell seal's blood flow to the abdominal organs is shut off and oxygen-rich blood is diverted to the eyes, brain, and spinal cord. Which of the following is the most likely reason for this adaptation?
    (A) To increase the number of red blood cells in the nervous system
    (B) To increase the amount of oxygen reaching the skeletomuscular system
    (C) To increase the amount of oxygen reaching the central nervous system
    (D) To increase the oxygen concentration in the lungs

GO ON TO THE NEXT PAGE.

Directions: This part B consists of questions requiring numeric answers. Calculate the correct answer for each question.

64. An experiment was conducted to observe the light-absorbing properties of chlorophylls and carotenoids using a spectrophotometer. The pigments were first extracted and dissolved in a solution. They were then illuminated with pure light of different wavelengths to detect which wavelengths were absorbed by the solution. The results are presented in the absorption spectrum below.

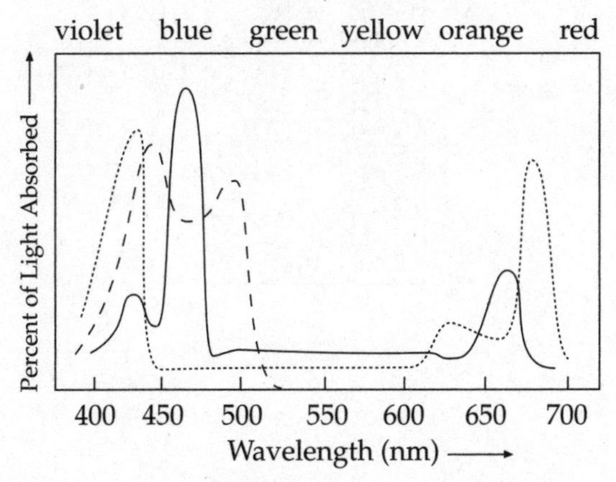

Absorption Spectrum for Green Plants

.... Chlorophyll *a*
——— Chlorophyll *b*
- - - - Carotenoids

At approximately what wavelength does chlorophyll *a* maximally absorb light?

65. A woman with blood genotype I^Ai and a man with blood genotype I^Bi have two children, both type AB. What is the probability that a third child will be blood type AB?

GO ON TO THE NEXT PAGE.

66. The trophic level efficiency of large herbivores such as elks is frequently only about 5 percent. In tons, what volume of plants would be required to maintain 24,000 lbs of elk?

67. If the genotype frequencies of an insect population are AA = 0.49, Aa = 0.42, and aa = 0.09, what is the gene frequency of the recessive allele?

GO ON TO THE NEXT PAGE.

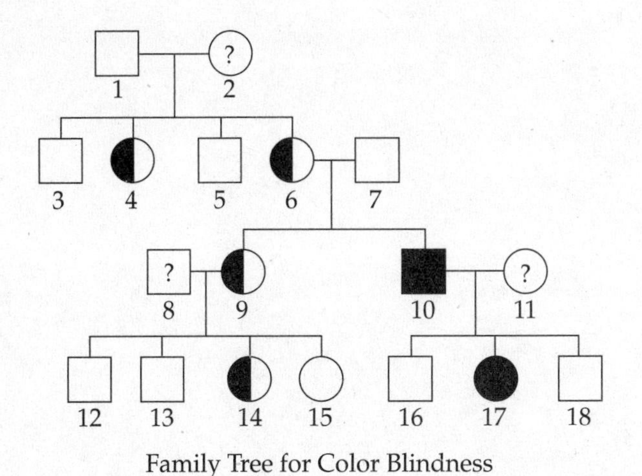

Family Tree for Color Blindness

68. Based on the pedigree above, what is the probability that a male child born to individuals 6 and 7 will be color-blind?

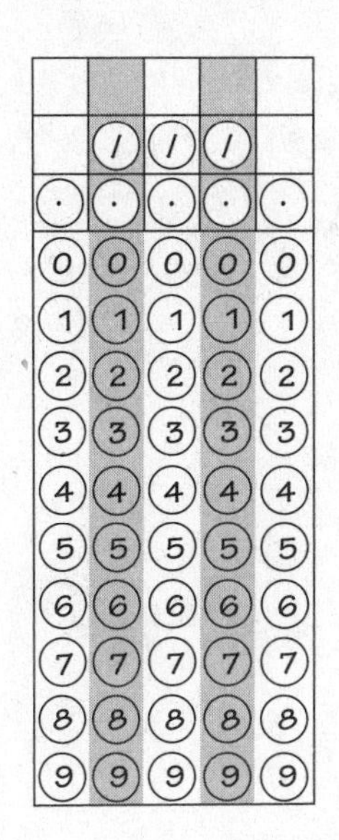

69. The loss of water by evaporation from the leaf openings is known as transpiration. The transpiration rates of various plants are shown below.

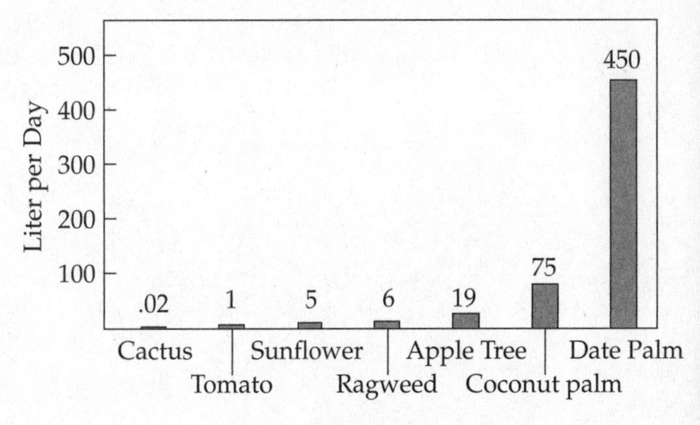

Transpiration Rates for Plants

How many liters of water per week are lost by a coconut palm?

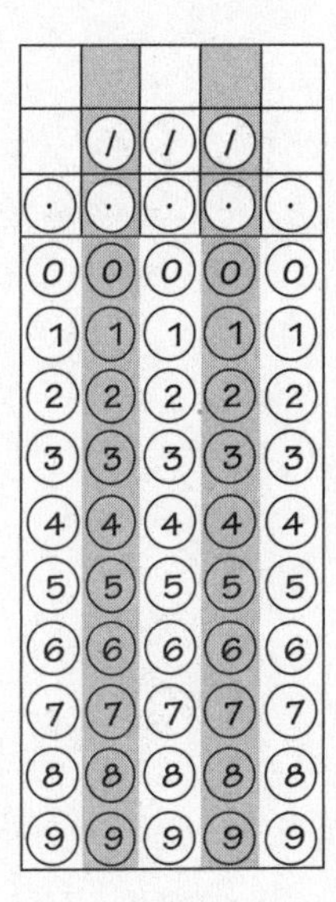

END OF SECTION I

This page intentionally left blank.

BIOLOGY
SECTION II
Planning time—10 minutes

Writing time—1 hour and 30 minutes

Directions: Questions 1 and 2 are long-form essay questions that should require about 20 minutes each to answer. Questions 3 through 8 are short free-response questions that should require about 6 minutes each to answer. Read each question carefully and write your response. Answers must be written out. Outline form is not acceptable. It is important that you read each question completely before you begin to write.

1. Cell size is limited by the surface area to volume ratio of the cell membrane.
   a. **Discuss** why cell size is limited by this ratio.
   b. **Describe** two adaptations that increase surface area in organisms.
   c. **Describe** the processes by which small polar and small nonpolar molecules can cross cell membranes according to their concentration gradients and give an example of each type of molecule.

2. Sickle-cell anemia is a genetic disorder caused by the abnormal gene for hemoglobin S. A single substitution occurs in which glutamic acid is substituted for valine in the sixth position of the hemoglobin molecule. This change reduces hemoglobin's ability to carry oxygen.
   a. **Discuss** the process by which mutation occurs in base substitution.
   b. Biologists used gel electrophoresis to initially identify the mutant gene. **Explain** how gel electrophoresis could be applied to the identification of the gene mutation. **Discuss** the use of restriction enzymes.
   c. Hemoglobin S is transmitted as a simple Mendelian allele. **Describe** the outcome if a female who does not carry the abnormal allele mates with a male homozygous for the disease. **Include** a Punnett square and phenotypic and genotypic ratios.

3. The cell membrane is an important structural feature of a nerve cell.
   a. **Discuss** what ions and concentration are associated with the resting state of a neuron.
   b. **Describe** the role of membranes in the conduction of a nerve impulse.

4. **Discuss** the Krebs cycle, the electron transport chain, and oxidative phosphorylation.
   a. **Explain** why these steps are considered aerobic processes.
   b. **Discuss** the location at which **each** stage occurs.

GO ON TO THE NEXT PAGE.

5. **Describe** three main differences between meiosis and mitosis.

6. **Define** homologous structures and give an example.

7. **Describe** the three ways that genetic information is transmitted laterally between bacteria.

8. **Describe** why viruses are typically not considered to be alive.

STOP

END OF EXAM

17

Practice Test 1 Answers and Explanations

ANSWER KEY

1. D
2. A
3. B
4. B
5. C
6. B
7. D
8. A
9. B
10. D
11. C
12. A
13. A
14. B
15. D
16. D
17. C
18. D
19. D
20. B
21. B
22. D
23. D
24. B
25. C
26. D
27. D
28. D
29. D
30. A
31. C
32. D
33. D
34. C
35. C
36. D
37. D
38. C
39. A
40. B
41. B
42. C
43. C
44. B
45. C
46. A
47. D
48. C
49. D
50. D
51. B
52. D
53. C
54. A
55. C
56. D
57. D
58. C
59. C
60. B
61. D
62. B
63. C
64. 425
65. 1/4 or 0.25
66. 240
67. 0.3
68. 1/2
69. 525

MULTIPLE-CHOICE ANSWERS AND EXPLANATIONS

1. **D** The resting potential depends on active transport (the Na^+/K^+ pump) and the selective permeability of the axon membrane to K^+ than to Na^+, which leads to a differential distribution of ions across the axonal membrane.

2. **A** The Krebs cycle occurs in the mitochondrial matrix. Don't forget to review the site of each stage of aerobic respiration. Glycolysis, the first step in aerobic respiration, occurs in the cytoplasm. The electron transport chain occurs along the inner mitochondrial membrane. Oxidative phosphorylation occurs as protons (H^+ ions) move from the intermembrane space to the mitochondrial matrix.

3. **B** A heterotroph obtains its energy from organic molecules. An autotroph obtains energy from sunlight utilizing pigments including chlorophyll (A) and uses CO_2 and water to make organic molecules. heterotrophs can obtain their energy from ingesting autotrophs, but can also consume other heterotrophs.

4. **B** In meiosis, the sister chromatids separate during the second metaphase of meiosis (Meiosis II) whereas the sister chromatids separate during metaphase of mitosis. (A), In meiosis, there are two rounds of cell division, whereas in mitosis, there is only one round of cell division. (C), Chromosomes are replicated during interphase in both meiosis and mitosis. (D), Spindle fibers form during prophase in both mitosis and meiosis.

5. **C** (A), (B), and (D), Amino acids are organic molecules that contain carbon, hydrogen, oxygen, and nitrogen. Don't forget to associate amino acids with nitrogen because of the amino group (NH_2).

6. **B** Unlike eukaryotes, prokaryotes (which include bacteria) do not contain membrane-bound organelles. (A) and (D), Bacteria contain circular double-stranded DNA, ribosomes, and a cell wall. (C), Bacterial cell membranes are made up of a bilipid layer with proteins interspersed.

7. **D** Populations can be described as the evolutionary unit because changes in the genetic makeup of populations can be measured over time. (A), Genetic changes occur only at the individual level. (B), Only under Hardy-Weinberg equilibrium does the gene pool remain fixed over time in a population. However, this statement does not explain why the population is the evolving unit. (C), This statement is true but does not address the question.

8. **A** Positive feedback occurs when a stimulus causes an increased response. Answer choices (B), (C), and (D) are examples of negative feedback.

9. **B** In order to determine the genotype of the parents, use the ratio of the offspring given in the question and work backward. The ratio of black-haired to white-haired guinea pigs is 3,1. The offspring were therefore BB, Bb, Bb, and bb. Now draw a Punnett square to figure out the genotype of the parents. The parents are Bb and Bb.

10. **D** The mean weight of the offspring in the next generation will be heavier than the mean weight of the original population because all the lighter horses in the original population died off. The normal distribution for weight will therefore shift to the heavier end (to the right of the graph). You can therefore eliminate answer choice (C) because the mean weight should increase. (A) and (B), The mean weight of the offspring could be heavier or lighter than their parents.

11. **C** Apoptosis (programmed cell death), hox and homeotic genes (genes that control differentiation), and inductive effects (a tissue affecting the differentiation of another tissue) play a role in cell differentiation. Operons are sets of multiple genes regulated by a single regulatory unit in bacteria.

12. **A** The primitive atmosphere lacked oxygen (O_2). It contained methane (CH_4), ammonia (NH_3), hydrogen (H_2).

13. **A** Villi and microvilli are fingerlike projections present in the small intestine that dramatically increase the surface area available for nutrient absorption.

14. **B** Pepsin works in the stomach (not the small intestine) to break down proteins to peptides. Complete digestion occurs in the small intestine. (A), Pancreatic lipase breaks down fats into three fatty acids and glycerol. (C), Pancreatic amylase breaks down carbohydrates into simple sugars. (D), Bile emulsifies lipids and makes them more accessible to lipase.

15. **D** Ribosomes are the site of protein synthesis. Therefore, the correct answer should start with ribosome. So eliminate answer choices (A) and (C). The polypeptide then moves through the rough ER to the Golgi apparatus, where it is modified and packaged into a vesicle. The vesicle then floats to the plasma membrane and is secreted.

16. **D** This statement is false because an individual with two identical alleles is said to be *homozygous* not heterozygous with respect to that gene. (A) and (B), Alleles are different forms of the same gene found on corresponding positions of homologous chromosomes. (C), More than two alleles can exist for a gene, but a person can have only two alleles for each trait. (Remember our discussion of blood groups in Chapter 10.)

17. **C** To make multiple copies of a plasmid (a small circular DNA), it should be inserted into a bacterium. (A), A plasmid would not replicate if it were inserted into a virus. (B), If a plasmid were treated with a restriction enzyme, it would be cut into smaller fragments. This would not give us cloned versions of the plasmid. (D), If the plasmid were run on a gel (using gel electrophoresis), this would only tell us the size of the plasmid.

18. **D** The least likely explanation for why mutations are low is that mutations produce variability in a gene pool. Any gene is bound to mutate. This produces a constant input of new genetic information in a gene pool. This answer choice doesn't give us any additional information about the rate of mutations. (A), Some mutations are subtle and cause only a slight decrease in reproductive output. (B), Some mutations are harmful and decrease the productive success of the individual. (C), Some mutations are deleterious and lead to total reproductive failure. The zygote fails to develop.

19. **D** The main challenge for organisms residing in hot arid environments is water loss. Long loops of Henle and storage of water in fatty tissues conserve water resources and large ears help disperse heat. Secretion is the elimination of water in the kidneys which would not be advantageous for an organism in a dry environment.

20. **B** Legume plants are able to live in nitrogen-poor soil because they obtain nitrogen from nitrogen-fixing bacteria. (A), (C), and (D), These plants cannot make their own proteins without nitrogen from nitrogen-fixing bacteria.

21. **B** Every cell in an organism has the same set of genes. Differences in the timing and expression levels of these genes leads to structural and functional differences.

22. **D** Deoxygenated blood from the vena cava enters the right atrium (2), then the right ventricle (1), and then enters the pulmonary arteries (4). The left atrium (5), left ventricle (6), and aorta (3) all carry oxygenated blood.

23. **D** The egg yolk provides food for the embryo. (A), The amnion protects the embryo. (B), The chorion is the outermost layer that surrounds the embryo. (C), The placenta provides nourishment for the embryo in mammals, not in birds.

24. **B** Normal cells can become cancerous when a virus invades the cell and takes over the replicative machinery. (A), A pilus forms between two bacteria. (C), The host's genome is not *converted* to the viral genome. (D), Spores are released by fungi, not viruses.

25. **C** The primary difference between bone and cartilage is that cartilage is flexible and lacks blood vessels. (A), They are both part of the skeletal system. (B), Bone and cartilage are connective tissues made up of collagen and calcium salts. (D), Both cartilage and bones secrete a matrix.

26. **D** If potential mates have similar breeding seasons they will most likely mate. Use common sense to eliminate the other answer choices. (A), If the organisms don't meet, they won't reproduce. (B) and (C), If the potential mates do not share the same behaviors (such as courtship rituals), they may not mate.

27. **D** There is no union of gametes in mitosis. (A) and (C), Asexual reproduction involves the production of two new cells with the same number of chromosomes as the parent cell. If the parent cell is diploid, then the daughter cells will be diploid. (B), The daughter cells are identical to the parent cell.

28. **D** Arteries are thick-walled vessels that carry blood *away* from the heart. Blood moves by contracting muscles. (A), (B), and (C) are all characteristics of veins. Veins are thin-walled vessels (with valves) that return blood to the heart.

29. **D** Cohesion, adhesion, and capillary action are special properties of water that are necessary for transpiration to occur. Water is a hydrophilic, polar molecule.

30. **A** Crossing-over (exchange of genetic material) occurs in prophase I of meiosis. (B), During metaphase I, the tetrads line up at the metaphase plate. (C), During anaphase, the tetrads separate. (D), During prophase II, the chromosomes become visible again.

31. **C** Crossing over and synapsis occur during meiosis, which produces haploid gametes. Separation of homologous chromosomes occurs during meiosis I, while separation of sister chromatids does not occur until meiosis II.

32. **D** Tropism describes the reaction of plants to a stimulus. Gravitropism specifies a reaction to gravity and negative specifies that the reaction is away from the stimulus.

33. **D** The net primary productivity is the energy producers have left for storage after their own energy needs have been met. Energy stored = (Total energy produced) – (energy used in cellular respiration). (A), It is not the energy available to producers. (B), It is not the total chemical energy in producers. That's the gross primary productivity. (C), It is not the biomass (total living material) among producers.

34. **C** (A) and (B), These organisms exhibit the same behavior because they were subjected to the same environmental conditions and similar habitats. This is an example of convergent evolution. (C), However, they are not genetically similar. (One is an insect and the other a bird.) (D), They are analogous. They exhibit the same function but are structurally different.

35. **C** A symbiotic relationship is a relationship among two organisms where both benefit. (A) describes parasitism and (B) describes commensalisms. (D) is an example of altruistic behavior.

36. **D** Beta cells secrete insulin. Destruction of beta cells in the pancreas will halt the production of insulin. Therefore, eliminate answer choices (A), (B) and (C). This will lead to an increase in blood glucose levels.

37. **D** The sympathetic division is active during emergency situations. This leads to a decrease in peristalsis in your gastrointestinal tract. (Your stomach shuts down.) Stimulation of the sympathetic nervous system leads to (A), pupils dilating, (B), peripheral blood vessels constricting, and (C), sweating.

38. **C** This is an example of mutualism. Both organisms benefit. (A), Parasitism is an example of a symbiotic relationship in which one organism benefits and the other is harmed. (B), Commensalism is when one organism benefits and the other is unaffected. (D), Endosymbiosis is the idea that some organelles originated as symbiotic prokaryotes that live inside larger cells.

39. **A** They both contain genetic material (DNA), a plasma membrane, and a cell wall. Use Process of Elimination. Unlike fungi, bacteria lack a definite nucleus. Therefore, eliminate (B). Bacteria are unicellular, whereas fungi are both unicellular and multicellular. Therefore, eliminate (C) and (D).

40. **B** Let's use Process of Elimination. The two hormones that are responsible for the maintenance of Ca^{2+} in the blood are parathyroid hormone and calcitonin. Therefore, eliminate answer choices (A) and (C). Now review the effects of these hormones. Parathyroid hormone increases Ca^{2+} in the blood while calcitonin decreases Ca^{2+} in the blood. A sustained decrease in circulating Ca^{2+} levels might be caused by decreased levels of parathyroid hormone.

41. **B** This question tests your understanding of what stage is responsible for the synthesis of new proteins for lactose utilization. The region of bacterial DNA that controls gene expression is the *lac* operon. Structural genes will be transcribed to produce enzymes, which produce an mRNA involved in digesting lactose.

42. **C** (A) and (B), Enzymes, which are proteins, are organic catalysts that speed up reactions without altering them. They are not consumed in the process. (D), The rate of reaction can be affected by the concentration of the substrate up to a point.

43. **C** DNA polymerase, not RNA polymerase, is the enzyme that causes the DNA strands to elongate. (A), DNA helicase unwinds the double helix. (B), DNA ligase seals the discontinuous Okazaki fragments. (D), DNA strands, in the presence of DNA polymerase, always grow in the 5' to 3' direction as complementary bases attach.

44. **B** A voltage change from +50 to –70 is called repolarization. (A), A voltage change from –70 to +50 is called depolarization. (C), A voltage change from –70 to –90 is called hyperpolarization. (D), An action potential is a traveling depolarized wave. It refers to the whole thing, from depolarization, repolarization, hyperpolarization, and back to a resting potential.

45. **C** Electrons passed down along the electron transport chain from one carrier to another lose energy and provide energy for making ATP. (A), Glucose is decomposed during glycolysis, but this process is not associated with energy given up by electrons. (B), Glucose is made during photosynthesis. (D), NADH is an are energy-rich molecule, which accepted electrons during the Krebs cycle.

46. **A** The oxygen released during the light reaction comes from the splitting of water. (Review the reaction for photosynthesis.) Therefore water must have originally contained the radioactive oxygen. (B), Carbon dioxide is involved in the dark reaction and produces glucose. (C), Glucose is the final product and would not be radioactive unless carbon dioxide was the radioactive material. (D), Nitrogen is not part of photosynthesis.

47. **D** Hydrophobic means "water fearing;" so water cannot be a hydrophobic solvent. Water is a polar solvent; it contains both negatively and positively charged ends that can form hydrogen bonds with other polar substances. Water can break up into H^+ and OH^- ions.

48. **C** Pheromones act as sex attractants, alarm signals, or territorial markers. Use Process of Elimination for this question. Auxins are plant hormones that promote growth. Hormones are chemical messengers that produce a specific effect on target cells within the same organism. Enzymes are catalysts that speed up reactions. Coenzymes are organic substances that assist enzymes in a chemical reaction.

49. **D** Homologous structures are organisms with the same structure but different functions. The forelegs of an insect and the forelimbs of a dog are not structurally similar. (One is an invertebrate and the other a vertebrate.) They do not share a common ancestor. However, both structures are used for movement. All of the other examples are vertebrates that are structurally similar.

50. **D** (I), Muscle contractions require calcium ions. (II), In order to have a muscle contraction you need energy—ATP. (III), Actin is one of the proteins involved in muscle contractions.

51. **B** Speciation occurred in the Galapagos finches as a result of the different environments on the islands. This is an example of divergent evolution. The finches were geographically isolated. (A), Convergent evolution is the evolution of similar structures in distantly related organisms. (C), Disruptive selection is selection that favors both extremes at the expense of the intermediates in a population. (D), Stabilizing selection is selection that favors the intermediates at the expense of the extreme phenotypes in a population.

52. **D** Mutations produce genetic variability. All of the other answer choices are forms of asexual reproduction.

53. **C** Secondary consumers feed on primary consumers. If you set up a pyramid of numbers, you'll see that the herrings belong to the third trophic level.

54. **A** The biomass is the total bulk of a particular living organism. The plankton population has both the largest biomass and the most energy.

55. **C** If the herring population decreases, this will lead to an increase in the number of crustaceans and a decrease in the plankton population. Reorder the organisms according to their trophic levels and determine which populations will increase and decrease accordingly.

56. **D** This question tests your ability to trace the neural pathway of a motor (effector) neuron. The nerve conduction will travel from the spinal cord (where interneurons are located) to the muscle.

57. **D** Because the brain is destroyed, it is not associated with the movement of the leg. (A), Reflex actions are automatic. (B) and (C), Both of these statements are true but are not supported by the experiment.

58. **C** Neurotransmitters are released from the axonal bulb of one neuron and diffuse across a synapse to activate a second neuron. The second neuron is called a postsynaptic neuron. A neurotransmitter can either excite or inhibit the postsynaptic neuron. (A), The myelin sheath speeds up the conduction in a neuron. (B) and (D), Both sodium and potassium channels open during an action potential. Neurotransmitters are not involved in actions related to the axon membrane. They do not force potassium ions to move against a concentration gradient.

59. **C** The DNA template strand is complementary to the mRNA strand. Using the mRNA strand, work backward to establish the sequence of the DNA strand. Don't forget that DNA strands do not contain uracil, so eliminate answer choices (B) and (D).

60. **B** Use the amino acid chart to determine the sequence after uracil is deleted. The deletion of uracil creates a frameshift.

61. **D** The mRNA is modified before it leaves the nucleus. It becomes smaller when introns (intervening sequences) are removed. (A) and (B), A poly (A) tail and a cap are added to the mRNA and would therefore increase the length of the mRNA. (C), Exons are the coding sequences that are kept by the mRNA.

62. **B** The Weddell seal stores twice as much oxygen as humans. Calculate the liters per kilograms weight for both the seal and man using the information at the bottom of the chart. The Weddell seal stores 0.058 liters/kilograms (25.9 liters/450 kilograms) compared to 0.028 liters/kilograms (1.95 liters/70 kilograms) in humans.

63. **C** The most plausible answer is that blood is redirected toward the central nervous system, which permits the seal to navigate for long durations. (A), The seal does not need to increase the number of red blood cells in the nervous system. (B), The seal does not need to increase the amount of oxygen to the skeletal system. (D), The diversion of blood does not increase the concentration of oxygen in the lungs.

64. **425** If you look at the absorption spectrum, you'll see that chlorophyll a has two peaks, one at 425 nm and 680 nm. Chlorophyll *a* maximally absorbs light at approximately 425 nm.

65. $\mathbf{\frac{1}{4}}$ or **0.25**

Make a Punnett square to determine the probability that the couple has a child with blood type AB. The probability is $\frac{1}{4}$ whether it's the first child or the third child.

66. **240** 24,000 lbs of elk is equivalent to 12 tons since there are 2,000 lbs per ton. With a 5% efficiency of transfer from their food source, that 12 tons represents 5% of the weight of the plants they would need to consume. 5/100 = 12/tons needed, making the answer 240.

67. **0.3** The frequency of the homozygous dominant genotype (AA) is 0.49. To find the dominant *allele* frequency, we can use the formula provided by the Hardy-Weinberg theory, $p^2 + 2pq + p^2 = 1$, where p represents the dominant allele and q represents the recessive allele. Because we know that p^2 represents the frequency of the homozygous dominant genotype, we can find the frequency of the dominant allele (p) by taking the square root of the frequency of the genotype. The square root of 0.49 is 0.7. Note that by using the formula $p + q = 1$, we can also determine the frequency of the recessive allele (q). It would be 0.3 ($0.7 + q = 1$).

68. $\mathbf{\frac{1}{2}}$ Draw a Punnett square for the couple and determine the probability of color-blindness for the boys. Individuals 6 and 7, (X^cX and XY) will produce two males, one X^cY and one XY. The probability of color-blindness is therefore $\frac{1}{2}$.

69. **525** Make sure you read the question carefully. You are asked to calculate the number of liters per week, not per day. The chart tells us that a coconut palm loses 75 liters a day, which would mean 525 liters a week (7 × 75 = 525).

FREE-RESPONSE ANSWERS AND EXPLANATIONS

In the following pages you'll find two types of aids for grading your essays, checklists, and written paragraphs. You'll recall from our discussion of the essays in Chapter 13 that ETS uses checklists just like these to grade your essays on the actual test. They're actually quite simple to use.

For each item you mentioned, give yourself the appropriate number of points (1 point, 2 points, etc.). Remember that you can only get a maximum of 10 points for each essay. As you evaluate your work, don't be kind to yourself just because you like your own essay, If the checklist mentions an explanation of structure and function and you failed to give both, then do not give yourself a point. ETS is very particular about this. You need to mention precisely the things they've listed, in the way they've listed them, in order to gain points. What if something you've mentioned doesn't appear on the list?

Provided you know it's valid, give yourself a point. If it was something you pulled out of your hat at the last minute, odds are it's not directly applicable to the question. However, because you might come up with details even more specific than those contained in the checklist, it's not unlikely that the example or structure you've cited is perfectly valid. If so, go ahead and give yourself the point. Remember that ETS hires college professors and high school teachers to read your essays, so they'll undoubtedly recognize any legitimate information you slip into the essay.

The second part of the answer key involves short paragraphs. These are not templates, ETS does not expect you to write this way. They are simply additional tools to help illustrate the things you need to squeeze into your essay in order to rack up the points. They explain in some detail how the various parts of the checklist relate to one another, and may give you an idea about how best to integrate them into your own essays come test-time.

If you find that you didn't do too well on the essay portion, go back and practice some essays from Chapter 13. You can use the chapters themselves as checklists. If you find it too difficult to grade your own essays, see if your teacher or a classmate will help you out. Good luck!

LONG-FORM FREE RESPONSE CHECKLIST 1

I. **Cell size ratio**—4 points maximum

(2 points each)

A higher ratio of surface area to volume allows for greater space for solutes to move in an out of cells.

Cells must maintain homeostasis and in order to do this must eliminate wastes, ingest nutrients and maintain osmotic and ion balances.

As a cell grows larger this ratio diminishes, limiting cell size.

II. **Adaptations**—4 points maximum

(2 points each)

Alveoli

Convoluted membranes in chloroplasts and mitochondria

Root hairs

Villi and microvilli

Endoplasmic reticulum

III. Transport across the membrane

(2 points each)

Small polar molecules cross by facilitated diffusion and require membrane channels.

Water is an example of a small polar molecule that crosses through channels called aquaporins

Ions are polar species that cross the cell membrane via protein channels

Small nonpolar molecules can freely diffuse across the lipid bilayer.

Seed/seedcoat

Examples of small nonpolar molecules include CO2, O2 and steroid hormones.

Long-Form Free Response 1

Cells need a certain surface area to volume ratio to exchange materials with the environment in order to maintain homeostasis. A high surface area to volume ratio allows cells to regulate ion concentrations, as well as take in nutrients and eliminate wastes. As a cell becomes larger, the ratio of surface area to volume decreases and if a cell grows too large it cannot carry out these functions and therefore will not survive. Organisms have adapted a variety of ways to increase surface area. One example is alveoli, small air sacs in the lungs that maximize surface area in order to allow for maximum respiration. Another example of increasing surface area is villi in the small intestine, which results in a large surface area for the absorption of nutrients.

Depending on polarity, molecules cross the cell membrane in different ways. Small nonpolar molecules, like carbon dioxide, can freely diffuse through the cell membrane. However, due to the hydrophobic nature of the interior of the cell membrane, small polar molecules need protein channels to allow selective diffusion into or out of the cell. One example of a polar molecule is water, which crosses the membrane through channels called aquaporins.

Long-Form Free Response Checklist 2

a. Type of Mutation—4 points maximum

(1 point each)

Base substitution (involves one nucleotide being replaced by another)

Change in DNA/codon of mRNA draws new tRNA molecule

New tRNA carries wrong amino acid, altering polypeptide

Change in polypeptide in turn alters hemoglobin protein

Distorted red blood cell cannot efficiently carry oxygen

b. Gel Electrophoresis—5 points maximum

How it works—3 points maximum

(1 point each)

Apparatus	DNA is put into wells of an agarose gel with a buffer
Electricity	Electrical potential (electrical charge moves fragments)
Charge	Negatively charged fragments move toward the positive pole
Size/Molecular weight	Smaller fragments move faster across the gel

How it can be applied to identify the mutant gene—2 points maximum **(1 point each)**

Compare the normal hemoglobin to the abnormal hemoglobin

Use restriction enzymes to cut the two hemoglobin proteins into several fragments

Normal hemoglobin can be used as a marker

Fragments of the two proteins should be identical except for the fragment containing the abnormal gene

c. **Punnett Square**—2 points maximum

A normal female who does not carry the X-linked allele mates with a male homozygous for the disease,

HH → hh → All the offspring are Hh (heterozygotes)

They are all carriers (sickle-cell trait)

LONG-FORM FREE RESPONSE 2

Sickle-cell anemia is a disease in which the red blood cells have an abnormal shape because of a base substitution. Base substitution involves an error in DNA replication in which one nucleotide is replaced by another nucleotide. This causes the codon of an mRNA to contain an incorrect base. The codon therefore matches up with the anticodon of a different tRNA. This tRNA carries a different amino acid. The change in the amino acid alters the polypeptide. The polypeptide, in turn, alters the hemoglobin protein. In this case, the distorted red blood cell cannot efficiently carry oxygen.

Biologists must have determined the nature of the hemoglobin mutation by comparing the normal hemoglobin gene to the abnormal hemoglobin gene, using the technique gel electrophoresis. Gel electrophoresis identifies the difference between two molecules by examining the different rates each molecule moves across the gel. Substances move across a gel according to their molecular weight. For example, smaller fragments move faster than larger fragments. The biologists must have placed fragments of the two genes on an agarose gel and used the normal hemoglobin gene as the marker—the source of comparison to the abnormal hemoglobin gene. (A restriction enzyme was used to cut the two DNA sequences into several fragments prior to loading the gel). The two DNA sequences should have been identical except for the fragment that contained the abnormal gene.

If a normal noncarrier female mates with a male who is homozygous for the disease, these are the results using a Punnett square,

	H	H
h	Hh	Hh
h	Hh	Hh

All of the offspring would be heterozygotes. For sickle-cell anemia, these offspring would be carriers.

Short-Form Free Response Checklist 3

a. **Ions associated with plasma membrane**—4 points maximum

(1 point each)

Ions	**Description**
Selectively permeable to K^+ ions	(Concentrated on the inside of the cell)
Impermeable to Na^+ ions	(Concentrated on the outside of the cell)
Build up of positively charged ions outside of the membrane	(Inside is negatively charged)
Sodium-potassium pump	(Maintains the ion gradient)

b. **Role of the membrane**—4 points maximum

(.5 point each)

Nerve impulses are electrochemical (associated with ion and electrical changes in plasma membrane)
Action potential (change in the membrane potential)
Resting stage (voltage charge is –70 millivolts)
Depolarization (Na^+ moves into the cell down its cell gradient)
Repolarization (K^+ ions move out of the cell down their cell gradient)
Sodium channels (membrane channel that is voltage-gated)
Potassium channels (membrane channel that is voltage-gated)
Sodium-potassium pump (membrane protein that maintains ion gradient)

Short-Form Free Response 3

A nerve cell is fundamentally similar in structure and function to other somatic cells. Like other cells, a nerve cell consists of a bilipid layer made up of phospholipids and proteins. This membrane is semipermeable; it regulates the passage of certain ions across its membrane. The cell body of a nerve cell also contains cellular structures such as a nucleus, which regulates the activities of the cell, and ribosomes, which make proteins.

The plasma membrane of a resting neuron is selectively permeable to K^+ ions and impermeable to Na^+ ions. Because of this selective permeability, the K^+ ions are concentrated on the inside of the cell and are able to slowly diffuse outward, while the Na^+ ions are concentrated on the outside of the cell. This leads to a build up of positively-charged ions outside of the neuronal membrane. The inside of the plasma membrane is therefore negatively charged. These chemical gradients are maintained by a sodium-potassium pump. This pump actively moves three Na^+ ions out of the cell for every two K^+ ions brought into the cell.

The neuronal membrane plays an important role in the conduction of a nerve impulse. Nerve impulses are electrochemical. This means, the forces that cause ions to move across a membrane are both a concentration gradient and an electrical gradient. The membrane is associated with two chemicals (Na^+ ions and K^+ ions) as well as a change in the voltage charge. When a nerve cell is undisturbed, the membrane is said to be in a resting stage. The membrane is polarized and the voltage charge is –70 millivolts. During a nerve impulse, there is a change in the membrane permeability, Na^+ ions rush into the cell, and the inside becomes more positively charged. The membrane is now said to be depolarized. Na^+ move into the cell down its concentration gradient. Next, the Na^+ channels close and K^+ channels open and K^+ ions move out of the cell. The cell is now said to be repolarized. The original ion concentration is reestablished by the sodium-potassium pump.

Short-Form Free Response Checklist 4

a. **Krebs cycle, the electron transport chain, oxidative phosphorylation as aerobic processes**—2 points maximum

They require oxygen
They cannot occur under anaerobic conditions

b. **Site of each step**—3 points maximum

Stage	Site
Krebs cycle	Mitochondrial matrix
Electron transport chain	Along the inner mitochondrial membrane
Oxidative phosphorylation	As hydrogens move from the intermembrane space to the mitochondrial matrix

Short-Form Free Response 4

a. The Krebs cycle, electron transport chain, and oxidative phosphorylation are all part of aerobic respiration. During aerobic respiration, glucose is completely converted to CO_2, ATP, and water. These steps are considered aerobic processes because they cannot occur under anaerobic conditions; they require oxygen. Glycolysis, on the other hand, can occur under both aerobic and anaerobic conditions.

b. These three stages of aerobic respiration occur in different parts of the mitochondria. The Krebs cycle occurs in the mitochondrial matrix. Two acetyl CoA enter the Krebs cycle and produce NADH, $FADH_2$, ATP, and CO_2. The products of the Krebs cycle (NADH and $FADH_2$) are sent to the electron transport chain. The electron transport chain occurs along the inner mitochondrial membrane. The final stage of aerobic respiration—oxidative phosphorylation—occurs as hydrogens move across from the intermembrane space to the mitochondrial matrix.

Short-Form Free Response Checklist 5

(1 point each)

Produces somatic cells versus produces gametes
One round of division versus two rounds of division
Meiosis only: recombination between homologous chromosomes

Short-Form Free Response 5

The difference between meiosis and mitosis arise because of the different goals of these cellular processes. First, mitosis is designed to reproduce copies of somatic cells while meiosis is designed to create gametes with greater variability. Second, in order to create the appropriate level of genetic information, meiosis requires two rounds of division whereas recreating somatic cells only requires one round. Third, in order to increase genetic variability, recombination can occur as part of meiosis between homologous chromosomes. Such mixing is not part of mitosis since the goal is to simply recreate the original cell.

Short-Form Free Response Checklist 6

(1 point each)
Definition of homologous structures
Link to divergent evolution
Relevant example (many possible)

Short-Form Free Response 6

Homologous structures are those with similar structures because they arose from the same ancestral source, but which now have different functions. These structures are often the product of divergent evolution, where members of a certain species have been divided by barriers such that a change in bodily structures was dictated by the forces of natural selection. An example of homologous structures would be the arm of a human and the front leg of a cat. Their functions are different in their current form, but the ancestral origin is shared.

Short-Form Free Response Checklist 7

(1 point each)
Transduction
Transformation
(2 points each)
Conjugation

Short-Form Free Response 7

There are three ways in which bacteria transmit genetic information laterally to introduce new phenotypes. These are transduction, transformation, and conjugation. Transduction is the transmission of genetic material from one bacteria to another via a lysogenic virus. Transformation is the uptake of naked DNA from the environment. Conjugation is the cell-to-cell transfer of genetic material in the form of plasmids across a pilus formed between two cells.

Short-Form Free Response Checklist 8

(1 point each)

No respiration

Not able to create own energy

Not able to reproduce independently

Not able to live outside host cells

Short-Form Free Response 8

Viruses are typically not considered to be alive because they are not able to engage in many of the functions associated with life. Though viruses must engage in processes like translation that require energy, they do not engage in any process to create that energy. In that same line, there is no respiration of any kind being done by viruses. Further, viruses have no existence outside of their host cells; no viral functions happen outside of the context of a cell. Due to that restriction, viruses are also not able to reproduce themselves. Being entirely reliant on a host cellular environment for reproduction is another reason to consider them as not alive.

18

Practice Test 2

AP® Biology Exam

SECTION I: Multiple-Choice Questions

DO NOT OPEN THIS BOOKLET UNTIL YOU ARE TOLD TO DO SO.

At a Glance

Total Time
1 hour and 30 minutes
Number of Questions
69
Writing Instrument
Pencil required

Instructions

Section I of this examination contains 69 multiple-choice questions. These are broken down into Part A (63 multiple-choice questions) and Part B (6 grid-in questions).

Indicate all of your answers to the multiple-choice questions on the answer sheet. No credit will be given for anything written in this exam booklet, but you may use the booklet for notes or scratch work. After you have decided which of the suggested answers is best, completely fill in the corresponding oval on the answer sheet. Give only one answer to each question. If you change an answer, be sure that the previous mark is erased completely. Here is a sample question and answer.

Sample Question	Sample Answer
Chicago is a	Ⓐ ● Ⓒ Ⓓ

(A) state
(B) city
(C) country
(D) continent

Use your time effectively, working as quickly as you can without losing accuracy. Do not spend too much time on any one question. Go on to other questions and come back to the ones you have not answered if you have time. It is not expected that everyone will know the answers to all the multiple-choice questions.

About Guessing

Many candidates wonder whether or not to guess the answers to questions about which they are not certain. Multiple choice scores are based on the number of questions answered correctly. Points are not deducted for incorrect answers, and no points are awarded for unanswered questions. Because points are not deducted for incorrect answers, you are encouraged to answer all multiple-choice questions. On any questions you do not know the answer to, you should eliminate as many choices as you can, and then select the best answer among the remaining choices.

This page intentionally left blank.

BIOLOGY

SECTION I

Time—1 hour and 30 minutes

Directions: Each of the questions or incomplete statements below is followed by four suggested answers or completions. Select the one that is best in each case and then fill in the corresponding oval on the answer sheet.

1. In general, animal cells differ from plant cells in that animal cells have

(A) a cell wall made of cellulose
(B) lysosomes
(C) large vacuoles that store water
(D) centrioles within centrosomes

2. A cell from the leaf of the aquatic plant *Elodea* was soaked in a 15 percent sugar solution, and its contents soon separated from the cell wall and formed a mass in the center of the cell. All of the following statements are true about this event EXCEPT:

(A) The vacuole lost water and became smaller.
(B) The space between the cell wall and the cell membrane expanded.
(C) The large vacuole contained a solution with much lower osmotic pressure than that of the sugar solution.
(D) The concentration of solutes in the extracellular environment is hypertonic with respect to the cell's interior.

3. A chemical agent is found to denature all enzymes in the synaptic cleft. What effect will this agent have on acetylcholine?

(A) Acetylcholine will not be released from the presynaptic membrane.
(B) Acetylcholine will not bind to receptor proteins on the postsynaptic membrane.
(C) Acetylcholine will not diffuse across the cleft to the postsynaptic membrane.
(D) Acetylcholine will not be degraded in the synaptic cleft.

4. The base composition of DNA varies from one species to another. Which of the following ratios would you expect to remain constant in the DNA?

(A) Cytosine : Adenine
(B) Pyrimidine : Purine
(C) Adenine : Guanine
(D) Guanine : Deoxyribose

5. In reptile eggs, the extraembryonic membrane that functions in excretion and respiration is the

(A) amnion
(B) chorion
(C) allantois
(D) yolk sac

6. Consider the following enzyme pathway:

$$A \xrightarrow{1} B \xrightarrow{2} C \xrightarrow{3} D \xrightarrow{4} E \xrightarrow{5} F$$
$$C \xrightarrow{6} X \xrightarrow{7} Y$$

An increase in substance F leads to the inhibition of enzyme 3. All of the following are results of the process EXCEPT

(A) an increase in substance X
(B) increased activity of enzyme 6
(C) decreased activity of enzyme 4
(D) increased activity of enzyme 5

7. The liver is a vital organ that performs all of the following functions EXCEPT

(A) storing amino acids that were absorbed in the capillaries of the small intestine
(B) detoxifying harmful substances such as alcohol or certain drugs
(C) synthesizing bile salts that emulsify lipids
(D) breaking down peptides into amino acids

GO ON TO THE NEXT PAGE.

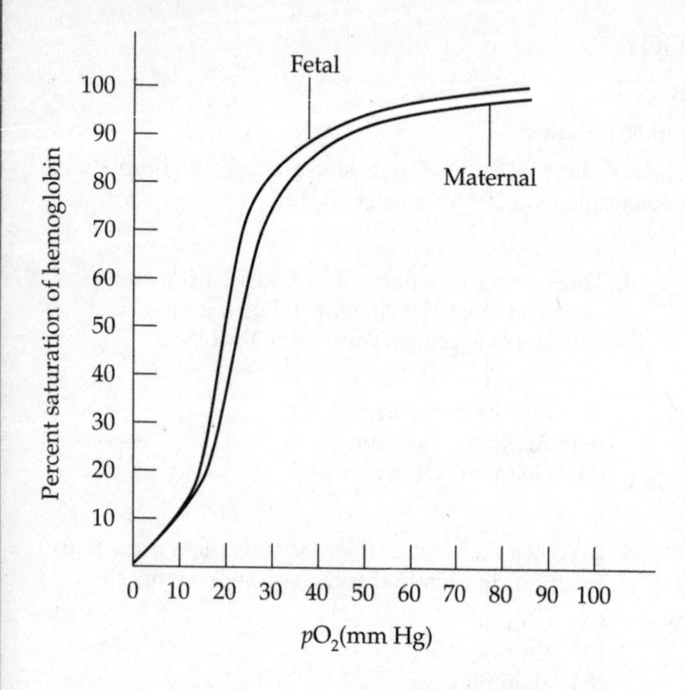

8. The graph above shows the oxygen dissociation curves of maternal hemoglobin and fetal hemoglobin. Based on the graph, it can be concluded that
   (A) fetal hemoglobin surrenders O_2 more readily than maternal hemoglobin
   (B) the dissociation curve of fetal hemoglobin is to the right of maternal hemoglobin
   (C) fetal hemoglobin has a higher affinity for O_2 than does maternal hemoglobin
   (D) fetal and maternal hemoglobin differ in structure

9. In minks, the gene for brown fur (B) is dominant over the gene for silver fur (b). Which set of genotypes represents a cross that could produce offspring with silver fur from parents that both have brown fur?
   (A) BB × BB
   (B) BB × Bb
   (C) Bb × Bb
   (D) Bb × bb

10. Hemoglobin is a molecule that binds to both O_2 and CO_2. There is an allosteric relationship between the concentrations of O_2 and CO_2. Hemoglobin's affinity for O_2
    (A) decreases as blood pH decreases
    (B) increases as H^+ concentration increases
    (C) increases in exercising muscle tissue
    (D) decreases as CO_2 concentration decreases

11. All viruses contain at least these two principal components:
    (A) DNA and proteins
    (B) nucleic acid and a capsid
    (C) DNA and cell membrane
    (D) RNA and cell wall

12. All of the following are differences between prokaryotes and eukaryotes EXCEPT
    (A) eukaryotes have linear chromosomes, while prokaryotes have circular chromosomes
    (B) eukaryotes possess double stranded DNA, while prokaryotes possess single stranded DNA
    (C) eukaryotes process their mRNA, while in prokaryotes transcription and translation occur simultaneously
    (D) eukaryotes contain membrane-bound organelles, prokaryotes do not

13. In humans, fertilization normally occurs in the
    (A) ovary
    (B) fallopian tube
    (C) uterus
    (D) placenta

14. The development of an egg without fertilization is known as
    (A) meiosis
    (B) parthenogenesis
    (C) embryogenesis
    (D) vegetative propagation

GO ON TO THE NEXT PAGE.

15. All of the following are examples of hydrolysis EXCEPT
    (A) conversion of fats to fatty acids and glycerol
    (B) conversion of proteins to amino acids
    (C) conversion of starch to simple sugars
    (D) conversion of pyruvic acid to acetyl CoA

16. In cells, which of the following can catalyze reactions involving hydrogen peroxide, provide cellular energy, and make proteins, in that order?
    (A) Peroxisomes, mitochondria, and ribosomes
    (B) Peroxisomes, mitochondria, and lysosomes
    (C) Peroxisomes, mitochondria, and Golgi apparatus
    (D) Lysosomes, chloroplasts, and ribosomes

17. All of the following play an important role in regulating respiration in humans EXCEPT
    (A) an increase in the amount of CO_2 in the blood
    (B) a decrease in the amount of O_2 in the blood
    (C) a decrease in the plasma pH level
    (D) strenuous exercise

18. The primary site of glucose reabsorption is the
    (A) glomerulus
    (B) proximal convoluted tubule
    (C) loop of Henle
    (D) collecting duct

Questions 19 and 20 refer to the graph.

The graph below shows the growth curve of a bacterial culture.

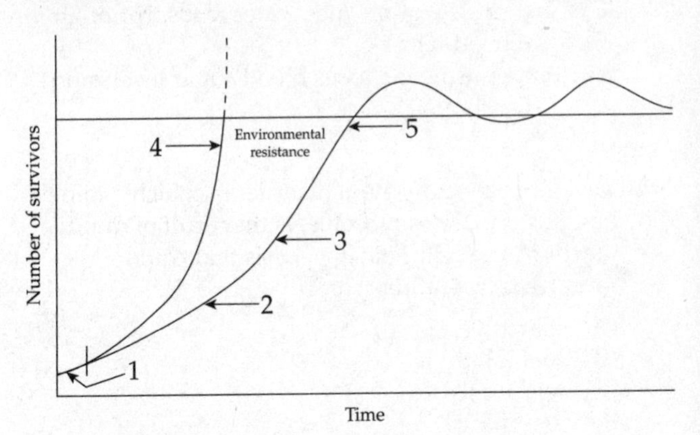

19. Which of the following represents the carrying capacity of the environment?
    (A) 2
    (B) 3
    (C) 4
    (D) 5

20. Which of the following shows the exponential growth curve of the population?
    (A) 1
    (B) 2
    (C) 3
    (D) 4

GO ON TO THE NEXT PAGE.

21. All of the following statements are true EXCEPT:
    (A) Thyroxine increases the rate of metabolism.
    (B) Insulin decreases storage of glycogen.
    (C) Vasopressin stimulates water reabsorption in the kidney.
    (D) Epinephrine increases blood sugar levels and heart rate.

22. Metafemale syndrome, a disorder in which a female has an extra X chromosome, is the result of nondisjunction. The failure in oogenesis that could produce this would occur in
    (A) metaphase I
    (B) metaphase II
    (C) telophase I
    (D) anaphase II

23. In plants, the tendency of climbing vines to twine their tendrils around a trellis is called
    (A) thigmotropism
    (B) hydrotropism
    (C) phototropism
    (D) geotropism

24. Females with Turner's syndrome have a high incidence of hemophilia, a recessive, X-linked trait. Based on this information, it can be inferred that females with this condition
    (A) have an extra X chromosome
    (B) have an extra Y chromosome
    (C) lack an X chromosome
    (D) have red blood cells that clump

25. When a retrovirus inserted its DNA into the middle of a bacterial gene, it altered the normal reading frame by one base pair. This type of mutation is called
    (A) duplication
    (B) translocation
    (C) inversion
    (D) frameshift mutation

26. High levels of estrogen from maturing follicles inhibit the release of gonadotropin releasing hormone (GnRH). Which of the following endocrine glands produces GnRH?
    (A) Anterior pituitary
    (B) Posterior pituitary
    (C) Hypothalamus
    (D) Pineal gland

27. The principle inorganic compound found in living things is
    (A) carbon
    (B) oxygen
    (C) water
    (D) glucose

28. Kangaroo rats are better able to concentrate urine than humans are. It would be expected that, compared to the nephrons of human kidneys, the nephrons of kangaroo-rat kidneys would have
    (A) thicker walls, which are impermeable to water
    (B) shorter loops of Henle
    (C) longer loops of Henle
    (D) shorter collecting ducts

GO ON TO THE NEXT PAGE.

29. All of the following are modes of asexual reproduction EXCEPT

(A) sporulation
(B) fission
(C) budding
(D) meiosis

30. The moist skin of earthworms, spiracles of grasshoppers, and the mucus membranes lining alveoli are all associated with the process of

(A) excretion
(B) respiration
(C) circulation
(D) digestion

31. Invertebrate immune systems possess which of the following?

(A) Cytotoxic T lymphocytes
(B) Phagocytes
(C) B cells
(D) Helper T cells

32. All of the following are examples of connective tissue EXCEPT

(A) ligaments
(B) muscle
(C) blood
(D) cartilage

33. Photosynthesis requires

(A) glucose, light, CO_2
(B) light, CO_2, water
(C) water, soil, O_2
(D) O_2, water, light

34. If a forest of fir, birch, and white spruce trees was devastated by fire, which of the following would most likely happen?

(A) Only animal life would continue to inhabit the region.
(B) Secondary succession would begin to occur.
(C) Only tough grasses would appear.
(D) The number of species would stabilize as the ecosystem matures.

35. Which of the following processes occur in the cytoplasm of an eukaryotic cell?

I. DNA replication

II. Transcription

III. Translation

(A) I only
(B) III only
(C) II and III only
(D) I, II, and III

36. Crossing-over during meiosis permits scientists to determine

(A) the chance for variation in zygotes
(B) the rate of mutations
(C) the distance between genes on a chromosome
(D) which traits are dominant or recessive

37. An animal cell that is permeable to water but not salts has an internal NaCl concentration of 10%. If placed in freshwater the cell will

(A) plasmolyze
(B) swell and eventually lyse
(C) endocytose water into a large central vacuole
(D) shrivel

GO ON TO THE NEXT PAGE.

38. Three distinct bird species, flicker, woodpecker, and elf owl, all inhabit a large cactus, *Cereus giganteus*, in the desert of Arizona. Since competition among these birds rarely occurs, the most likely explanation for this phenomenon is that these birds
    (A) have a short supply of resources
    (B) have different ecological niches
    (C) do not live together long
    (D) are unable to breed

39. Lampreys attach to the skin of lake trout and absorb nutrients from its body. This relationship is an example of
    (A) commensalism
    (B) parasitism
    (C) mutualism
    (D) gravitropism

40. The nucleotide sequence of a DNA molecule is 5'-C-A-T-3'. A mRNA molecule with a complementary codon is transcribed from the DNA in the process of protein synthesis a tRNA pairs with a mRNA codon. What is the nucleotide sequence of the tRNA anticodon?
    (A) 5'-G-T-A-3'
    (B) 5'-G-U-A-3'
    (C) 5'-C-A-U-3'
    (D) 5-'U-A-C-3'

41. Viruses are considered an exception to the cell theory because they
    (A) are not independent organisms
    (B) have only a few genes
    (C) move about via their tails
    (D) have evolved from ancestral protists

42. All of the following organs in the digestive system secrete digestive enzymes EXCEPT the
    (A) mouth
    (B) stomach
    (C) gall bladder
    (D) small intestine

43. Memory loss would most likely be due to a malfunction of which part of the brain?
    (A) Medulla
    (B) Cerebellum
    (C) Cerebrum
    (D) Pons

GO ON TO THE NEXT PAGE.

44. The sequence of amino acids in hemoglobin molecules of humans is more similar to the hemoglobin of chimpanzees than it is to the hemoglobin of dogs. This similarity suggests that

(A) humans and dogs are more closely related than humans and chimpanzees
(B) humans and chimpanzees are more closely related than humans and dogs
(C) humans are related to chimpanzees but not to dogs
(D) humans and chimpanzees are closely analogous

45. According to the heterotroph hypothesis, which event had to occur before oxygen filled the atmosphere?

(A) Heterotrophs had to remove carbon dioxide from the air.
(B) Autotrophs, which make their own food, had to evolve.
(C) Heterotrophs had to evolve.
(D) Autotrophs had to convert atmospheric nitrogen to nitrate.

46. Two individuals, one with type B blood and one with type AB blood have a child. The probability that the child has type O blood is

(A) 0%
(B) 25%
(C) 50%
(D) 100%

Questions 47–48 refer to the bar graph, which shows the relative biomass of four different populations of a particular food pyramid.

Relative Biomass

Population A ▬▬
Population B ▬▬▬▬▬▬▬
Population C ▬▬▬▬
Population D ▬

47. The largest amount of energy is available to

(A) population A
(B) population B
(C) population C
(D) population D

48. Which of the following would be the most likely result if there was an increase in the number of organisms in population C?

(A) The biomass of population D will remain the same.
(B) The biomass of population B will decrease.
(C) The biomass of population C will steadily increase.
(D) The food source available to population C would increase.

GO ON TO THE NEXT PAGE.

Questions 49–52 refer to the following illustration and information.

The cell cycle is a series of events in the life of a dividing eukaryotic cell. It consists of four stages: G_1, S, G_2, and M. The duration of the cell cycle varies from one species to another, and from one cell type to another. The G_1 phase varies the most. For example, embryonic cells can pass through the G_1 phase so quickly that it hardly exists, whereas neurons are arrested in the cell cycle and do not divide.

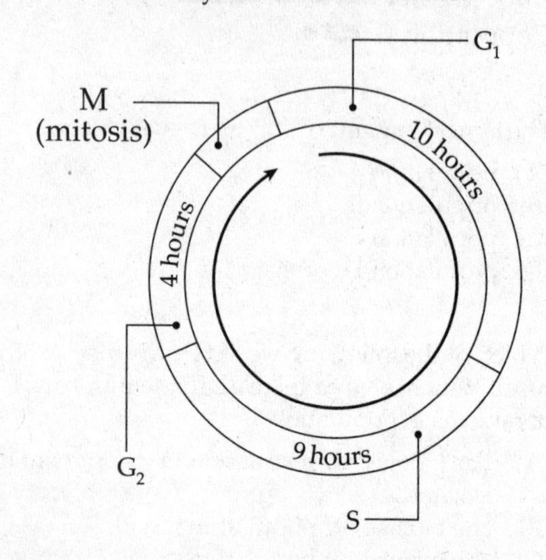

49. During which phase do chromosomes replicate?
    (A) G_1
    (B) S
    (C) G_2
    (D) M

50. In mammalian cells, the first sign of prophase is the
    (A) appearance of chromosomes
    (B) separation of chromatids
    (C) disappearance of the nuclear membrane
    (D) replication of chromosomes

51. Mitosis occurs in all of the following types of cells EXCEPT
    (A) epidermal cells
    (B) hair cells
    (C) red blood cells
    (D) pancreatic cells

52. Since neurons are destined never to divide again, what conclusion can be made?
    (A) These cells will go through cell division.
    (B) These cells will be permanently arrested in the G_1 phase.
    (C) These cells will be permanently arrested in the G_2 phase.
    (D) These cells will quickly enter the S-phase.

GO ON TO THE NEXT PAGE.

Questions 53–56 refer to the graphs, which show the permeability of ions during an action potential in a ventricular contractile cardiac fiber. The action potential of cardiac muscle fibers resembles that of skeletal muscles.

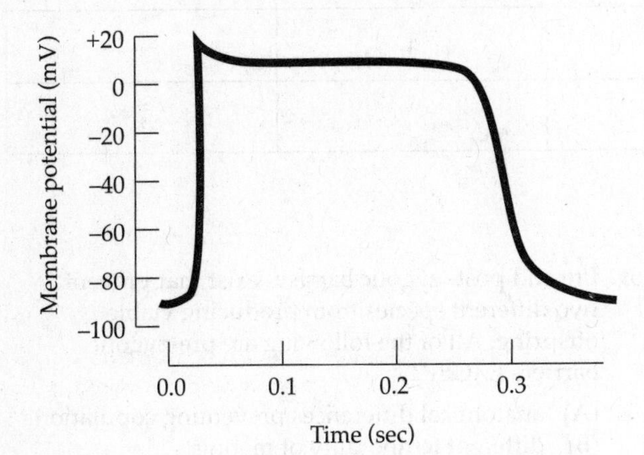

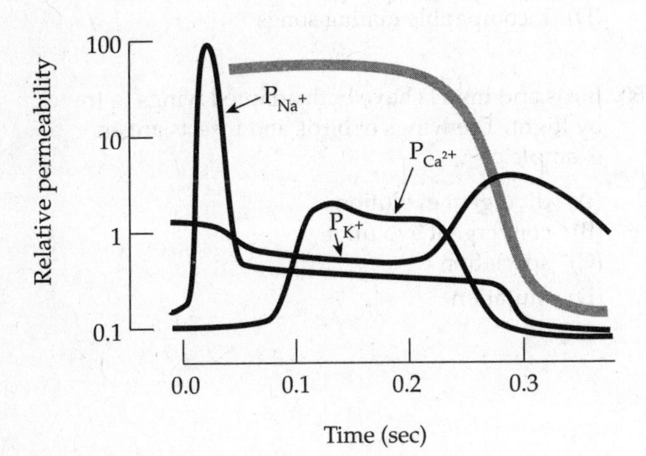

53. Based on the graph, the resting membrane potential of the muscle fibers is close to
 (A) –90 mV
 (B) –70 mV
 (C) 0 mV
 (D) +70 mV

54. Which of the following statements is true concerning the initial phase of depolarization?
 (A) Voltage-gated K^+ channels open in the plasma membrane.
 (B) The concentration of Ca^{2+} ions within the plasma membrane becomes more negative.
 (C) The membrane potential stays close to –40 mV.
 (D) The permeability of the sarcolemma to Na^+ ions increases.

55. In cardiac fibers, the duration of an action potential is approximately
 (A) 0.10 secs
 (B) 0.20 secs
 (C) 0.25 secs
 (D) 0.30 secs

56. One major difference between the action potential of cardiac muscle fibers and the action potential of skeletal muscle fibers is that in cardiac muscle fibers
 (A) the membrane is permeable to Na^+, not K^+
 (B) voltage-gated K+ channels open during depolarization, not repolarization
 (C) depolarization is prolonged compared to that in skeletal muscle fibers
 (D) the refractory period is shorter than that of skeletal muscle fibers

GO ON TO THE NEXT PAGE.

Questions 57–60 refer to the data below concerning the general animal body plan of five organisms.

Characteristic	Sea anemone	Hagfish	Eel	Salamander
Vertebral column		+	+	+
Jaws	+		+	+
Walking legs				+

Note: + indicates a feature presence in an organism.

57. The two most closely related organisms are
    (A) sea anemone and hagfish
    (B) eel and salamander
    (C) hagfish and eel
    (D) sea anemone and salamander

58. The correct order of evolution for the traits above is
    (A) jaws – vertebral column – walking legs
    (B) walking legs - jaws – vertebral column
    (C) jaws - walking legs – vertebral column
    (D) vertebral column – jaws – walking legs

59. Pre and post- zygotic barriers exist that prevent two different species from producing viable offspring. All of the following are pre-zygotic barriers EXCEPT
    (A) anatomical differences preventing copulation
    (B) different temporality of mating
    (C) sterility of offspring
    (D) incompatible mating songs

60. Birds and insects have both adapted wings to travel by flight. The wings of birds and insects are an example of
    (A) divergent evolution
    (B) convergent evolution
    (C) speciation
    (D) mutation

GO ON TO THE NEXT PAGE.

Questions 61–63 refer to the synthetic pathway of a pyrimidine, cytidine 5′ triphosphate, CTP. This pathway begins with the condensation of two small molecules by the enzyme, aspartate transcarbamylase (ATCase).

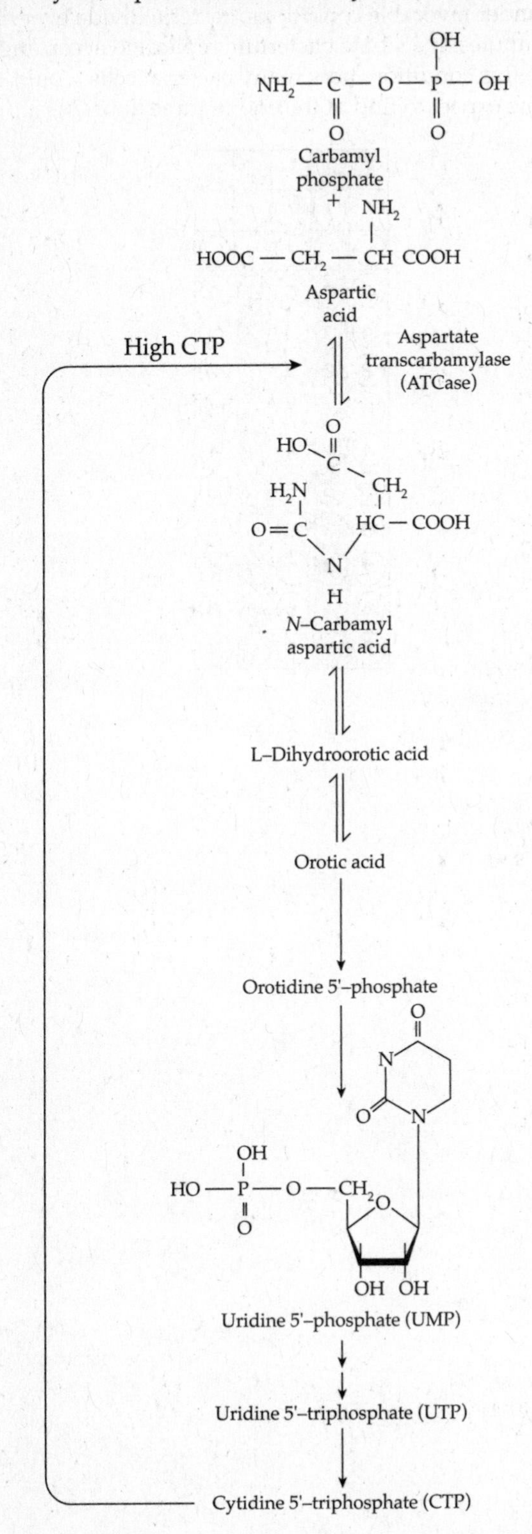

Regulation of CTP biosynthesis

61. Which of the following is true when the level of CTP is low in a cell?
    (A) CTP is converted to ATCase
    (B) The metabolic traffic down the pathway increases
    (C) ATCase is inhibited, which slows down CTP synthesis
    (D) The final product of the pathway is reduced

62. This enzymatic phenomenon is an example of
    (A) transcription
    (B) feedback inhibition
    (C) dehydration synthesis
    (D) photosynthesis

63. The biosynthesis of cytidine 5′-triphosphate requires
    (A) a ribose sugar, a phosphate group, and a nitrogen base
    (B) a deoxyribose sugar, a phosphate group, and a nitrogen base
    (C) a ribose sugar, phosphate groups, and a nitrogen base
    (D) a deoxyribose sugar, phosphate groups, and a nitrogen base

GO ON TO THE NEXT PAGE.

Directions: This part B consists of questions requiring numeric answers. Calculate the correct answer for each question.

64. In a diploid organism with the genotype AaBbCCDDEE, how many genetically distinct kinds of gametes would be produced?

65. Under favorable conditions, bacteria divide every 20 minutes. If a single bacterium replicated according to this condition, how many bacterial cells would one expect to find at the end of three hours?

GO ON TO THE NEXT PAGE.

66. In snapdragon plants that display intermediate dominance, the allele C^R produces red flowers and C^W produces white flowers. If a homozygous red-flowered snapdragon is crossed with a homozygous white-flowered snapdragon, what will the percentage of pink offspring be?

67. Translation is an energy-intensive process. Approximately how many ATPs are required to synthesize a protein containing 115 amino acids?

GO ON TO THE NEXT PAGE.

Question 68 refers to the following experiment.

A group of 100 *Daphnia*, small crustaceans known as water fleas, were placed in one of three culture jars of different sizes to determine their reproductive rate. The graph below shows the average number of offspring produced per female each day in each jar of pond water.

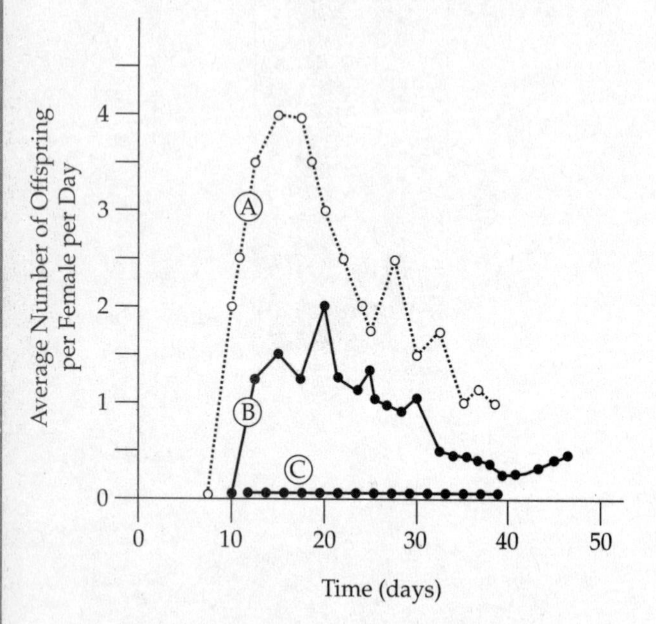

Key: Ⓐ Water fleas in a 1-liter jar of pond water
Ⓑ Water fleas in a 0.5-liter jar of pond water
Ⓒ Water fleas in a 0.25-liter jar of pond water

68. What is the total number of offspring produced in the 0.5-liter jar on the twentieth day, assuming all survive?

GO ON TO THE NEXT PAGE.

69. On average, there is a 90 percent reduction of productivity for each trophic level. Based on this information, 10,000 pounds of grass should be able to support how many pounds of crickets?

	/	/	/	
.	.	.	.	.
0	0	0	0	0
1	1	1	1	1
2	2	2	2	2
3	3	3	3	3
4	4	4	4	4
5	5	5	5	5
6	6	6	6	6
7	7	7	7	7
8	8	8	8	8
9	9	9	9	9

END OF SECTION I

GO ON TO THE NEXT PAGE.

BIOLOGY

SECTION II

Planning time—10 minutes

Writing time—1 hour and 30 minutes

Directions: Questions 1 and 2 are long-form essay questions that should require about 20 minutes each to answer. Questions 3 through 8 are short free-response questions that should require about 6 minutes each to answer. Read each question carefully and write your response. Answers must be written out. Outline form is not acceptable. It is important that you read each question completely before you begin to write.

1. Chlorophyll is one of a class of pigments that absorbs light energy in photosynthesis.
   a. **Relate** the structure of chlorophyll to its function.
   b. **Design** an experiment to investigate the influence of sunlight on the activity of chlorophyll.
   c. **Describe** what information concerning the structure of chlorophyll could be inferred from your experiment.

2. Over the course of early evolution, organisms had to develop various methods to regulate internal fluids and excrete wastes. **Discuss** the problems faced by **three** organisms and how these problems were solved. In your discussion include structural adaptations and their functional significance.

3. **Describe** the chemical nature of genes. Name two types of gene mutations that could occur during replication.

4. Select **one** of the following three pairs of hormones and discuss the concept of negative feedback.
   a. Thyroid-stimulating hormone (TSH) and thyroxine
   b. Parathyroid hormone and calcitonin
   c. ACTH and cortisol

5. **Describe** why fermentation is a less efficient way to produce energy than aerobic respiration.

6. **Define** analogous structures and give an example.

7. **Describe** symbiosis and give an example involving humans.

8. **Define** the role of ADH (anti-diuretic hormone) and aldosterone in the regulation of blood pressure.

STOP

END OF EXAM

19

Practice Test 2 Answers and Explanations

ANSWER KEY

1. D
2. D
3. D
4. B
5. B
6. D
7. D
8. C
9. C
10. A
11. B
12. A
13. B
14. B
15. D
16. A
17. B
18. B
19. D
20. D
21. B
22. D
23. A
24. C
25. D
26. C
27. C
28. C
29. D
30. B
31. B
32. B
33. B
34. B
35. B
36. C
37. B
38. B
39. B
40. C
41. A
42. C
43. C
44. B
45. B
46. A
47. B
48. B
49. B
50. A
51. C
52. B
53. A
54. D
55. D
56. C
57. B
58. D
59. C
60. B
61. B
62. B
63. C
64. 4
65. 512
66. 100
67. 460
68. 200
69. 1,000

MULTIPLE-CHOICE ANSWERS AND EXPLANATIONS

1. **D** Animal cells have centrioles. Both animal and plant cells have an endoplasmic reticulum, membrane-bound organelles, and lysosomes. Only plants have a cell wall made of cellulose and large vacuoles.

2. **D** If the contents of the cell separated from the cell wall, then water was moving *out* of the cell. (B), This would cause the space between the cell wall and the cell membrane to expand. This also means that the concentration of solutes in the extracellular environment would therefore be *hypotonic* with respect to the cell's interior. (C), Because the vacuole is within the cell, it contains a solution with a much lower osmotic pressure than that of the sugar solution. (A), Lastly, because the fluid in the cell was hypertonic to the sugar solution, fluid was moving out of the vacuole and caused it to become smaller.

3. **D** This question tests your ability to associate what happens when enzymes are denatured and what would happen in the synaptic cleft. Acetylcholinesterase is an enzyme that degrades acetylcholine in the synaptic cleft. (A), (B) and (C), If acetylcholinesterase is denatured, acetylcholine will still be released from the presynaptic membrane and continue to diffuse across the synaptic cleft and bind to the postsynaptic membrane because acetylcholine is not degraded.

4. **B** The ratio of purines to pyrimidines should be constant because purines always bind with pyrimidines, no matter which ones they may be.

5. **B** Antibodies are only produced by plasma B-cells. Helper T cells (D) activate B cells, cytotoxic T cells (C) kill infected cells and phagocytes (B) are involved in non-specific immune responses

6. **D** If substance F leads to the inhibition of enzyme 3, then substances D and E and enzymes 3, 4, and 5 will be affected. The activity of enzyme 5 will be decreased, not increased.

7. **D** The liver does not break down peptides into amino acids. It performs all of the following functions, (A), stores amino acids absorbed in the capillaries, (B), detoxifies harmful substances, and (C), makes bile.

8. **C** Based on the graph, fetal hemoglobin has a higher affinity for oxygen than maternal hemoglobin. (A), Fetal hemoglobin does not give up oxygen more readily than maternal hemoglobin. (B), The dissociation curve of fetal hemoglobin is to the left of the maternal hemoglobin. (D), Fetal hemoglobin and maternal hemoglobin are different structurally, but you can't tell this from the graph.

9. **C** The set of genotypes that represents a cross that could produce offspring with silver fur from parents that both have brown fur is Bb and Bb. Complete a Punnett square for this question. In order for the offspring to have silver fur, both parents must have the silver allele.

10. **A** (B) and (D), Hemoglobin's affinity for O_2 decreases as the concentration of H^+ increases (or the pH decreases) and as the concentration of CO_2 increases (or the concentration of HCO_3^- increases). (C), Hemoglobin's affinity for oxygen in tissue muscles does not increase during exercise.

11. **B** Viruses are made up of nucleic acid surrounded by a protein coat called a capsid. They do not contain a cell wall, proteins, or cell membrane.

12. **A** Both prokaryotes and eukaryotes possess double-stranded DNA. Answer choices (B), (C), and (D) are correct differences between prokaryotes and eukaryotes.

13. **B** In humans, fertilization normally occurs in the fallopian tube. (A), The ovary is the female gonad that contains the eggs. (C), The uterus is the organ that houses the developing embryo. (D), The placenta is the structure that nourishes the embryo.

14. **B** The development of an egg without fertilization is known as parthenogenesis. Parthenogenesis is a form of asexual reproduction found in insects and lizards. (A), Meiosis is a form of sexual reproduction that produces gametes. (C), Embryogenesis refers to the early stages of embryo development. (D), Vegetative propagation is a form of asexual reproduction by which plants produce identical offsprings from stem, leaves, or roots.

15. **D** All of the choices are examples of hydrolysis except the conversion of pyruvic acid to glucose. Hydrolysis is the breaking of a covalent bond by adding water. In all of the correct examples, complex compounds are broken down to simpler compounds. The conversion of pyruvic acid to acetyl CoA is an example of decarboxylation—a carboxyl group is removed as carbon dioxide and the 2-carbon fragment is oxidized.

16. **A** Peroxisomes catalyze reactions that produce hydrogen peroxide, ribosomes are involved in protein synthesis, and mitochondria contain enzymes involved in cellular respiration. (B) and (D), Lysosomes are the sites of degredation; they contain hydrolytic enzymes, but do not produce hydrogen peroxide. (C), The Golgi apparatus sorts and packages substances that are destined to be secreted out of the cell. Interestingly, O_2 concentration generally doesn't play an important role in regulating respiration.

17. **B** Use process of elimination. The respiratory rate in humans will be affected by an increase in the amount of CO_2, a drop in pH levels, which is the same as an increase in hydrogen ion levels. Strenuous exercising will also modify the respiratory rate. O_2 concentration generally does not play an important role in regulating respiration.

18. **B** The primary site of glucose reabsorption is the proximal convoluted tubule. (A), The glomerulus is a tuft of capillaries that filters fluid into the Bowman's capsule. (C), The loop of Henle is the site of salt reabsorption. (D), The collecting duct is the site in which urine is concentrated.

19. **D** The carrying capacity is the maximum number of organisms of a given species that can be maintained in a given environment. Once a population reaches its carrying capacity, the number of organisms will fluctuate around it.

20. **D** During the exponential growth phase of a population, the size doubles during each time interval. This part of the graph looks like a parabola.

21. **B** Insulin decreases the level of blood glucose by increasing the storage of glycogen in muscles. The other hormones are correctly paired with their respective function.

22. **D** The failure in oogenesis that could produce this syndrome would occur in anaphase I or anaphase II. Anaphase refers to the stage of meiosis in which chromatids separate from each other. If the chromosomes or chromatids fail to separate during anaphase, one egg cell will contain two X chromosomes, instead of one.

23. **A** The tendency of climbing vines to twine their tendrils around a trellis is called thigmotropism. Thigmotropism refers to growth stimulated by contact with an object. (B), Hydrotropism refers to growth of a plant toward water. (C), Phototropism refers to growth toward light. (D), Geotropism refers to growth toward or against gravity.

24. **C** Females with Turner's syndrome lack an X chromosome. If females with this syndrome have a high rate of hemophilia, they must not have the second X to mask the expression of the disease.

25. **D** This type of mutation is called frameshift mutation. The insertion of DNA leads to a change in the normal reading frame by one base pair. The other answer choices refer to chromosomal aberrations. (A), Duplication is when an extra copy of a chromosome segment is introduced. (B), Translocation is when a segment of a chromosome moves to another chromosome. (C), Inversion is when a segment of a chromosome is inserted in the reverse orientation.

26. **C** The hypothalamus releases GnRH, which stimulates the release of FSH. The key word in this question is "releasing." The organ that produces releasing or inhibiting factors is the hypothalamus. (A), The anterior pituitary is the master gland that secretes several hormones. (B), The posterior pituitary secretes oxytocin and vasopressin. (D), The pineal gland is a gland at the base of the brain that secretes melatonin and helps regulate circadian rhythms.

27. **C** The principle inorganic compound found in living things is water. Water is a necessary component for life.

28. **C** The loop of Henle is responsible for the concentration of urine. The longer the loop of Henle, the more water would be reabsorbed, which would make the filtrate more concentrated. (D), If the loop of Henle were longer, the collecting ducts may or may not also be longer. (A), If the walls of the nephrons were thicker and impermeable, water would not be conserved.

29. **D** All of the following are modes of asexual reproduction except meiosis, which is the production of gametes for sexual reproduction. (A), Sporulation is a form of asexual reproduction in which spores are produced. (B), Fission is the equal division of a bacterial cell. (C), Budding is a form of asexual reproduction in yeasts in which small cells grow from a parent cell.

30. **B** Respiration is the transport of oxygen into an organism and carbon dioxide out of an organism.

31. **B** Invertebrates lack specific immune responses including B cells and T cells. Phagocytes is the only answer choice that describes non-specific immune responses.

32. **B** All of the choices are examples of connective tissue except muscle. Connective tissue connects and supports other tissues. Ligaments, blood, cartilage, and bone are all connective tissues.

33. **B** Photosynthesis requires light, carbon dioxide and water and produces oxygen and glucose.

34. **B** When a forest of trees is devastated by fire, secondary succession is most likely to occur. Secondary succession refers to ecological succession in a disturbed community. (A), Plants and animals will continue to inhabit the region once the community is reestablished. (C), Tough grasses are not pioneers. (D), It cannot be determined if the number of species will be stabilized.

35. **B** Translation, the synthesis of proteins from mRNA, occurs in the cytoplasm. (I), DNA replication occurs in the nucleus. (II), Transcription, the synthesis of RNA from DNA, occurs in the nucleus.

36. **C** Crossing-over permits scientists to determine chromosome mapping. Chromosome mapping is a detailed map of all the genes on a chromosome. The frequency of crossing-over between any two alleles is proportional to the distance between them. The farther apart the two linked alleles are on a chromosome, the more often the chromosome will break between them. Crossing-over does not tell us about the chance of variation in zygotes, the rate of mutations, or whether the traits are dominant, recessive, or masked.

37. **B** In this scenario the inside of the cell is hypertonic to the outside environment, this will cause water to move into the cell and this can cause the cell to lyse.

38. **B** The most likely explanation for this phenomenon is that these birds have different ecological niches. An ecological niche is the position or function of an organism or population in its environment. (A), We do not know if there is a short supply of resources. (C), We do not know how long the bird species live together. (D), The breeding patterns of the bird species does not explain the lack of competition.

39. **B** This relationship is an example of parasitism. Parasitism is a form of symbiosis in which one organism benefits and the other is harmed. (A), Commensalism is a form of symbiosis in which one organism benefits and the other is unaffected. (C), Mutualism is a form of symbiosis in which both organisms benefit. (D), Gravitropism is the growth of a plant toward or away from gravity.

40. **C** If the nucleotide sequence of a DNA molecule is 5'-C-A-T-3', then the transcribed DNA strand (mRNA) would be 3'-G-U-A-5'. The nucleotide sequence of the tRNA codon would be 5'-C-A-U-3'.

41. **A** Viruses are considered an exception to the cell theory because they are not independent organisms. They can only survive by invading a host. (B), Viruses have either DNA or RNA. (C), Not all viruses have a tail. (D), Viruses did not evolve from ancestral protists.

42. **C** All of the listed organs secrete digestive enzymes except the gall bladder. The gall bladder stores and secretes bile produced by the liver, which is an emulsifier, not an enzyme. (A), The mouth secretes salivary amylase. (B), The stomach secretes pepsin. (D), The small intestine secretes protease.

43. **C** Memory loss would most likely be due to a malfunction of the cerebrum. The cerebrum controls all voluntary activities and receives and interprets sensory information. (A), The medulla controls involuntary actions such as breathing. (B), The cerebellum coordinates muscle activity and controls balance. (D), The pons is a mass of nerve fibers running across the surface of the mammalian brain.

44. **B** The similarity suggests that humans and chimpanzees are more closely related than humans and dogs. Because these two organisms share similar amino acid sequences, they must share more recent common ancestors than with the dog.

45. **B** The event that had to occur before oxygen filled the atmosphere was that autotrophs, which make their own food and give off oxygen, had to evolve.

46. **A** There is no way for these two parents to produce a type O child. The only genotype that will result in type O blood is two "O" alleles, one from each parent, since the allele for O blood is recessive to the alleles for A or B blood. The AB parent only has alleles for A and B blood, so it is impossible for these two individuals to produce a child with type O blood.

47. **B** The largest amount of energy is available to producers. Population B is most likely composed of producers, because they have the largest biomass.

48. **B** An increase in the number of organisms in population C would most likely lead to a decrease in the biomass of B because population B is the food source for population C. Make a pyramid based on the biomasses given. If population C increases, population B will decrease. (A) and (C), We cannot necessarily predict what will happen to the biomass of populations that are above population C. (D), The food source available to population C would most likely decrease, not increase.

49. **B** Chromosomes replicate during interphase, the S phase. (A) and (C), During G_1 and G_2, the cell makes protein and performs other metabolic duties.

50. **A** The first sign of prophase in mammalian cells is the appearance of chromosomes.

51. **C** Mitosis occurs in all of the following type of cells except mature red blood cells. Mature red blood cells are short-lived and do not divide.

52. **B** Because neurons are not capable of dividing, it is reasonable to conclude that these cells will not complete the G_1 phase. This is a reading comprehension question. The passage states that cells that do not divide are arrested at the G_1 phase. (A), These cells will not be committed to go through cell division. (C) and (D), These cells will not enter the G_2 or S-phase.

53. **A** According to the graph, the resting membrane potential of the muscle fiber is close to –90mV.

54. **D** Refer to the second graph about the membrane permeability of ions during a muscle contraction. During depolarization, the membrane is permeable to Na^+. (A), The voltage-gated K^+ channels do not open until after depolarization. (B), The concentration of Ca^{2+} does not become more negative. (C), The membrane potential changes from –90 mV to +20 mV.

55. **D** Refer to both graphs in the passage. In cardiac fibers, the duration of an action potential—a neuronal impulse—is approximately 0.3 seconds.

56. **C** In cardiac muscle fibers, depolarization is prolonged compared to that in skeletal muscle fibers. (A), The membrane is permeable to both Na^+ and K^+. (B), In cardiac muscle fibers, voltage-gated K^+ channels open during repolarization. (D), The refractory period is longer in cardiac muscle fibers compared to skeletal muscle fibers.

57. **B** The two most closely related organisms are the two with the most shared derived characteristics.

58. **D** Shared derived characteristics are newly evolved traits that are shared with every group on a phylogenic tree except for one. Vertebral columns are present in every group except for the sea anemone so it must have evolved first. Walking legs are only found in the salamander, indicating that it most likely evolved most recently.

59. **C** Pre-zygotic barriers to reproduction are those that prevent fertilization (Answer choices (A), (B), and (D)). Answer choice (C) is an example of a post-zygotic barrier to reproduction.

60. **B** Convergent evolution occurs when two organisms that are not closely related independently evolve similar traits, like the wings of insects and birds. Divergent evolution occurs when two closely related individuals become more different over time and can lead to speciation.

61. **B** When the level of CTP is low in a cell, the metabolic traffic down the pathway increases. This pathway is controlled by feedback inhibition. The final product of the pathway inhibits the activity of the first enzyme. When the supply of CTP is low, the pathway will continue to produce CTP.

62. **B** This enzymatic phenomenon is an example of feedback inhibition. Feedback inhibition is the metabolic regulation in which high levels of an enzymatic pathway's final product inhibit the activity of its rate-limiting enzyme. (A), Transcription is the production of RNA from DNA. (C), Dehydration synthesis is the formation of a covalent bond by the removal of water. (D), In photosynthesis, radiant energy is converted to chemical energy.

63. **C** The biosynthesis of cytidine 5'-triphate requires a nitrogenous base, three phosphates, and the sugar ribose. Pyrimidines are a class of nitrogenous bases with a single ring structure. The sugar they contain is ribose (shown in the pathway).

64. **4** There are four genetically distinct kinds of gametes that could be produced, ABCDE, AbCDE, aBCDE, and abCDE. Notice that the only alleles that vary are A and B ($2^2 = 4$).

65. **512** If bacteria divide every 20 minutes, you would produce 512 bacterial cells. One method would be to use the equation 2^x, where x equals the number of 20 minute intervals in three hours; $2^9 = 512$.

66. **100** Because snapdragon plants display intermediate dominance, the heterozygous phenotype is affected by the alleles of both homozygotes. If a homozygous red plant is crossed with a homozygous white plant, all offspring would be heterozygous and pink.

67. **460** To calculate an estimate of how many ATPs are used in the synthesis of a protein, the number of amino acids in the protein should be multiplied by four. Since the question states there are 115 amino acids, the answer is therefore 460.

68. **200** The graph shows the average number of offspring per female per day. Because there are 100 females in the $\frac{1}{2}$ liter, the total number of offspring on the twentieth day would be 100 females times 2 offspring per day, which equals 200 offspring.

69. **1,000**

A 90 percent reduction of productivity in the grass is a reduction of 9,000 pounds. That means the grass should be able to support 1,000 pounds of crickets.

FREE-RESPONSE ANSWERS AND EXPLANATIONS

Long-Form Free Response 1

Chlorophyll is the principal light-harnessing pigment found in the thylakoid membrane of chloroplasts. It is a molecule that's made up of a large ring structure composed of smaller rings (pyrroles). In the center of the large porphyrin ring is a magnesium atom surrounded by four nitrogen atoms. The small rings consist of many alternating single and double bonds. These alternating bonds allow electrons in the porphyrin ring and magnesium atoms to move around freely. Alternating double and single bonds are commonly found in molecules that strongly absorb visible light. Photosynthesis occurs when sunlight activates chlorophyll by exciting their electrons. The chlorophyll molecule has a long hydrophobic hydrocarbon tail. This shape allows many chlorophyll molecules to be grouped together like a stack of saucers.

The following experiment could determine the influence of sunlight on chlorophyll. First, extract chlorophyll from the leaf by boiling it gently in a dissolved solution. Then pass light of a known wavelength through the solution and measure it, using a spectrophotometer. The wavelength of the entering light can be varied to see which wavelength is most absorbed by the solution. You could then plot the data on a graph to form an absorption spectrum. The absorption spectrum will show two peaks, which represent the type of light absorbed by the pigment. The valley would represent the light that's reflected.

Based on this experiment, it would be inferred that certain wavelengths, such as orange-red and violet-blue, are strongly absorbed by chlorophyll, whereas green and yellow hues are least absorbed, and are reflected. Chlorophyll gives plants their characteristic green color because it reflects yellow-green light.

Long-Form Free Response 2

Because the amount of intracellular water is critical, one major challenge that various organisms faced was how they would regulate body water and get rid of wastes. Waste products include carbon dioxide, salts, and nitrogenous wastes. This issue is especially important because nitrogenous wastes are highly toxic to the body. Over the course of evolution, organisms developed various structural adaptations to deal with this problem.

Insects developed excretory organs called Malpighian tubules. These long, slender tubules take up water and salt, concentrate the waste, and empty it into the intestine. The waste product, uric acid, is excreted as a dry pellet through the anus. Insects also have a hard, dry cuticle that has a waxy outer layer; this aids in preventing water loss.

Organisms that live in the sea, such as marine fishes, had to cope with the gradual increase in the saltiness of the water. Seawater is hypertonic relative to their body fluids, and marine fish had to protect their body cells from water loss because without a specific mechanism for this, they would become dehydrated even though they're surrounded by water. They excrete concentrated urine that is isotonic with their body fluids. They also eliminate excess salt by actively pumping it out through their gills.

Because humans are terrestrial organisms, they must conserve plenty of water. Humans also must get rid of nitrogenous wastes, and their major excretory organ is the kidney. Each kidney is made up of millions of function units called nephrons. As the filtrate travels along the nephron, glucose, amino acids, and salts are retained by the body, and the rest of the fluid is concentrated into urine. The skin is another important organ that gets rid of wastes. It contains sweat glands that help to maintain an optimal salt balance in the body.

Short-Form Free Response 3

A gene is a heredity unit located at a specific locus along a chromosome. Genes are made up of DNA, and DNA is made up of repeating subunits of nucleotides. A nucleotide consists of three parts, a five-carbon sugar, a phosphate group, and a nitrogenous base.

A gene mutation is a change in the sequence of base pairs in a DNA. It results from defects in the sequence of bases. Mutations that involve a single base change in the DNA sequence are called point mutations. These are base substitutions involving a single DNA nucleotide being replaced by another nucleotide. Another type of mutation is a frameshift mutation, caused by an insertion or deletion of bases in the DNA sequence. This causes a shift in the reading frame of DNA.

Short-Form Free Response 4

Hormones are chemical messengers that travel through the blood and act on target cells in the body. Hormone secretion is regulated by a negative feedback system, which is the basis of hormonal regulation. In part, negative feedback loops mean that an excess of a hormone in the bloodstream temporarily shuts down the production of that hormone.

The anterior pituitary gland secretes a hormone called thyroid-stimulating hormone (TSH). TSH stimulates the thyroid gland to secrete its hormone; thyroxine. Thyroxine regulates the basal metabolic rate in most body tissues. If the blood level of thyroxine rises above normal, it will suppress the secretion of TSH which, in turn, will lower the blood level of thyroxine. Both responses are negative feedback mechanisms.

Two hormones play a critical role in the regulation of calcium, parathyroid hormone and calcitonin. Parathyroid hormone, which is secreted from the parathyroid glands, increases blood calcium levels. It triggers the bones to release the calcium stored inside them. Calcitonin acts as an antagonist of parathyroid hormone. It decreases blood calcium levels. When blood calcium levels are above normal, the thyroid gland secretes calcitonin. This mechanism helps to maintain homeostasis.

ACTH is another hormone that's released by the anterior pituitary. It stimulates the adrenal cortex to secrete a number of steroid hormones, including cortisol. Cortisol increases the blood's concentration of glucose and helps the body adapt to stress. When cortisol levels reach a peak level in the bloodstream, the anterior pituitary is temporarily prevented from producing ACTH. This, in turn, shuts down the production of cortisol. Once the level of cortisol falls below normal, the adrenal cortex resumes production of cortisol.

Short-Form Free Response 5

Fermentation occurs when oxygen is not available to act as the final electron acceptor. When this does not occur, then NADH and $FADH_2$ are not able to be regenerated into NAD^+ and FAD, respectively. Instead, the only step in respiration that is occurring is glycolysis, thus generating its end product of pyruvate. Pyruvate can act as the final electron acceptor and allow NAD^+ to be regenerated, but the only energy that has been produced is the net of two ATP from glycolysis. Since not all the energy can be mined from the carbons when processed in fermentation, the process is less efficient.

Short-Form Free Response 6

Analogous structures are ones which have similar functions, but whose underlying structures are very different. They are often the product of convergent evolution, where differing organisms ended up in the same environment and thus subjected to the same forces of natural selection. An example of analogous structures would be the wing of bird and the wing of a bee. Both are used for flight, but their components are completely different.

Short-Form Free Response 7

Symbiosis describes a situation in which two organisms co-exist, in which at least one benefits by gaining food, shelter, transportation or some other requirement for existence. If the host provides a benefit, but remains unimpacted, then it is commensalistic, but if both organisms benefit then it is mutualistic. An example involving humans would be the bacterial flora of the skin or of the intestine. Many of these bacteria provide a protective or digestive benefit to humans while some utilize the body for food while also not doing any harm.

Short-Form Free Response 8

When blood pressure drops, aldosterone is released and allows more sodium to be reabsorbed into the blood stream at the distal convoluted tubule of the nephron of the kidney. When the concentration of the blood goes up, it is possible for it to take up more fluid so when aldosterone acts on the kidney, it is followed by anti-diuretic hormone. ADH creates water channels which then allow water to pass from the collecting duct into the circulatory system; this water is following the sodium that has already passed into the blood. By increasing the volume in the circulatory system, the body's blood pressure is raised.

ABOUT THE AUTHOR

Kim Magloire received a degree in Biology and Science in Human Affairs from Princeton University. She received her Master's degree in Epidemiology from Columbia University and is completing her doctorate in Epidemiology. In her many years of teaching, Kim has prepared hundreds of students for the AP Biology Exam as well as for the SAT, MCAT, GMAT, LSAT, GRE, and other science SAT Subject Tests.

Kim is the founder of SciTech, an educational company dedicated to helping students achieve academic excellence in science, math, and technology. Founded in 1996, SciTech offers a number of programs that help students make the "science and math connection." She has received numerous awards for her innovative educational programs in New York City, including the 1999 National "Community Leader of the Year" award from The Quaker Oats Company. She resides in New York City with her husband and daughter.

ANSWER SHEETS

Completely darken bubbles with a No. 2 pencil. If you make a mistake, be sure to erase mark completely. Erase all stray marks.

1. YOUR NAME: ______________________________
(Print) Last First M.I.

SIGNATURE: ______________________________ **DATE:** ____/____/____

HOME ADDRESS: ______________________________
(Print) Number and Street

____
City State Zip Code

PHONE NO. : ______________________________
(Print)

IMPORTANT: Please fill in these boxes exactly as shown on the back cover of your test book.

2. TEST FORM

3. TEST CODE

0 1 2 3 4 5 6 7 8 9
A B C D E F G
0 1 2 3 4 5 6 7 8 9

4. REGISTRATION NUMBER

0 1 2 3 4 5 6 7 8 9

5. YOUR NAME

First 4 letters of last name | FIRST INIT | MID INIT

A B C D E F G H I J K L M N O P Q R S T U V W X Y Z

6. DATE OF BIRTH

Month	Day	Year
JAN		
FEB		
MAR	0 0	0 0
APR	1 1	1 1
MAY	2 2	2 2
JUN	3 3	3 3
JUL	4	4 4
AUG	5	5 5
SEP	6	6 6
OCT	7	7 7
NOV	8	8 8
DEC	9	9 9

7. SEX

MALE
FEMALE

The Princeton Review

FORM NO. 00001-PR

Section 1

Start with number 1 for each new section.
If a section has fewer questions than answer spaces, leave the extra answer spaces blank.

1. A B C D
2. A B C D
3. A B C D
4. A B C D
5. A B C D
6. A B C D
7. A B C D
8. A B C D
9. A B C D
10. A B C D
11. A B C D
12. A B C D
13. A B C D
14. A B C D
15. A B C D
16. A B C D
17. A B C D
18. A B C D
19. A B C D
20. A B C D
21. A B C D
22. A B C D
23. A B C D
24. A B C D
25. A B C D
26. A B C D
27. A B C D
28. A B C D
29. A B C D
30. A B C D
31. A B C D
32. A B C D
33. A B C D
34. A B C D
35. A B C D
36. A B C D
37. A B C D
38. A B C D
39. A B C D
40. A B C D
41. A B C D
42. A B C D
43. A B C D
44. A B C D
45. A B C D
46. A B C D
47. A B C D
48. A B C D
49. A B C D
50. A B C D
51. A B C D
52. A B C D
53. A B C D
54. A B C D
55. A B C D
56. A B C D
57. A B C D
58. A B C D
59. A B C D
60. A B C D
61. A B C D
62. A B C D
63. A B C D
64. A B C D
65. A B C D
66. A B C D
67. A B C D
68. A B C D
69. A B C D
70. A B C D
71. A B C D
72. A B C D
73. A B C D
74. A B C D
75. A B C D
76. A B C D
77. A B C D
78. A B C D
79. A B C D
80. A B C D
81. A B C D
82. A B C D
83. A B C D
84. A B C D
85. A B C D
86. A B C D
87. A B C D
88. A B C D
89. A B C D
90. A B C D
91. A B C D
92. A B C D
93. A B C D
94. A B C D
95. A B C D
96. A B C D
97. A B C D
98. A B C D
99. A B C D
100. A B C D
101. A B C D
102. A B C D
103. A B C D
104. A B C D
105. A B C D
106. A B C D
107. A B C D
108. A B C D
109. A B C D
110. A B C D
111. A B C D
112. A B C D
113. A B C D
114. A B C D
115. A B C D
116. A B C D
117. A B C D
118. A B C D
119. A B C D
120. A B C D

Completely darken bubbles with a No. 2 pencil. If you make a mistake, be sure to erase mark completely. Erase all stray marks.

1. YOUR NAME: ______________________________
(Print) Last First M.I.

SIGNATURE: ______________________________ **DATE:** ___/___/___

HOME ADDRESS: ______________________________
(Print) Number and Street

____
City State Zip Code

PHONE NO. : ______________________________
(Print)

IMPORTANT: Please fill in these boxes exactly as shown on the back cover of your test book.

2. TEST FORM

3. TEST CODE

0 1 2 3 4 5 6 7 8 9
A B C D E F G
0 1 2 3 4 5 6 7 8 9

4. REGISTRATION NUMBER

0 1 2 3 4 5 6 7 8 9

5. YOUR NAME

First 4 letters of last name | FIRST INIT | MID INIT

A B C D E F G H I J K L M N O P Q R S T U V W X Y Z

6. DATE OF BIRTH

Month	Day	Year
JAN, FEB, MAR, APR, MAY, JUN, JUL, AUG, SEP, OCT, NOV, DEC	0 1 2 3 / 0 1 2 3 4 5 6 7 8 9	0 1 2 3 4 5 6 7 8 9 / 0 1 2 3 4 5 6 7 8 9

7. SEX

MALE
FEMALE

The Princeton Review

FORM NO. 00001-PR

Section 1

Start with number 1 for each new section.
If a section has fewer questions than answer spaces, leave the extra answer spaces blank.

1. A B C D
2. A B C D
3. A B C D
4. A B C D
5. A B C D
6. A B C D
7. A B C D
8. A B C D
9. A B C D
10. A B C D
11. A B C D
12. A B C D
13. A B C D
14. A B C D
15. A B C D
16. A B C D
17. A B C D
18. A B C D
19. A B C D
20. A B C D
21. A B C D
22. A B C D
23. A B C D
24. A B C D
25. A B C D
26. A B C D
27. A B C D
28. A B C D
29. A B C D
30. A B C D
31. A B C D
32. A B C D
33. A B C D
34. A B C D
35. A B C D
36. A B C D
37. A B C D
38. A B C D
39. A B C D
40. A B C D
41. A B C D
42. A B C D
43. A B C D
44. A B C D
45. A B C D
46. A B C D
47. A B C D
48. A B C D
49. A B C D
50. A B C D
51. A B C D
52. A B C D
53. A B C D
54. A B C D
55. A B C D
56. A B C D
57. A B C D
58. A B C D
59. A B C D
60. A B C D
61. A B C D
62. A B C D
63. A B C D
64. A B C D
65. A B C D
66. A B C D
67. A B C D
68. A B C D
69. A B C D
70. A B C D
71. A B C D
72. A B C D
73. A B C D
74. A B C D
75. A B C D
76. A B C D
77. A B C D
78. A B C D
79. A B C D
80. A B C D
81. A B C D
82. A B C D
83. A B C D
84. A B C D
85. A B C D
86. A B C D
87. A B C D
88. A B C D
89. A B C D
90. A B C D
91. A B C D
92. A B C D
93. A B C D
94. A B C D
95. A B C D
96. A B C D
97. A B C D
98. A B C D
99. A B C D
100. A B C D
101. A B C D
102. A B C D
103. A B C D
104. A B C D
105. A B C D
106. A B C D
107. A B C D
108. A B C D
109. A B C D
110. A B C D
111. A B C D
112. A B C D
113. A B C D
114. A B C D
115. A B C D
116. A B C D
117. A B C D
118. A B C D
119. A B C D
120. A B C D

NOTES

NOTES

NOTES